thelightningpress.com

First Edition
w/SMARTupdate 1
(CYBER1-1)

Cyberspace Operations & Electronic Warfare

Multi-Domain Guide to Offensive/Defensive CEMA and CO

The Lightning Press
Norman M Wade

The Lightning Press

2227 Arrowhead Blvd.
Lakeland, FL 33813
24-hour Order/Voicemail: 1-800-997-8827
E-mail: SMARTbooks@TheLightningPress.com

www.TheLightningPress.com

(CYBER1-1) The Cyberspace Operations & Electronic Warfare SMARTbook (w/SMARTupdate 1*)
Multi-Domain Guide to Offensive/Defensive CEMA and CO

* SMARTupdate 1 to CYBER1 (Aug '21) updates the first printing of the CYBER1 SMARTbook (Oct '19) by incorporating new material from FM 3-12 (Aug '21), ATP 3-12.3 (Jul '19), ATP 6-02.70 (Oct '19), JP 3-85 (May '20) and adding a new section on Cyberspace IPB (ATP 2-01.3). An asterisk marks changed pages. (Learn more at www.thelightningpress.com/smartupdates/)

Copyright © 2021 The Lightning Press

ISBN: 978-1-935886-71-6

All Rights Reserved

No part of this book may be reproduced or utilized in any form or other means, electronic or mechanical, including photocopying, recording or by any information storage and retrieval systems, without permission in writing by the publisher. Inquiries should be addressed to The Lightning Press.

Notice of Liability

The information in this SMARTbook and quick reference guide is distributed on an "As Is" basis, without warranty. While every precaution has been taken to ensure the reliability and accuracy of all data and contents, neither the author nor The Lightning Press shall have any liability to any person or entity with respect to liability, loss, or damage caused directly or indirectly by the contents of this book. If there is a discrepancy, refer to the source document. This SMARTbook does not contain classified or sensitive information restricted from public release. "The views presented in this publication are those of the author and do not necessarily represent the views of the Department of Defense or its components."

SMARTbook is a trademark of The Lightning Press.

Credits: Cover image licensed from Shutterstock.com. All other photos courtesy Dept. of the Army and/or Dept. of Defense and credited individually where applicable.

Printed and bound in the United States of America.

View, download FREE samples and purchase online:
www.TheLightningPress.com

(CYBER1-1) Notes to Reader

The Cyberspace Operations & Electronic Warfare SMARTbook

United States armed forces operate in an increasingly **network-based world**. The proliferation of information technologies is changing the way humans interact with each other and their environment, including interactions during military operations. This broad and rapidly changing operational environment requires that today's armed forces must operate in cyberspace and leverage an **electromagnetic spectrum** that is increasingly competitive, congested, and contested.

Cyberspace is a global domain within the information environment consisting of the interdependent network of information technology infrastructures and resident data, including the Internet, telecommunications networks, computer systems, and embedded processors and controllers. Operations in cyberspace contribute to gaining a significant operational advantage for achieving military objectives.

Cyber electromagnetic activities (CEMA) are activities leveraged to seize, retain, and exploit an advantage over adversaries and enemies in both cyberspace and the electromagnetic spectrum, while simultaneously denying and degrading adversary and enemy use of the same and protecting the mission command system.

Cyberspace operations (CO) are the employment of cyberspace capabilities where the primary purpose is to achieve objectives in or through cyberspace. Cyberspace operations consist of three functions: offensive cyberspace operations, defensive cyberspace operations, and Department of Defense information network operations.

Electromagnetic Warfare (EW) is military action involving the use of electromagnetic and directed energy to control the electromagnetic spectrum or to attack the enemy. EW consists of three functions: electromagnetic attack, electromagnetic protection, and electromagnetic support.

Spectrum management operations (SMO) are the interrelated functions of spectrum management, frequency assignment, host-nation coordination, and policy that enable the planning, management, and execution of operations within the electromagnetic operational environment during all phases of military operations.

Department of Defense information network (DODIN) operations are operations to secure, configure, operate, extend, maintain, and sustain DOD cyberspace.

Cybersecurity incorporates actions taken to protect, monitor, analyze, detect, and respond to unauthorized activity on DOD information systems and computer networks.

SMARTbooks - DIME is our DOMAIN!

SMARTbooks: Reference Essentials for the Instruments of National Power (D-I-M-E: Diplomatic, Informational, Military, Economic)! Recognized as a "whole of government" doctrinal reference standard by military, national security and government professionals around the world, SMARTbooks comprise a comprehensive professional library.

SMARTbooks can be used as quick reference guides during actual operations, as study guides at education and professional development courses, and as lesson plans and checklists in support of training. Visit **www.TheLightningPress.com**!

(CYBER1-1) References

The following references were used in part to compile "CYBER1: The Cyberspace Operations and Electronic Warfare SMARTbook." All military references used to compile SMARTbooks are in the public domain and are available to the general public through official public websites and designated as approved for public release with unlimited distribution. The SMARTbooks do not contain ITAR-controlled technical data, classified, or other sensitive material restricted from public release. SMARTbooks are reference books that address general military principles, fundamentals and concepts rather than technical data or equipment operating procedures.

* SMARTupdate 1 to CYBER1 (Aug '21) updates the first printing of the CYBER1 SMARTbook (Oct '19) by incorporating new material from FM 3-12 (Aug '21), ATP 3-12.3 (Jul '19), ATP 6-02.70 (Oct '19), JP 3-85 (May '20) and adding a new section on Cyberspace IPB (ATP 2-01.3). An asterisk marks changed pages. (Learn more at www.thelightningpress.com/smartupdates/)

Joint Publications

JP 3-0	Oct 2018	Joint Operations (w/Change 1)
JP 3-12	Jun 2019	Cyberspace Operations
JP 3-13.1	Feb 2012	Electronic Warfare
JP 3-13	Nov 2014	Information Operations (with Change 1)
JP 3-85*	May 2020	Joint Electromagnetic Spectrum Management Operations

Field Manuals (FMs) and Training Circulars (TCs)

FM 3-0	Dec 2017	Operations (with Change 1)
FM 3-12*	Aug 2021	Cyberspace Operations and Electromagnetic Warfare
FM 6-0	Apr 2016	Commander and Staff Organization and Operations (w/change 2*)

Army Tactics, Techniques and Procedures (ATPs/ATTPs)

ATP 3-12.3*	Jul 2019	Electronic Warfare Techniques
ATP 6-02.70*	Oct 2019	Techniques for Spectrum Management Operations
ATP 6-02.71	Apr 2019	Techniques for Department of Defense Information Network Operations
ATP 2-01.3*	Jan 2020	Intelligence Preparation of the Battlefield (w/Change 1)

Other Publications

CSL (USAWC)	June 2017	Strategic Cyberspace Operations Guide
PAM 525-3-1	Dec 2018	The U.S. Army in Multi-Domain Operations 2028

* Denotes new/updated reference since first printing.

(CYBER1-1) Table of Contents

Introduction (Threat/COE/Info)

Introduction/Overview 0-1
    Cyberspace 0-1
    Cyberspace Operations (CO) 0-1
    Electromagnetic Warfare (EW) *(Note about change in terms)* 0-1
  I. The Global Cyber Threat 0-2
  II. Cyber Operations against the U.S. (2010-2015) 0-4
  III. Contemporary Operational Environment 0-6
    A. Critical Variables 0-6
    B. Today's Operational Environment 0-6
    C. Anticipated Operational Environments 0-7
    D. The Multi-Domain Extended Battlefield 0-8
        - Information Environment Operations (IEO) 0-9
  IV. The Information Environment 0-10
  V. Information as a Joint Function 0-10
        - Information Function Activities 0-12
        - Joint Force Capabilities, Operations, and Activities for
          Leveraging Information 0-14
  VI. Information Operations (IO) 0-11
  CYBER1-1 Overview (How This Book is Organized) 0-16

Joint Cyberspace Operations

I. Joint Cyberspace Operations 1-1
  I. Introduction 1-1
    A. The Impact of Cyberspace on Joint Operations 1-1
    B. Viewing Cyberspace Based on Location and Ownership 1-6
    C. DOD Cyberspace (DODIN) 1-6
    D. Connectivity and Access 1-7
  II. The Nature of Cyberspace 1-2
    Cyberspace Layer Model 1-2
        - Physical Network Layer 1-3
        - Logical Network Layer 1-3
        - Cyber-Personal Layer 1-3
    The Services' Cyberspace Doctrine 1-4

Table of Contents-1

III. The Operational Environment 1-7
　　　　　A. Key Terrain 1-8
　　　　　B. The Information Environment 1-8
　　　IV. Integrating Cyberspace Operations with Other Operations 1-9
　　　IV. Cyberspace Operations Forces
　　　　　A. United States Cyber Command (USCYBERCOM) 1-10
　　　　　B. Cyber Mission Force (CMF) 1-10
　　　　　　　- Cyber National Mission Force (CNMF) 1-1
　　　　　　　- Cyber Protection Force (CPF) 1-10
　　　　　　　- Cyber Combat Mission Force (CCMF) 1-10
　　　　　C. USCYBERCOM Subordinate Command Elements 1-10
　　　　　D. Other Cyberspace Forces and Staff 1-11
　　　VI. Challenges to the Joint Force's Use of Cyberspace 1-12
　　　　　A. Threats 1-12
　　　　　B. Anonymity and Difficulties with Attribution 1-13
　　　　　C. Geography Challenges 1-13
　　　　　D. Technology Challenges 1-13
　　　　　E. Private Industry and Public Infrastructure 1-14

II. Cyberspace Operations Core Activities 1-15
　　　Cyberspace Operations (CO) 1-15
　　　Cyberspace-Enabled Activities 1-15
　　　I. Cyberspace Missions 1-15
　　　　　　　- Military Operations In and Through Cyberspace 1-16
　　　　　　　- National Intelligence Operations In and Through Cyberspace 1-17
　　　　　　　- DOD Ordinary Business Operations In and Through Cyberspace 1-17
　　　　　A. DODIN Operations 1-18
　　　　　B. Offensive Cyberspace Operations (OCO) 1-18
　　　　　C. Defensive Cyberspace Operations (DCO) 1-19
　　　　　　　- Defensive Cyberspace Operations Internal Defensive Measures (DCO-IDM) ..1-19
　　　　　　　- Defensive Cyberspace Operations Response Action (DCO-RA) 1-19
　　　　　　　- Defense of Non-DOD Cyberspace 1-19
　　　II. Cyberspace Actions 1-20
　　　　　A. Cyberspace Security 1-20
　　　　　B. Cyberspace Defense 1-21
　　　　　C. Cyberspace Exploitation 1-21
　　　　　D. Cyberspace Attack 1-21
　　　III. Assignment of Cyberspace Forces to CO 1-23
　　　IV. The Joint Functions and Cyberspace Operations 1-24

III. Authorities, Roles, & Responsibilities 1-29
　　　I. Introduction 1-29
　　　II. Authorities 1-30
　　　　　　　- United States Code 1-31
　　　III. Roles and Responsibilities 1-30
　　　IV. Legal Considerations 1-38

IV. Planning, Coordination, Execution & Assessment 1-39
　　　I. Joint Planning Process (JPP) and Cyberspace Operations 1-39
　　　II. Cyberspace Operations Planning Considerations 1-39
　　　III. Intelligence and Operational Analytic Support 1-43
　　　IV. Targeting 1-46
　　　　　　　- Targeting In and Through Cyberspace 1-47
　　　V. Command and Control (C2) of Cyberspace Forces 1-48
　　　VI. Synchronization of Cyberspace Operations 1-52
　　　VII. Assessment of Cyberspace Operations 1-55

V. Interorganizational & Multinational 1-57

2-Table of Contents

Chap 2: Cyberspace Operations

I. Cyberspace and the Electromagnetic Spectrum 2-1*
- I. Cyberspace and the Electromagnetic Spectrum (EMS) 2-2*
  - Cyberspace Operations & Electromagnetic Warfare (EW) Logic Chart 2-3*
  - A. Operational Environment (OE) Overview 2-4*
    - Operational Initiative 2-4*
    - The Multi-Domain Extended Battlefield 2-4*
  - B. Cyberspace Domain 2-6*
    - Physical Network Layer 2-6*
    - Logical Network Layer 2-6*
    - Cyber-Persona Layer 2-6*
  - C. Operational & Mission Variables 2-8*
- III. Trends and Characteristics 2-10*
  - A. Congested Environments 2-10*
  - B. Contested Environments 2-10*
  - C. Threats 2-10*
  - D. Hazards 2-11*
- III. Core Competencies & Fundamentals 2-12*
  - A. Core Competencies 2-12*
  - B. Fundamental Principles 2-12*
- IV. Contributions to the Warfighting Functions 2-14*
- V. Conflict and Competition 2-16*
  - A. Competition Continuum 2-16*
  - B. Multi-Domain Extended Battlefield 2-16*
- C. Positions of Relative Advantage (in Cyberspace and the EMS) 2-16*

II. Cyberspace Operations 2-17*
- Electromagnetic Spectrum Superiority 2-17*
- I. Cyberspace Operations 2-17*
  - Cyberspace Operations (Missions & Actions) Overview 2-19*
  - A. Department of Defense Information Network Operations (DODIN) ... 2-18*
  - B. Defensive Cyberspace Operations (DCO) 2-18*
    - Defensive Cyberspace Operations Internal Defensive Measures (DCO-IDM) . 2-20*
    - Defensive Cyberspace Operations Response Action (DCO-RA) 2-20*
  - C. Offensive Cyberspace Operations (OCO) 2-20*
- II. Cyberspace Actions 2-21*
  - A. Cyberspace Security 2-21*
  - B. Cyberspace Defense 2-21*
  - C. Cyberspace Exploitation 2-21*
  - D. Cyberspace Attack 2-22*
- III. Interrelationship with Other Operations 2-23*
  - A. Intelligence Operations 2-23*
  - B. Space Operations 2-24*
  - C. Information Operations (IO) 2-25*

III. Army Organizations & Command and Control 2-27*
- I. United States Army Cyber Command 2-27*
- II. Army Information Warfare Operations Center 2-27*
- III. Cyberspace Electromagnetic Activities at Corps and Below 2-28*
  - A. Commander's Role 2-28*
  - B. Cyberspace Electromagnetic Activities (CEMA) Section 2-29*

Table of Contents-3 *

- Cyber Electromagnetic Warfare Officer (CEWO).................................. 2-30*
- Cyber Warfare Officer or Cyber-Operations Officer 2-31*
- Electromagnetic Warfare Technician (EWT).. 2-31*
- Electromagnetic Warfare Sergeant Major or NCOIC........................ 2-31*
- Electromagnetic Warfare Noncommissioned Officer (EW NCO)....... 2-31*
- Cyberspace Electromagnetic Activities Spectrum Manager 2-31*
  C. Cyberspace Electromagnetic Activities (CEMA) Working Group.......... 2-29*
IV. Staff and Support at Corps and Below 2-32*
  A. Assistant Chief of Staff, Intelligence.. 2-32*
  B. Assistant Chief of Staff, Signal.. 2-33*
  C. G-6 or S-6 Spectrum Manager 2-34*
  D. Information Operations Officer or Representative 2-34*
  E. Fires Support Element 2-34*
  F. Staff Judge Advocate.. 2-34*
V. Electromagnetic Warfare (EW) Organizations... 2-36*
  Electromagnetic Warfare (EW) Platoon 2-36*
  Intelligence, Information, Cyber, EW, & Space (I2CEWS) 2-36*

IV. Integration through the Operations Process....................... 2-37*
I. The Operations Process.. 2-37*
  A. Planning... 2-38*
  B. Preparation 2-38*
  C. Execution.. 2-39*
  D. Assessment 2-39*
II. Integrating Processes 2-40*
  A. Intelligence Preparation of the Battlefield (IPB) 2-40*
  B. Information Collection... 2-40*
  C. Targeting... 2-41*
  D. Risk Management... 2-41*
  E. Knowledge Management 2-41*
III. Risks In Cyberspace and the EMS 2-42*
  A. Operational Risks.. 2-42*
  B. Technical Risks 2-42*
  C. Policy Risks 2-43*
  D. Operations Security Risks 2-43*

Chap 3: Electromagnetic Warfare (EW)

I. Electromagnetic Warfare (EW) 3-1*
I. Electromagnetic Warfare (EW) *(Note about change in terms)* 3-1*
  A. Electromagnetic Attack (EA) 3-2*
    - Offensive EA.. 3-2*
    - Defensive EA... 3-2*
    - Electromagnetic Attack (EA) Effects 3-2*
    - Electromagnetic Attack (EA) Tasks 3-3*
  B. Electromagnetic Protection (EP).. 3-6*
    - Electromagnetic Protection Tasks 3-6*
  C. Electromagnetic Support (ES) 3-8*
    - Electromagnetic Support Tasks 3-8*
    - Electromagnetic Support (ES) Actions... 3-9*
    * Electromagnetic Warfare Reprogramming 3-8*

III. Spectrum Management 3-10*
　　　　- Electromagnetic Interference (EMI) 3-10*
　　　　- Frequency Interference Resolution 3-10*
　　　　- Spectrum Management Operations (SMO)........................ 3-10*
　　　　- Electromagnetic Warfare Coordination 3-10*

II. EW Key Personnel... 3-11*
　　I. Electronic Warfare Personnel 3-11*
　　II. Theater Army, Corps, Division and Brigade............................ 3-11*
　　　　A. Cyber Electronic Warfare Officer (CEWO)......................... 3-12*
　　　　B. Electronic Warfare Technician.. 3-12*
　　　　C. Electronic Warfare Noncommissioned Officer 3-13*
　　　　D. Spectrum Manager 3-13*
　　　　E. Battalion Electronic Warfare Personnel 3-13*
　　　　F. Company CREW Specialists 3-16*
　　　　G. Electronic Warfare Control Authority 3-16*
　　III. Staff Members and Electronic Warfare 3-14*

III. EW Preparation & Execution 3-17*
　　I. Electronic Warfare Preparation.. 3-17*
　　II. Integration of Electronic Warfare and Signals Intelligence 3-18*
　　　　Deconflicting the Electromagnetic Spectrum............................ 3-19*
　　　　A. Distinctions Between Electronic Warfare and Signals Intelligence 3-18*
　　　　B. Sensing Activity Distinctions 3-18*
　　III. Electronic Warfare Execution 3-20*

IV(a). Electronic Attack Techniques 3-21*
　　I. Planning Electronic Attack 3-21*
　　　　A. Electronic Attack Effects 3-21*
　　　　B. Electronic Attack (EA) Considerations............................... 3-22*
　　II. Preparing Electronic Attack... 3-24*
　　　　- Electronic Attack Requests (EARFs)................................... 3-25*
　　　　- Electronic Attack Considerations... 3-24*
　　III. Executing Electronic Attack 3-24*
　　　　Close Air Support (CAS)... 3-24*
　　　　A. Airborne Electronic Attack 3-26*
　　　　B. Defensive Electronic Attack 3-28*
　　　　　　- Counter Radio-Controlled Improvised Device (CREW)..... 3-28*
　　IV. Electronic Attack Techniques in Large Scale Combat Operations 3-28*

IV(b). Electronic Protection Techniques................................. 3-29*
　　I. Planning Electronic Protection.. 3-29*
　　　　Electronic Protection Considerations 3-30*
　　II. Electromagnetic Interference.. 3-31*
　　　　A. Recognizing Electromagnetic Jamming.............................. 3-31*
　　　　B. Remedial Electronic Protection Techniques....................... 3-32*
　　　　C. Concealment.. 3-32*
　　　　D. Threat Electronic Attack on Friendly Command Nodes....... 3-32*
　　　　E. Electromagnetic Interference (EMI) Battle Drill.................. 3-33*
　　III. Staff Electronic Protection Responsibilities 3-34*
　　IV. Equipment and Communications Enhancements................... 3-34*

IV(c). Electronic Warfare Support Techniques........................ 3-35*
　　I. Planning Electronic Warfare Support.. 3-35*
　　　　A. Electronic Reconnaissance... 3-35*
　　　　B. Electronic Warfare Support Considerations....................... 3-35*
　　II. Preparing Electronic Warfare Support..................................... 3-35*
　　　　A. Electromagnetic Environment (EME) Survey..................... 3-36*
　　　　B. Direction Finding (DF).. 3-36*

Chap 4: Cyberspace & EW (CEMA) Planning

IPB Cyberspace Considerations 4-a*
   Intelligence Preparation of the Battlefield (IPB) 4-a*
   Step 1 — Define the Operational Environment 4-b*
      A. Step 1 Cyberspace Considerations 4-b*
      B. Cyber-Centric Activities and Outputs for Step 1 .... 4-d*
   Step 2 — Describe Environmental Effects on Operations 4-e*
      A. Step 2 Cyberspace Considerations 4-e*
      B. Cyber-Centric Activities & Outputs for Step 2 .... 4-f*
         Threat Overlay 4-f*
         Threat Description Table 4-f*
         Modified Combined Obstacle Overlay 4-g*
         Terrain Effects Matrix 4-h*
         Weather, Light, and Illumination Charts or Tables 4-h*
         Civil Considerations Data Files, Overlays, and Assessments................ 4-h*
   Step 3 — Evaluate the Threat 4-i*
      A. Step 3 Cyberspace Considerations 4-i*
      B. Cyber-Centric Activities and Outputs for Step 3 .... 4-j*
         Threat Characteristics 4-k*
         Threat Model 4-l*
           - Cyber Kill Chain 4-l*
           - Threat Tactics, Options, and Peculiarities 4-m*
           - High-Value Targets 4-m*
         Threat Capabilities 4-k*
   Step 4 — Determine Threat Courses of Action 4-n*
      A. Step 4 Cyberspace Considerations 4-n*
      B. Cyber-Centric Activities and Outputs for Step 4 .... 4-n*
         Threat Situation Template 4-o*
         Event Template 4-p*
         Event Matrix 4-q*

I(a). Cyberspace (CEMA) Operations Planning 4-1
   I. Army Design Methodology 4-2
   II. The Military Decision-Making Process (MDMP) 4-2
      - Step 1: Receipt of Mission 4-2
      - Step 2: Mission Analysis 4-3
      - Step 3: Course of Action Development 4-4
      - Step 4: Course of Action Analysis 4-5
      - Step 5: Course of Action Comparison 4-6
      - Step 6: Course of Action Approval 4-7
      - Step 7: Orders Production, Dissemination, and Transition 4-8

I(b). Cyber Effects Request Format (CERF) 4-9
   I. Requesting Cyberspace Effects 4-9
      - Cyber Effects Request Format (CERF) 4-11
   II. Cyber Effects Request Format Preparation 4-12

II(a). Electronic Warfare Planning 4-15*
   I. Electronic Warfare Contributions to the Military Decision-Making Process 4-15*
   II. Electronic Warfare Planning Considerations 4-15*
      A. Planning Factors 4-15*
      Electronic Warfare Running Estimate 4-16*

* **6-Table of Contents**

    III. Staff Contributions to EW Planning 4-19*
        EW Contributions to the Staff 4-20*
        Joint Restricted Frequency List (JRFL) 4-22*
    III. Electronic Warfare Configurations 4-23*
    V. EW Employment Considerations 4-24*
    VI. Electronic Warfare Assessment 4-26*

II(b). Electromagnetic Attack Request 4-27*
    I. Electromagnetic Attack Request 4-27*
    II. Airborne Electromagnetic Attack Support 4-28*

III. Targeting (D3A) 4-29*
    Targeting Methodology 4-30*
    Targeting Crosswalk 4-31*
    I. Decide 4-32*
    II. Detect 4-33*
    III Deliver 4-33*
    IV. Assess 4-33*
    Considerations When Targeting 4-34*

IV. Cyberspace (CEMA) in Operations Orders 4-35*
    - ANNEX C–OPERATIONS (G-5 OR G-3 [S-3]) 4-35*
    - ANNEX H–SIGNAL (G-6 [S-6]) 4-35*
    - Appendix 12 (Cyberspace Electromagnetic Activities) to Annex C 4-35*
    (Operations) to Operations Plans and Orders
    - Appendix 12 to Annex C (Sample Format) 4-36*

V. Cyberspace Integration into Joint Planning (JPP) 4-41
    I. Cyberspace Planning Integration 4-42
    II. Cyberspace Planning and the JPP 4-42
        A. Initiation 4-42
        B. Mission Analysis 4-42
        C. Course of Action (COA) Development 4-43
        D. COA Analysis, Comparison, and Approval 4-43
        E. Plan or Order Development 4-43
    IV. Cyberspace-Related Intelligence Requirements (IRs) 4-44
    V. Information Operations (IO) 4-44
    VI. Planning Insights 4-44

VI. Integrating / Coordinating Functions of IO 4-45
    I. Information Operations and the Information-Influence 4-45
    Relational Framework
    II. The Information Operations Staff and Information Operations Cell 4-46
    III. Relationships and Integration 4-46
        - Commander's Communication Synchronization (CCS) 4-46
        A. Strategic Communication (SC) 4-46
        B. Joint Interagency Coordination Group (JIACG) 4-47
        C. Public Affairs (PA) 4-48
        D. Civil-Military Operations (CMO) 4-48
        E. Cyberspace Operations 4-48
        F. Information Assurance (IA) 4-49
        G. Space Operations 4-49
        H. Military Information Support Operations (MISO) 4-49
        I. Intelligence 4-49
        J. Military Deception (MILDEC) 4-49
        K. Operations Security (OPSEC) 4-50
        L. Special Technical Operations (STO) 4-50
        M. Joint Electromagnetic Spectrum Operations (JEMSO) 4-50
        N. Key Leader Engagement (KLE) 4-50

Chap 5: Spectrum Management Operations (SMO/JEMSO)

I. Spectrum Management Operations (SMO/JEMSO) 5-1*
    I. Electromagnetic Spectrum Operations (EMSO) 5-1*
        Electromagnetic Operational Environment (EMOE) 5-2*
            - Electromagnetic Spectrum (EMS) 5-2*
            - Electromagnetic Environment (EME) 5-2*
    II. Spectrum Management Operations (SMO) 5-4*
        A. Objective of Spectrum Management Operations 5-5*
        B. Spectrum Management Operations Core Functions 5-5*
    III. Joint Electromagnetic Spectrum Operations (JEMSO) 5-5*
        A. JEMSO Actions 5-6*
            - Exploitation 5-6*
            - Electronic Attack (EA) 5-6*
            - Protect 5-6*
            - Manage 5-7*
        B. Electromagnetic Environmental Effects (E3) 5-8*
            - HERP 5-8*
            - HERO 5-8*
            - HERF 5-8*
            - Electromagnetic Pulse (EMP) 5-8*
            - High-Altitude Electromagnetic Pulse (HEMP) 5-8*

II. Spectrum Management 5-9*
    Frequency Interference Resolution 5-9*
    I. Key SMO inputs to the MDMP 5-10*
    II. SMO Support to the Warfighting Functions 5-12*
    II. The Common Operational Picture (COP) 5-14*
        - Live Spectrum Analysis 5-14*
        - Movement of Forces to a New Location 5-14*

III. Planning Joint EMS Operations (JEMSO) 5-15*
    Planning Process 5-15*
    JEMSMO Cell Actions and Outputs as Part of Joint Planning 5-17*
    Information (Planning Considerations) 5-18*
    I. Electromagnetic Order of Battle (EOB) 5-15*
    II. EMOE Estimate 5-16*
    III. JEMSO Staff Estimate 5-20*
        - EMS Superiority Approach 5-20*
        - Determine Friendly EMS-Use Requirements 5-20*
    IV. JEMSO Appendix to Annex C 5-20*

* **8-Table of Contents**

Chap 6: Dept of Defense Info Network (DODIN) Ops

I. Department of Defense Information Network (DODIN) 6-1
- I. Department of Defense Information Network (DODIN) Operations 6-1
- II. Department of Defense Information Network Operations 6-2
  in Army Networks (DODIN-A)
- III. DODIN Critical Tasks 6-4
- IV. DODIN across the Operational Phases 6-5
  - Phase 0—Shape 6-6
  - Phase I—Deter 6-7
  - Phase II—Seize Initiative 6-8
  - Phase III—Dominate 6-8
  - Phase IV—Stabilize 6-8
  - Phase V—Enable Civil Authority 6-8

II. DODIN Roles & Responsibilities 6-9
- I. Global Level 6-9
  - A. United States Cyber Command (USCYBERCOM) 6-9
  - B. Defense Information Systems Agency (DISA) 6-9
  - C. Chief Information Officer/G-6 6-10
  - D. United States Army Cyber Command (ARCYBER) 6-10
    - 1st Information Operations Command 6-10
    - U.S. Army Network Enterprise Technology Command (NETCOM) 6-10
  - E. Global Department of Defense Information Network Operations 6-12
    - Army Cyber Operations and Integration Center (ACOIC) 6-12
    - Army Enterprise Service Desk 6-12
    - Functional Network Operations and Security Centers (NOSC) 6-13
- II. Theater Level 6-13
  - A. Geographic Combatant Commander (GCC) 6-14
    - Combatant Command J-6 6-14
    - Joint Cyberspace Center (JCC) 6-14
    - Theater Network Operations Control Center (TNCC) 6-14
  - B. Enterprise Operations Center 6-17
  - C. Joint Task Force 6-18
  - D. Theater Army G-6 6-18
  - E. Signal Command (Theater) SC(T) 6-18
  - F. Theater Tactical Signal Brigade 6-19
  - G. Strategic Signal Brigade 6-19
  - H. Theater DODIN Operations Organizations 6-20
  - I. Installation-Level DODIN Operations Infrastructure 6-23
- III. Corps and Below Units 6-24
  - A. Corps and Division 6-24
  - B. Brigade Combat Team and Multifunctional Support Brigade 6-30

III. DODIN Network Operations Components 6-35
- I. DODIN Operations Operational Construct 6-36
- II. DODIN Enterprise Management 6-36
  - A. Functional Services 6-36
  - B. Critical Capabilities 6-38
  - C. Enabled Effects 6-38
  - D. Objective 6-38
- III. Enterprise Management Activities 6-40

Table of Contents-9

Chap 7: Cybersecurity

I. Cybersecurity Fundamentals 7-1
    I. Cybersecurity Fundamental Attributes 7-2
    II. Cybersecurity Risk Management 7-2
        - Risk Management Framework 7-6
    III. Cybersecurity Principles 7-3
    IV. Enabled Effects 7-5
    V. Operational Resilience 7-8
    VI. Cybersecurity Integration and Interoperability 7-8
    VII. Cyberspace Defense 7-10
    VIII. Cybersecurity Performance 7-12

II. Cybersecurity Functions 7-13
    Cybersecurity Functional Services 7-15
    I. Identify 7-13
        A. Identify Mission-Critical Assets 7-13
        B. Identify Laws, Regulations, and Policies 7-14
        C. Identify Threat Activities 7-14
            - Tools of Cyber Attacks 7-18
        D. Identify Vulnerabilities 7-14
    II. Protect Function 7-21
    III. Detect Function 7-23
    IV. Respond Function 7-24
    V. Recover Function 7-24

III. Protection, Detection, & Reaction 7-25
    I. Protection 7-25
        - Cyber Attacks 7-26
        - Information Systems Security 7-26
        - Protection Levels 7-27
        - Mitigating Insider Threats 7-27
    II. Detection 7-29
    III. Reaction 7-29
    Information Assurance Vulnerability Management (IAVM) 7-30
    Scanning and Remediation 7-30
    Continuity of Operations 7-30

Chap 8: Acronyms & Glossary

I. Acronyms and Abbreviations 8-1
II. Glossary 8-3

Introduction (Threat/COE/Info)

United States armed forces operate in an increasingly network-based world. The proliferation of information technologies is changing the way humans interact with each other and their environment, including interactions during military operations. This broad and rapidly changing operational environment requires that today's armed forces must operate in cyberspace and leverage an electromagnetic spectrum that is increasingly competitive, congested, and contested.

Cyberspace

Cyberspace reaches across geographic and geopolitical boundaries and is integrated with the operation of critical infrastructures, as well as the conduct of commerce, governance, and national defense activities. Access to the Internet and other areas of cyberspace provides users operational reach and the opportunity to compromise the integrity of critical infrastructures in direct and indirect ways without a physical presence. The prosperity and security of our nation are significantly enhanced by our use of cyberspace, yet these same developments have led to increased exposure of vulnerabilities and a critical dependence on cyberspace, for the US in general and the joint force in particular.

See pp. 1-1 to 1-6 and 2-1 to 2-16

Cyberspace Operations (CO)

Cyberspace Operations (CO) are the employment of cyberspace capabilities where the primary purpose is to achieve objectives in or through cyberspace. CO comprise the military, national intelligence, and ordinary business operations of DOD in and through cyberspace. Although commanders need awareness of the potential impact of the other types of DOD CO on their operations, the military component of CO is the only one guided by joint doctrine and is the focus of this publication. CCDRs and Services use CO to create effects in and through cyberspace in support of military objectives. Military operations in cyberspace are organized into missions executed through a combination of specific actions that contribute to achieving a commander's objective.

See pp. 1-15 and 2-17.

Electromagnetic Warfare (EW)*

Electromagnetic Warfare (EW) is military action involving the use of electromagnetic and directed energy to control the electromagnetic spectrum or to attack the enemy. EW consists of three functions: electromagnetic attack, electromagnetic protection, and electromagnetic support.

> *Editor's Note: In keeping with doctrinal terminology changes in JP 3-85, Joint Electromagnetic Spectrum Operations (May '20) and FM 3-12, Cyberspace Operations and Electromagnetic Warfare (Aug '21), the term "electronic warfare (EW)" has been updated to "electromagnetic warfare (EW)". Likewise, the EW divisions have been updated as "electromagnetic attack (EA), electromagnetic protection (EP), and electromagnetic support (ES)." For purposes of the CYBER1 SMARTbook, EW/EA/EP/ES acronyms and terms will remain the same as presented in the original cited and dated source -- for example, ATP 3-12.3, Electronic Warfare Techniques (Jul '19). Readers should anticipate that as those specific references are updated/revised, so will the terms.*

I. The Global Cyber Threat

Ref: Daniel R. Coats, Director Of National Intelligence, Statement for the Record, Worldwide Threat Assessment of the Us Intelligence Community (Jan 29, 2019).

> Our adversaries and strategic competitors will increasingly use cyber capabilities—including cyber espionage, attack, and influence—to seek political, economic, and military advantage over the United States and its allies and partners. China, Russia, Iran, and North Korea increasingly use cyber operations to threaten both minds and machines in an expanding number of ways—to steal information, to influence our citizens, or to disrupt critical infrastructure.
>
> At present, China and Russia pose the greatest espionage and cyber attack threats, but we anticipate that all our adversaries and strategic competitors will increasingly build and integrate cyber espionage, attack, and influence capabilities into their efforts to influence US policies and advance their own national security interests. In the last decade, our adversaries and strategic competitors have developed and experimented with a growing capability to shape and alter the information and systems on which we rely. For years, they have conducted cyber espionage to collect intelligence and targeted our critical infrastructure to hold it at risk. They are now becoming more adept at using social media to alter how we think, behave, and decide. As we connect and integrate billions of new digital devices into our lives and business processes, adversaries and strategic competitors almost certainly will gain greater insight into and access to our protected information.

China

China presents a persistent cyber espionage threat and a growing attack threat to our core military and critical infrastructure systems. China remains the most active strategic competitor responsible for cyber espionage against the US Government, corporations, and allies. It is improving its cyber attack capabilities and altering information online, shaping Chinese views and potentially the views of US citizens—an issue we discuss in greater detail in the Online Influence Operations and Election Interference section of this report.

- Beijing will authorize cyber espionage against key US technology sectors when doing so addresses a significant national security or economic goal not achievable through other means. We are also concerned about the potential for Chinese intelligence and security services to use Chinese information technology firms as routine and systemic espionage platforms against the United States and allies.
- China has the ability to launch cyber attacks that cause localized, temporary disruptive effects on critical infrastructure—such as disruption of a natural gas pipeline for days to weeks—in the United States.

Russia

We assess that Russia poses a cyber espionage, influence, and attack threat to the United States and our allies. Moscow continues to be a highly capable and effective adversary, integrating cyber espionage, attack, and influence operations to achieve its political and military objectives. Moscow is now staging cyber attack assets to allow it to disrupt or damage US civilian and military infrastructure during a crisis and poses a significant cyber influence threat—an issue discussed in the Online Influence Operations and Election Interference section of this report.

- Russian intelligence and security services will continue targeting US information systems, as well as the networks of our NATO and Five Eyes partners, for technical information, military plans, and insight into our governments' policies.
- Russia has the ability to execute cyber attacks in the United States that generate localized, temporary disruptive effects on critical infrastructure—such as disrupting an electrical distribution network for at least a few hours—similar to those demonstrated in Ukraine in 2015 and 2016. Moscow is mapping our critical infrastructure with the long-term goal of being able to cause substantial damage.

Iran

Iran continues to present a cyber espionage and attack threat. Iran uses increasingly sophisticated cyber techniques to conduct espionage; it is also attempting to deploy cyber attack capabilities that would enable attacks against critical infrastructure in the United States and allied countries. Tehran also uses social media platforms to target US and allied audiences, an issue discussed in the Online Influence Operations and Election Interference section of this report.

- Iranian cyber actors are targeting US Government officials, government organizations, and companies to gain intelligence and position themselves for future cyber operations.
- Iran has been preparing for cyber attacks against the United States and our allies. It is capable of causing localized, temporary disruptive effects—such as disrupting a large company's corporate networks for days to weeks—similar to its data deletion attacks against dozens of Saudi governmental and private-sector networks in late 2016 and early 2017.

North Korea

North Korea poses a significant cyber threat to financial institutions, remains a cyber espionage threat, and retains the ability to conduct disruptive cyber attacks. North Korea continues to use cyber capabilities to steal from financial institutions to generate revenue. Pyongyang's cybercrime operations include attempts to steal more than $1.1 billion from financial institutions across the world—including a successful cyber heist of an estimated $81 million from the New York Federal Reserve account of Bangladesh's central bank.

Nonstate and Unattributed Actors

Foreign cyber criminals will continue to conduct for-profit, cyber-enabled theft and extortion against US networks. We anticipate that financially motivated cyber criminals very likely will expand their targets in the United States in the next few years. Their actions could increasingly disrupt US critical infrastructure in the health care, financial, government, and emergency service sectors, based on the patterns of activities against these sectors in the last few years.

Terrorists could obtain and disclose compromising or personally identifiable information through cyber operations, and they may use such disclosures to coerce, extort, or to inspire and enable physical attacks against their victims. Terrorist groups could cause some disruptive effects—defacing websites or executing denial-of-service attacks against poorly protected networks—with little to no warning.

The growing availability and use of publicly and commercially available cyber tools is increasing the overall volume of unattributed cyber activity around the world. The use of these tools increases the risk of misattributions and misdirected responses by both governments and the private sector.

See pp. 7-16 to 7-17 for discussion of cyber threat activities and pp. 7-18 to 7-19 for cyber attack tools. See also p. 2-24 for discussion of threats in cyberspace from FM 3-12.

II. Cyber Operations against the U.S. (2010-2015)

Ref: U.S. Army War College Strategic Cyberspace Operations Guide (Jun '16), pp. 8-13.

Although there have been hundreds, if not thousands, of cyber operations against the U.S. over the past five years, the following list includes those operations acknowledged by the U.S. Government:

2010
- **Insider** – Army PFC Manning was found not guilty of the most serious charge of knowingly aiding the enemy, but was convicted on 20 other specifications related to the misappropriation of hundreds of thousands of intelligence documents sent to WikiLeaks. Prosecutors alleged that Manning downloaded some 470,000 SIGACTS (from Iraq and Afghanistan) from the SIPRNET.

2011
- **Iran** – DDOS attacks on the U.S. financial sector. A group sponsored by Iran's Islamic Revolutionary Guard Corps – for conducting a coordinated campaign of distributed denial of service (DDoS) attacks against 46 major companies, primarily in the U.S. financial sector (2011-2013). These attacks, which occurred on more than 176 days, disabled victim bank websites, prevented customers from accessing their accounts online, and collectively cost the banks tens of millions of dollars in remediation costs as they worked to neutralize and mitigate the attacks on their servers.
- **Syria** – Two Syrian hackers charged with targeting Internet sites—in the U.S. and abroad—on behalf of the Syrian Electronic Army (SEA), a group of hackers that supports the regime of Syrian President Bashar al-Assad. The affected sites—which included computer systems in the Executive Office of the President in 2011 and a U.S. Marine Corps recruitment website in 2013. They collected usernames and passwords that gave them the ability to deface websites, redirect domains to sites controlled by the conspirators, steal e-mail, and hijack social media accounts. To obtain the login information they used a technique called "spear-phishing."

2012
- **China** – A Chinese national pleaded guilty to participating in a years-long conspiracy to hack into the computer networks of major U.S. defense contractors to steal military technical data (C-17 strategic transport aircraft and certain fighter jets) and send the stolen data to China.

2013
- **Iran** – An Iranian hacker obtained unauthorized access into the Supervisory Control and Data Acquisition (SCADA) systems of the Bowman Dam, located in Rye, NY. This allowed him to repeatedly obtain information regarding the status and operation of the dam, including information about the water levels and temperature, and the status of the sluice gate, which is responsible for controlling water levels and flow rates.
- **China** – Members of PRC's Third Department of the General Staff Department of the People's Liberation Army (3PLA), Second Bureau, Third Office, Military Unit Cover Designator (MUCD) 61398 charged with conspiracy to penetrate the computer networks of six American companies while those companies were engaged in negotiations or joint ventures or were pursuing legal action with, or against, state-owned enterprises in China. They then used their illegal access to allegedly steal proprietary information including, for instance, e-mail exchanges among company employees and trade secrets related to technical specifications for nuclear plant designs.

- **Insider**—Edward J. Snowden, was charged with violations of: Unauthorized Disclosure of National Defense Information; Unauthorized Disclosure of Classified Communication; and Theft of Government Property.
- **Unattributed** – Hackers penetrated U.S. Army Corps of Engineers (USACE) database about the nation's 85,000 dams. That data included their location, condition and potential for fatalities if the dams were to be breached.

2014
- **Iran** – Computer security experts reported that members of an Iranian organization were responsible for computer operations targeting U.S. military, transportation, public utility, and other critical infrastructure networks.
- **North Korea** – Conducted a cyber attack on Sony Pictures Entertainment, which stole corporate information and introduced hard drive erasing malware into the company's network infrastructure, according to the FBI.
- **China** – The U.S. company, Community Health Systems, informed the Securities and Exchange Commission that it believed hackers "originating from China" had stolen personally identifiable information on 4.5 million individuals.
- **Unattributed** – JP Morgan Chase suffered a hacking intrusion.
- **Syria** – A member of the SEA is suspected of being responsible for a series of cyber extortion schemes targeting a variety of American and international companies.
- **Unattributed** – A data breach at Home Depot exposed information from 56 million credit/debit cards and 53 million customer email addresses.
- **Iran** – Iranian actors have been implicated in the February 2014 cyber attack on the Las Vegas Sands casino company.

2015
- **Unattributed** – In June 2015, a Pentagon spokesman acknowledged that an element of the army.mil service provider's content was compromised. After this came to their attention, the Army took appropriate preventive measures to ensure there was no breach of Army data by taking down the website temporarily. Later, the Syrian Electronic Army (SEA) claimed responsibility for defacing the army.mil website.
- **Unattributed** – the Office of Personnel Management (OPM) discovered that a number of its systems were compromised. These systems included those that contain information related to the background investigations of current, former, and prospective federal government employees, as well as other individuals for whom a federal background investigation was conducted. OPM announced the compromise resulted in 21.5 million personal records being stolen. The Chinese government announced that it arrested a handful of hackers it says were connected to the breach of Office of Personnel Management's database.
- **Russia** – Cyber actors are developing means to remotely access industrial control systems (ICS) used to manage critical infrastructures. Unknown Russian actors successfully compromised the product supply chains of at least three ICS vendors so that customers downloaded malicious software ("malware") designed to facilitate exploitation directly from the vendors' websites along with legitimate software updates.
- **Insider** – A former U.S. Nuclear Regulatory Commission employee pleads guilty to attempted spear-phishing cyber-attack on Department of Energy computers to compromise, exploit and damage U.S. government computer systems that contained sensitive nuclear weapon-related information with the intent of allowing foreign nations to gain access to that information or to damage essential systems.
- **Unattributed** – A "group of hackers" was responsible for an intrusion into an unclassified network maintained by the Joint Staff.

III. Contemporary Operational Environment (COE)

The DOD officially defines an operational environment (OE) as "a composite of the conditions, circumstances, and influences that affect the employment of military forces and bear on the decisions of the unit commander" (JP 1-02). The contemporary operational environment (COE) is the operational environment that exists today and for the clearly foreseeable future.

A. Critical Variables

Any OE, in the real world or in the training environment, can be defined in terms of eleven critical variables. While these variables can be useful in describing the overall (strategic) environment, they are most useful in defining the nature of specific OEs. Each of these "conditions, circumstances, and influences" and their possible combinations will vary according to the specific situation. In this sense, they are "variables." These variables are interrelated and sometimes overlap. Different variables will be more or less important in different situations. Each OE is different, because the content of the variables is different. Only by studying and understanding these variables -- and incorporating them into its training -- will the U.S. Army be able to keep adversaries from using them against it or to find ways to use them to its own advantage.

Critical variables of COE include:
- Nature and Stability of the State
- Regional and Global Relationships
- Economics
- Sociological Demographics
- Information
- Physical Environment
- Technology
- External Organizations
- National Will
- Time
- Military Capabilities

B. Today's Operational Environment

Today's operational environment presents threats to the Army and joint force that are significantly more dangerous in terms of capability and magnitude than those we faced in Iraq and Afghanistan. Major regional powers like Russia, China, Iran, and North Korea are actively seeking to gain strategic positional advantage.

The proliferation of advanced technologies; adversary emphasis on force training, modernization, and professionalization; the rise of revisionist, revanchist, and extremist ideologies; and the ever increasing speed of human interaction makes large-scale ground combat more lethal, and more likely, than it has been in a generation. As the Army and the joint force focused on counter-insurgency and counter-terrorism at the expense of other capabilities, our adversaries watched, learned, adapted, modernized and devised strategies that put us at a position of relative disadvantage in places where we may be required to fight.

The Army and joint force must adapt and prepare for large-scale combat operations in highly contested, lethal environments where enemies employ potent long range fires and other capabilities that rival or surpass our own. The risk of inaction is great; the less prepared we are to meet these challenges, the greater the likelihood for conflict with those who seek windows of opportunity to exploit. The reduction of friendly, forward-stationed forces, significant reductions in capability and capacity across the entire joint force, and the pace of modernization make it imperative that we do everything possible to prepare for worst-case scenarios.

C. Anticipated Operational Environments

Ref: FM 3-0, Operations (Oct '17), pp. 1-4 to 1-6.

> "No matter how clearly one thinks, it is impossible to anticipate precisely the character of future conflict. The key is to not be so far off the mark that it becomes impossible to adjust once that character is revealed."
>
> - Sir Michael Howard

Factors that affect operations extend far beyond the boundaries of a commander's assigned AO. As such, commanders, supported by their staffs, seek to develop and maintain an understanding of their OE. An operational environment is a composite of the conditions, circumstances, and influences that affect the employment of capabilities and bear on the decisions of the commander (JP 3-0). An OE encompasses physical areas of the air, land, maritime, space, and **cyberspace domains; as well as the information environment (which includes cyberspace); the electromagnetic spectrum (EMS)**, and other factors. Included within these are adversary, enemy, friendly, and neutral actors that are relevant to a specific operation.

Commanders and staffs analyze an OE using the eight operational variables: political, military, economic, social, information, infrastructure, physical environment, and time (known as PMESII-PT).

How the many entities behave and interact with each other within an OE is difficult to discern and always results in differing circumstances. No two OEs are the same. In addition, an OE is not static; it continually evolves. This evolution results from opposing forces and actors interacting and their abilities to learn and adapt. The complex and dynamic nature of an OE makes determining the relationship between cause and effect difficult and contributes to the uncertain nature of military operations.

Trends

Several trends will continue to affect future OEs. The competition for resources, water access, declining birthrates in traditionally allied nations, and disenfranchised groups in many nations contribute to the likelihood of future conflict. Populations will continue to migrate across borders and to urban areas in search of the employment and services urban areas offer. Adversarial use of ubiquitous media platforms to disperse misinformation, propaganda, and malign narratives enables adversaries to shape an OE to their advantage and foment dissention, unrest, violence, or at the very least, uncertainty.

Proliferating technologies will continue to present challenges for the joint force. Unmanned systems are becoming more capable and common. Relatively inexpensive and pervasive anti-tank guided missiles and advanced rocket propelled grenades can defeat modern armored vehicles. Sensors and sensing technology are becoming commonplace. Adversaries have long-range precision strike capabilities that outrange and outnumber U.S. systems. Advanced integrated air-defense systems can neutralize friendly air power, or they can make air operations too costly to conduct. Anti-ship missiles working in concert with an IADS can disrupt access to the coastlines and ports necessary for Army forces to enter an AO. Adversary cyberspace and space control capabilities can disrupt friendly information systems and degrade C2 across the joint force. Use of WMD and the constant pursuit of the materials, expertise, and technology to employ WMD will increase in the future. Both state and non-state actors continue to develop WMD programs to gain advantage against the United States and its allies. These trends mean that adversaries can contest U.S. dominance in the air, land, maritime, space, and **cyberspace domains**.

See following pages (pp. 0-8 to 0-9) for an overview and discussion of the multi-domain extended battlefield.

D. The Multi-Domain Extended Battlefield

Ref: FM 3-0, Operations (Oct '17), pp. 1-6 to 1-8.

The interrelationship of the air, land, maritime, space, and the information environment (including cyberspace) requires a cross-domain understanding of an OE. Commanders and staffs must understand friendly and enemy capabilities that reside in each domain. From this understanding, commanders can better identify windows of opportunity during operations to converge capabilities for best effect. Since many friendly capabilities are not organic to Army forces, commanders and staffs plan, coordinate for, and integrate joint and other unified action partner capabilities in a multi-domain approach to operations.

A **multi-domain approach** to operations is not new. Army forces have effectively integrated capabilities and synchronized actions in the air, land, and maritime domains for decades. Rapid and continued advances in technology and the military application of new technologies to the space domain, the EMS, and the information environment (particularly cyberspace) require special consideration in planning and converging effects from across all domains.

Refer to TRADOC PAM 525-3-1, The U.S. Army in Multi-Domain Operations (Dec '18), for further information.

Space Domain

The space domain is the space environment, space assets, and terrestrial resources required to access and operate in, to, or through the space environment (FM 3-14). Space is a physical domain like land, sea, and air within which military activities are conducted. Proliferation of advanced space technology provides more widespread access to space-enabled technologies than in the past. Adversaries have developed their own systems, while commercially available systems allow almost universal access to some level of space enabled capability with military applications. Army forces must be prepared to operate in a denied, degraded and disrupted space operational environment (D3SOE).

Refer to FM 3-14 for doctrine on Army space operations.

Information Environment

The information environment is the aggregate of individuals, organizations, and systems that collect, process, disseminate, or act on information (JP 3-13). The information environment is not separate or distinct from the OE but is inextricably part of it. Any activity that occurs in the information environment simultaneously occurs in and affects one or more of the physical domains. Most threat forces recognize the importance of the information environment and emphasize information warfare as part of their strategic and operational methods.

The information environment is comprised of three dimensions: physical, informational, and cognitive. The physical dimension includes the connective infrastructure that supports the transmission, reception, and storage of information.

Across the globe, information is increasingly available in near-real time. The ability to access this information, from anywhere, at any time, broadens and accelerates human interaction across multiple levels, including person to person, person to organization, person to government, and government to government. Social media, in particular, enables the swift mobilization of people and resources around ideas and causes, even before they are fully understood. Disinformation and propaganda create malign narratives that can propagate quickly and instill an array of emotions and behaviors from anarchy to focused violence. From a military standpoint, information enables decision making, leadership, and combat power; it is also key to seizing, gaining, and retaining the initiative, and to consolidating gains in an OE. Army commanders conduct information operations to affect the information environment.

See following pages (pp. 0-10 to 0-11) for further discussion of the information environment, information as a joint function, and information operations.

Cyberspace and the Electromagnetic Spectrum

Cyberspace is a global domain within the information environment consisting of interdependent networks of information technology infrastructures and resident data, including the Internet, telecommunications networks, computer systems, and embedded processors and controllers.

Ref: FM 3-0 (Oct '17), fig. 1-2. Cyberspace in the multi-domain extended battlefield.

Cyberspace is an extensive and complex global network of wired and wireless links connecting nodes that permeate every domain. Networks cross geographic and political boundaries connecting individuals, organizations, and systems around the world. Cyberspace is socially enabling, allowing interactivity among individuals, groups, organizations, and nation-states.

Information Environment Operations (IEO)

The Joint Force seizes the initiative in competition by actively engaging in the information space across domains (to include cyberspace) and the EMS. The theater army converges Army actions and messaging in support of the Joint Force Commander's IEO, though all echelons engage in the information space in support of policy and commander's intent. To accomplish this mission, subordinate echelons must be enabled with access to intelligence, cyberspace, and EMS capabilities; appropriate authorities and permissions normally reserved for conflict or at higher echelons; and policy guidance expressed as intent rather than narrow, restrictive directives. This allows forward presence forces to aggressively take tailored actions and employ messages to counter and expose inconsistencies in the adversary's information warfare operations. The Army primarily contributes to the strategic narrative, however, by reinforcing the resolve and commitment of the U.S. to its partner and demonstrating its capabilities as a credible deterrent to conflict. *See also p. 1-9.*

TRADOC Pamphlet 525-3-1, *The U.S. Army in Multi-Domain Operations (Dec '18).*

(Threat/COE/Info) Introduction 0-9

IV. The Information Environment

The information environment is the aggregate of individuals, organizations, and systems that collect, process, disseminate, or act on information. This environment consists of three interrelated dimensions which continuously interact with individuals, organizations, and systems. These dimensions are the physical, informational, and cognitive. The JFC's operational environment is the composite of the conditions, circumstances, and influences that affect employment of capabilities and bear on the decisions of the commander.

The Information Environment

1 The Physical Dimension

2 The Informational Dimension

3 The Cognitive Dimension

See pp. 2-14 to 2-15 for discussion of the information environment from FM 3-12.

V. Information as a Joint Function

Ref: JP 3-0, Joint Operations, w/Chg 1 (Oct '18), pp. III-17 to III-27.

All military activities produce **information**. Informational aspects are the features and details of military activities observers interpret and use to assign meaning and gain understanding. Those aspects affect the perceptions and attitudes that drive behavior and decision making. The JFC leverages informational aspects of military activities to gain an advantage; failing to leverage those aspects may cede this advantage to others. Leveraging the informational aspects of military activities ultimately affects strategic outcomes.

The **information function** encompasses the management and application of information and its deliberate integration with other joint functions to change or maintain perceptions, attitudes, and other elements that drive desired behaviors and to support human and automated decision making. The information function helps commanders and staffs understand and leverage the pervasive nature of information, its military uses, and its application during all military operations. This function provides JFCs the ability to integrate the generation and preservation of friendly information while leveraging the inherent informational aspects of military activities to achieve the commander's objectives and attain the end state.

Information Function Activities

The information function includes activities that facilitate the JFC's understanding of the role of information in the OE, facilitate the JFC's ability to leverage information to affect behavior, and support human and automated decision making.

See following pages (pp. 0-12 to 0-15) for an overview and further discussion.

VI. Information Operations (IO)

Ref: JP 3-0, Joint Operations, w/Chg 1 (Oct '18), pp. III-17 to III-22.

All military activities produce **information**. Informational aspects are the features and details of military activities observers interpret and use to assign meaning and gain understanding. Those aspects affect the perceptions and attitudes that drive behavior and decision making. The JFC leverages informational aspects of military activities to gain an advantage; failing to leverage those aspects may cede this advantage to others. Leveraging the informational aspects of military activities ultimately affects strategic outcomes.

The **information function** encompasses the management and application of information and its deliberate integration with other joint functions to change or maintain perceptions, attitudes, and other elements that drive desired behaviors and to support human and automated decision making.

The **instruments of national power** (diplomatic, informational, military, and economic) provide leaders in the US with the means and ways of dealing with crises around the world. Employing these means in the information environment requires the ability to securely transmit, receive, store, and process information in near real time. The nation's state and non-state adversaries are equally aware of the significance of this new technology, and will use information-related capabilities (IRCs) to gain advantages in the information environment, just as they would use more traditional military technologies to gain advantages in other operational environments. As the strategic environment continues to change, so does information operations (IO).

Regardless of its mission, the joint force considers the likely impact of all operations on **relevant actor** perceptions, attitudes, and other drivers of behavior. The JFC then plans and conducts every operation in ways that **create desired effects** that include maintaining or inducing relevant actor behaviors. These ways may include the timing, duration, scope, scale, and even visibility of an operation; the deliberately planned presence, posture, or profile of assigned or attached forces in an area; the use of signature management in deception operations; the conduct of activities and operations to similarly impact behavioral drivers; and the **employment of specialized capabilities** -- e.g., key-leader engagements (KLE), cyberspace operations (CO), military information support operations (MISO), electronic warfare (EW), and civil affairs (CA) -- to reinforce the JFC's efforts.

Inform activities involve the release of accurate information to domestic and international audiences to put joint operations in context; facilitate informed perceptions about military operations; and counter adversarial misinformation, disinformation, and propaganda. Inform activities help to assure the trust and confidence of the US population, allies, and partners and to deter and dissuade adversaries and enemies.

The joint force **attacks and exploits information, information networks, and systems** to affect the ability of relevant actors to leverage information in support of their own objectives. This includes the manipulation, modification, or destruction of information or disruption of the flow of information for the purpose of gaining a position of military advantage. This also includes targeting the credibility of information.

Refer to INFO1: The Information Operations & Capabilities SMARTbook (Guide to Information Operations & the IRCs). See following pages (pp. 0-11a to 0-11b) for an overview of this companion book to the CYBER1 SMARTbook.

See pp. 4-45 to 4-50 for discussion of the the integrating/coordinating functions of information operations (IO) and pp. 4-51 to 4-54 for related discussion of IO planning.

INFO1: The Information Operations & Capabilities SMARTbook

Guide to Information Operations & the IRCs

Over the past two decades, information operations (IO) has gone through a number of doctrinal evolutions, explained, in part, by the rapidly changing nature of information, its flow, processing, dissemination, impact and, in particular, its military employment. INFO1: The Information Operations & Capabilities SMARTbook examines the most current doctrinal references available and charts a path to emerging doctrine.

FM 3-13   ATP 3-13.1   JP 3-13 (Chg 1)   JP 3-0 (Chg 1)

Plus more than a dozen primary references on the IRCs and more!

INFO1 chapters and topics include information operations (IO defined and described), information in joint operations (joint IO), information-related capabilities (PA, CA, MILDEC, MISO, OPSEC, CO, EW, Space, STO), information planning (information environment analysis, IPB, MDMP, JPP), information preparation, information execution (IO working group, IO weighted efforts and enabling activities, intel support), fires & targeting, and information assessment.

Chap 1: Information Operations (Defined & Described)

Information is a resource. As a resource, it must be obtained, developed, refined, distributed, and protected. The **information element of combat power** is integral to optimizing combat power, particularly given the increasing relevance of operations in and through the information environment to achieve decisive outcomes.

Information Operations (IO) is the integrated employment, during military operations, of information-related capabilities in concert with other lines of operation to influence, disrupt, corrupt, or usurp the decision-making of adversaries and potential adversaries while protecting our own. The purpose of IO is to **create effects in and through the information environment** that provide commanders decisive advantage over enemies and adversaries.

Chap 2: Information in Joint Operations

The joint force commander (JFC) **leverages informational aspects of military activities to gain an advantage**; failing to leverage those aspects may cede this advantage to others. Leveraging the informational aspects of military activities ultimately affects strategic outcomes. The joint force **attacks and exploits information, information networks, and systems to affect the ability of relevant actors to leverage information** in support of their own objectives. This includes the manipulation, modification, or destruction of information or disruption of the flow of information for the purpose of gaining a position of military advantage. This also includes targeting the credibility of information.

Chap 3: Information-Related Capabilities (IRCs)

An **information-related capability (IRC)** is a tool, technique, or activity employed within a dimension of the information environment that can be used to create effects and operationally desirable conditions. IO brings together information-related capabilities (IRCs) at a specific time and in a coherent fashion to create effects in and through the information environment that advance the ability to deliver operational advantage to the commander.

All unit operations, activities, and actions affect the information environment. Even if they primarily affect the physical dimension, they nonetheless also affect the informational and cognitive dimensions. For this reason, whether or not they are routinely considered an IRC, a wide variety of unit functions and activities can be adapted for the purposes of conducting information operations or serve as enablers to its planning, execution, and assessment.

Chap 4: Information Planning

Planning is the art and science of understanding a situation, envisioning a desired future, and laying out effective ways of bringing that future about. Commanders, supported by their staffs, ensure IO is fully integrated into the plan, starting with Army design methodology (ADM) and progressing through the military decisionmaking process (MDMP). The focal point for IO planning is the IO officer (or designated representative for IO). However, the entire staff contributes to planning products that describe and depict how IO supports the commander's intent and concept of operations.

Chap 5: Information Planning

Preparation consists of those activities performed by units and Soldiers to improve their ability to execute an operation. Preparation creates conditions that improve friendly force opportunities for success. Because many IO objectives and IRC tasks require long lead times to create desired effects, preparation for IO often starts earlier than for other types of operations. Initial preparation for specific IRCs and IO units (such as 1st IO Command or a Theater IO Group) may begin during peacetime.

Chap 6: Information Execution

Execution of IO includes IRCs executing the synchronization plan and the commander and staff monitoring and assessing their activities relative to the plan and adjusting these efforts, as necessary. The primary mechanism for monitoring and assessing IRC activities is the **IO working group.** There are two variations of the IO working group. The first monitors and assesses ongoing planned operations and convenes on a routine, recurring basis. The second monitors and assesses unplanned or crisis situations and convenes on an as-needed basis.

Chap 7: Fires & Targeting

The **fires warfighting function** is the related tasks and systems that **create and converge effects in all domains** against the threat to enable actions across the range of military operations. These tasks and systems create **lethal and nonlethal effects** delivered from both Army and Joint forces, as well as other unified action partners.

Targeting is the process of selecting and prioritizing targets and matching the appropriate response to them, considering operational requirements and capabilities (JP 3-0). IO is integrated into the targeting cycle to produce effects in and through the information environment that support objectives.

Chap 8: Information Assessment

Assessment precedes and guides the other activities of the operations process. It is also part of targeting. In short, assessment occurs at all levels and within all operations and has a role in any process or activity. The purpose of assessment is to improve the commander's decision making and make operations more effective. Assessment is a key component of the commander's decision cycle, helping to determine the results of unit actions in the context of overall mission objectives.

(Threat/COE/Info) Introduction 0-11b *

Information Function Activities
Ref: JP 3-0, Joint Operations, w/Chg 1 (Oct '18), pp. III-17 to III-22.

The information function includes activities that facilitate the JFC's understanding of the role of information in the OE, facilitate the JFC's ability to leverage information to affect behavior, and support human and automated decision making.

1. Understand Information in the Operational Environment (OE)

In conjunction with activities under the intelligence joint function, this activity facilitates the JFC's understanding of the pervasive nature of information in the OE, its impact on relevant actors, and its effect on military operations. It includes determining relevant actor perceptions, attitudes, and decision-making processes and requires an appreciation of their culture, history, and narratives, as well as knowledge of the means, context, and established patterns of their communication.

Information affects the perceptions and attitudes that drive the behavior and decision making of humans and automated systems. In order to affect behavior, the JFC must understand the perceptions, attitudes, and decision-making processes of humans and automated systems. These processes reflect the aggregate of social, cultural, and technical attributes that act upon and impact knowledge, understanding, beliefs, world views, and actions.

The human and automated systems whose behavior the JFC wants to affect are referred to as relevant actors. Relevant actors may include any individuals, groups, and populations, or any automated systems, the behavior of which has the potential to substantially help or hinder the success of a particular campaign, operation, or tactical action. For the purpose of military activities intended to inform audiences, relevant actors may include US audiences; however, US audiences are not considered targets for influence.

See pp. 0-6 to 0-9 for related discussion of the operational environment.

Language, Regional, and Cultural Expertise

Language skills, regional knowledge, and cultural awareness enable effective joint operations. Deployed joint forces should understand and effectively communicate with HN populations; local and national government officials; multinational partners; national, regional, and international media; and other key stakeholders, including NGOs. This capability includes knowledge about the human aspects of the OE and the skills associated with communicating with foreign audiences. Knowledge about the human aspects of the OE is derived from the analysis of national, regional, and local culture, economy, politics, religion, and customs. Consequently, commanders should integrate training and capabilities for foreign language and regional expertise in contingency, campaign, and supporting plans and provide for them in support of daily operations and activities. Commanders should place particular emphasis on foreign language proficiency in technical areas identified as key to mission accomplishment.

For specific planning guidance and procedures regarding language and regional expertise, refer to CJCSI 3126.01, Language, Regional Expertise, and Culture (LREC) Capability Identification, Planning, and Sourcing.

2. Leverage Information to Affect Behavior

Tasks aligned under this activity apply the JFC's understanding of the impact information has on perceptions, attitudes, and decision-making processes to affect the behaviors of relevant actors in ways favorable to joint force objectives.

Influence Relevant Actors
Regardless of its mission, the joint force considers the likely impact of all operations on relevant actor perceptions, attitudes, and other drivers of behavior. The JFC then plans and conducts every operation in ways that create desired effects that include maintaining or inducing relevant actor behaviors. These ways may include the timing, duration, scope, scale, and even visibility of an operation; the deliberately planned presence, posture, or profile of assigned or attached forces in an area; the use of signature management in deception operations; the conduct of activities and operations to similarly impact behavioral drivers; and the employment of specialized capabilities (e.g., KLE, CO, military information support operations [MISO], EW, CA) to reinforce the JFC's efforts. Since some relevant actors will be located outside of the JFC's OA, coordination, planning, and synchronization of activities with other commands or mission partners is vital.

Inform Domestic, International, and Internal Audiences
Inform activities involve the release of accurate information to domestic and international audiences to put joint operations in context; facilitate informed perceptions about military operations; and counter adversarial misinformation, disinformation, and propaganda. Inform activities help to assure the trust and confidence of the US population, allies, and partners and to deter and dissuade adversaries and enemies.

Attack and Exploit Information, Information Networks, and Systems
The joint force attacks and exploits information, information networks, and systems to affect the ability of relevant actors to leverage information in support of their own objectives. This includes the manipulation, modification, or destruction of information or disruption of the flow of information for the purpose of gaining a position of military advantage. This also includes targeting the credibility of information.

3. Support Human and Automated Decision Making
The management aspect of the information joint function includes activities that facilitate shared understanding across the joint force and that protect friendly information, information networks, and systems to ensure the availability of timely, accurate, and relevant information necessary for JFC decision making.

Facilitating Shared Understanding
Facilitating shared understanding is related to building shared understanding in the C2 joint function. Where building shared understanding is an element of C2 and focuses on purpose (i.e., the commander's objective), facilitating shared understanding is concerned with process (i.e., the methods). Key components of facilitating understanding are collaboration, KS, and IM.

Protecting Friendly Information
Information Networks, and Systems. The information function reinforces the protection function and focuses on protecting friendly information, information networks, and systems. This aspect of the information function includes the preservation of friendly information across the staff and the joint force and any information shared with allies and partners. These activities reinforce the requirement to assure the flow of information important to the joint force, both by protecting the information and by assessing and mitigating risks to that information. The preservation of information includes both passive and active measures to prevent and mitigate adversary collection, manipulation, and destruction of friendly information, to include attempts to undermine the credibility of friendly information.

Joint Force Capabilities, Operations, and Activities for Leveraging Information

Ref: JP 3-0, Joint Operations, w/Chg 1 (Oct '18), pp.) III-17 to III-26.

In addition to planning all operations to benefit from the inherent informational aspects of physical power and influence relevant actors, the JFC also has additional means with which to leverage information in support of enduring outcomes. The following capabilities, operations, and activities may reinforce the actions of assigned or attached forces, support LOOs or LOEs, or constitute the primary activity in a LOE.

See pp. 4-45 to 4-50 for related discussion of the the integrating/coordinating functions of information operations (IO) from JP 3-13.

Key Leader Engagement (KLE)
Most operations require commanders and other leaders to engage key local and regional leaders to affect their attitudes and gain their support. Building relationships to the point of effective engagement and influence usually takes time. Language, regional expertise, and culture knowledge and skills are keys to successfully communicate with and, therefore, manage KLE. Commanders can be challenged to identify key leaders, develop messages, establish dialogue, and determine other ways and means of delivery, especially in societies where interpersonal relationships are paramount.

Public Affairs (PA)
PA contributes to the achievement of military objectives by countering incorrect information and propaganda through the dissemination of accurate information. PA personnel advise the JFCs on the possible direct and indirect effects of joint force actions on public perceptions, attitudes, and beliefs, and work to formulate and deliver timely and culturally attuned messages.

Civil-Military Operations (CMO)
CMO are activities that establish, maintain, influence, or exploit relationships between military forces and indigenous populations and institutions with the objective to reestablish or maintain stability in a region or HN. During all military operations, CMO can coordinate the integration of military and nonmilitary instruments of national power. CA support CMO by conducting military engagement and humanitarian and civic assistance to influence the populations of the HN and other PNs in the OA.

Military Deception (MILDEC)
Commanders conduct MILDEC to mislead enemy decision makers and commanders and cause them to take or not take specific actions. The intent is to cause enemy commanders to form inaccurate impressions about friendly force dispositions, capabilities, vulnerabilities, and intentions; misuse their intelligence collection assets; and fail to employ their combat or support units to best advantage. As executed by JFCs, MILDEC targets enemy leaders and decision makers through the manipulation of their intelligence collection, analysis, and dissemination systems. MILDEC depends on intelligence to identify deception targets, assist in developing credible stories, identify and orient on appropriate receivers (the readers of the story), and assess the effectiveness of the deception effort. Deception requires a thorough knowledge of the enemy and their decision-making processes.

Military Information Support Operations (MISO)
Psychological operations forces conduct MISO to convey selected information and indicators to foreign audiences to influence their emotions, motives, and objective reasoning and ultimately induce or reinforce foreign attitudes and behavior favorable to the origina-

tor's objectives. Psychological operations forces devise and execute psychological actions and craft persuasive messages using a variety of audio, visual, and audiovisual products, which can then be delivered to both targets and audiences.

Operations Security (OPSEC)

OPSEC uses a process to preserve friendly essential secrecy by identifying, controlling, and protecting critical information and indicators that would allow adversaries to identify and exploit friendly vulnerabilities. The purpose of OPSEC is to reduce vulnerabilities of US and multinational forces to adversary exploitation, and it applies to all activities that prepare, sustain, or employ forces.

Combat Camera (COMCAM)

Imagery is one of the most powerful tools available for informing internal and domestic audiences and for influencing foreign audiences. COMCAM forces provide imagery support in the form of a directed imagery capability to the JFC across the range of military operations. COMCAM imagery supports capabilities that use imagery for their products and efforts, including MISO, MILDEC, PA, and CMO, and provides critical operational documentation for sensitive site exploitation, legal and evidentiary requirements, and imagery for battle damage assessment/MOE analysis, as well as operational documentation and imagery for narrative development during foreign humanitarian assistance (FHA) operations and NEOs.

Space Operations

Space operations support joint operations throughout the OE by providing information in the form of ISR; missile warning; environment monitoring; satellite communications; and space-based positioning, navigation, and timing (PNT). Space operations also integrate offensive and defensive activities to achieve and maintain space superiority.

Special Technical Operations (STO)

STO should be deconflicted and synchronized with other activities. Detailed information related to STO and its contribution to joint force operations can be obtained from the STO planners at CCMD or Service component HQ.

*CYBERSPACE OPERATIONS (CO)

CO include the missions of OCO, DCO, and DODIN operations. These missions include the use of technical capabilities in cyberspace and cyberspace as a medium to leverage information in and through cyberspace.

See chap. 1 and 2 (CO), chap. 6 (DODIN), and chap. 7 (Cybersecurity).

*ELECTRONIC WARFARE (EW)

EW is the military action ultimately responsible for securing and maintaining freedom of action in the EMS for friendly forces while exploiting or denying it to adversaries. EW is an enabler for other activities that communicate or maneuver through the EMS, such as MISO, PA, or CO.

See chap. 3, Electronic Warfare.

Refer to INFO1: The Information Operations & Capabilities SMARTbook (Guide to Information Operations & the IRCs). INFO1 chapters and topics include information operations (IO defined and described), information in joint operations (joint IO), information-related capabilities (PA, CA, MILDEC, MISO, OPSEC, CO, EW, Space, STO), information planning (information environment analysis, IPB, MDMP, JPP), information preparation, information execution (IO working group, IO weighted efforts and enabling activities, intel support), fires & targeting, and information assessment.

(Threat/COE/Info) Introduction 0-15

CYBER1: The Cyberspace Operations & Electronic Warfare SMARTbook (Chapters)

Chap 1: Joint Cyberspace Operations (CO)
Most aspects of joint operations rely in part on cyberspace, which is the domain within the information environment that consists of the interdependent network of information technology (IT) infrastructures and resident data. It includes the Internet, telecommunications networks, computer systems, and embedded processors and controllers. Cyberspace operations (CO) is the employment of cyberspace capabilities where the primary purpose is to achieve objectives in or through cyberspace.

Chap 2: Cyberspace Operations (OCO/DCO/DODIN)
Army cyberspace operations range from defensive to offensive. These operations establish and maintain secure communications, detect and deter threats in cyberspace to the DODIN, analyze incidents when they occur, react to incidents, and then recover and adapt while supporting Army and joint forces from strategic to tactical levels while simultaneously denying adversaries effective use of cyberspace and the electromagnetic spectrum (EMS). Cyberspace missions include DODIN operations, defensive CO, and offensive CO.

Chap 3: Electromagnetic Warfare (EW) Operations
Electromagnetic Warfare (EW) is military action involving the use of electromagnetic and directed energy to control the electromagnetic spectrum or to attack the enemy. EW consists of three functions: electromagnetic attack, electromagnetic protection, and electromagnetic support.

Chap 4: Cyber & EW (CEMA) Planning
The cyberspace planner is the subject matter expert to create effects in cyberspace and the EMS, with considerations from the CEMA section. Involving the cyberspace planner early in development of the commander's vision and planning allows for synchronization and integration with missions, functions, and tasks.

Chap 5: Spectrum Management Operations (SMO/JEMSO)
Spectrum management operations are the interrelated functions of spectrum management, frequency assignment, host nation coordination, and policy that together enable the planning, management, and execution of operations within the electromagnetic operational environment during all phases of military operations JEMSO include all activities in military operations to successfully plan and execute joint or multinational operations in order to control the electromagnetic operational environment (EMOE).

Chap 6: DoD Info Network (DODIN) Operations
Department of Defense information network (DODIN) operations are operations to secure, configure, operate, extend, maintain, and sustain Department of Defense cyberspace to create and preserve the confidentiality, availability, and integrity of the Department of Defense information network. DODIN operations are one of the three cyberspace missions.

Chap 7: Cybersecurity
The Army depends on reliable networks and systems to access critical information and supporting information services to accomplish their missions. Threats to the DODIN exploit the increased complexity and connectivity of Army information systems and place Army forces at risk. Like other operational risks, cyberspace risks affect mission accomplishment. Robust cybersecurity measures prevent adversaries from accessing the DODIN through known vulnerabilities.

Chap 8: Acronyms & Glossary
A combined glossary lists acronyms and terms with Army, multi-Service, or joint definitions, and other selected terms.

I. Joint Cyberspace Operations

Ref: JP 3-12, Cyberspace Operations (Jun '18), chap. I.

> "... the United States (US) Department of Defense (DOD) is responsible for defending the US homeland and US interests from attack, including attacks that may occur in cyberspace. ... the DOD seeks to deter attacks and defend the US against any adversary that seeks to harm US national interests during times of peace, crisis, or conflict. To this end, the DOD has developed capabilities for cyberspace operations and is integrating those capabilities into the full array of tools that the US government uses to defend US national interests..."
>
> - The Department of Defense Cyber Strategy, April 2015

I. Introduction

Most aspects of joint operations rely in part on **cyberspace**, which is the domain within the information environment that consists of the interdependent network of information technology (IT) infrastructures and resident data. It includes the Internet, telecommunications networks, computer systems, and embedded processors and controllers. Cyberspace operations (CO) is the employment of cyberspace capabilities where the primary purpose is to achieve objectives in or through cyberspace.

See following page (p. 1-2) for discussion of the nature of cyberspace.

JP 3-12 focuses on military operations in and through cyberspace; explains the relationships and responsibilities of the Joint Staff (JS), combatant commands (CCMDs), United States Cyber Command (USCYBERCOM), the Service cyberspace component (SCC) commands, and combat support agencies (CSAs); and establishes a framework for the employment of cyberspace forces and capabilities. Cyberspace forces are those personnel whose primary duty assignment is to a CO mission.

A. The Impact of Cyberspace on Joint Operations

Cyberspace capabilities provide opportunities for the US military, its allies, and partner nations (PNs) to gain and maintain continuing advantages in the operational environment (OE) and enable the nation's economic and physical security. Cyberspace reaches across geographic and geopolitical boundaries and is integrated with the operation of critical infrastructures, as well as the conduct of commerce, governance, and national defense activities. Access to the Internet and other areas of cyberspace provides users operational reach and the opportunity to compromise the integrity of critical infrastructures in direct and indirect ways without a physical presence. The prosperity and security of our nation are significantly enhanced by our use of cyberspace, yet these same developments have led to increased exposure of vulnerabilities and a critical dependence on cyberspace, for the US in general and the joint force in particular.

Although it is possible for CO to produce stand-alone tactical, operational, or strategic effects and thereby achieve objectives, commanders integrate most CO with other operations to create coordinated and synchronized effects required to support mission accomplishment.

II. The Nature of Cyberspace

Ref: JP 3-12, Cyberspace Operations (Jun '18), pp. I-2 to I-4.

Cyberspace, while part of the information environment, is dependent on the air, land, maritime, and space physical domains. Much as operations in the physical domains rely on physical infrastructure created to take advantage of naturally occurring features, operations in cyberspace rely on networked, stand-alone, and platform-embedded IT infrastructure, in addition to the data that resides on and is transmitted through these components to enable military operations in a man-made domain. CO use links and nodes located in the physical domains and perform logical functions to create effects first in cyberspace and then, as needed, in the physical domains. Actions in cyberspace, through carefully controlled cascading effects, can enable freedom of action for activities in the physical domains. Likewise, activities in the physical domains can create effects in and through cyberspace by affecting the electromagnetic spectrum (EMS) or the physical infrastructure. The relationship between space and cyberspace is unique in that virtually all space operations depend on cyberspace, and a critical portion of cyberspace bandwidth can only be provided via space operations, which provide a key global connectivity option for CO. These interrelationships are important considerations during planning. While domains are useful constructs for visualizing and characterizing the physical environment in which operations are conducted (i.e., the operational area [OA]), the use of the term "domain" is not meant to imply or mandate exclusivity, primacy, or command and control (C2) in any domain.

Cyberspace Layer Model

To assist in the planning and execution of CO, cyberspace can be described in terms of three interrelated layers: physical network, logical network, and cyber-persona. Each layer represents a different focus from which CO may be planned, conducted, and assessed.

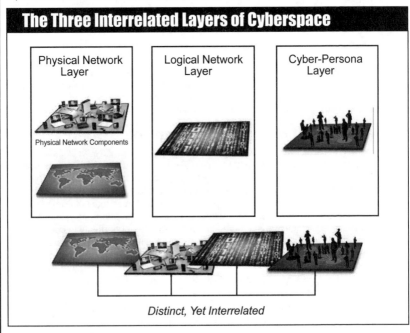

Ref: JP 3-12, Cyberspace Operations (Jun '18), fig. I-1. The Three Interrelated Layers of Cyberspace.

Physical Network Layer

The physical network layer consists of the IT devices and infrastructure in the physical domains that provide storage, transport, and processing of information within cyberspace, to include data repositories and the connections that transfer data between network components. The physical network components include the hardware and infrastructure (e.g., computing devices, storage devices, network devices, and wired and wireless links). Components of the physical network layer require physical security measures to protect them from physical damage or unauthorized physical access, which may be leveraged to gain logical access. The physical network layer is the first point of reference CO use to determine geographic location and appropriate legal framework. While geopolitical boundaries can easily and quickly be crossed in cyberspace, there are still sovereignty issues tied to the physical domains. Every physical component of cyberspace is owned by a public or private entity, which can control or restrict access to their components.

Logical Network Layer

The logical network layer consists of those elements of the network related to one another in a way that is abstracted from the physical network, based on the logic programming (code) that drives network components (i.e., the relationships are not necessarily tied to a specific physical link or node, but to their ability to be addressed logically and exchange or process data). Individual links and nodes are represented in the logical layer but so are various distributed elements of cyberspace, including data, applications, and network processes not tied to a single node. An example is the Joint Knowledge Online Website, which exists on multiple servers in multiple locations in the physical domains but is represented as a single URL [uniform resource locator] on the World Wide Web. More complex examples of the logical layer are the DOD's NIPRNET and SIPRNET, global, multi-segment networks that can be thought of as a single network only in the logical sense. For targeting purposes, planners may know the logical location of some targets, such as virtual machines and operating systems, that allow multiple servers or other network functions with separate IP addresses to reside on one physical computer, without knowing their geographic location. Logical layer targets can only be engaged with a cyberspace capability: a device or computer program including any combination of software, firmware, or hardware, designed to create an effect in or through cyberspace.

Cyber-Persona Layer

The cyber-persona layer is a view of cyberspace created by abstracting data from the logical network layer using the rules that apply in the logical network layer to develop descriptions of digital representations of an actor or entity identity in cyberspace (cyber-persona). The cyber-persona layer consists of network or IT user accounts, whether human or automated, and their relationships to one another. Cyber-personas may relate directly to an actual person or entity, incorporating some personal or organizational data (e.g., e-mail and IP addresses, Web pages, phone numbers, Web forum log-ins, etc). One individual may create and maintain multiple cyber-personas through use of multiple identifiers in cyberspace, such as separate work and personal e-mail addresses, and different identities on different Web forums, chat rooms, and social networking sites, which may vary in the degree to which they are factually accurate. Conversely, a single cyber-persona can have multiple users, such as multiple hackers using the same malicious software (malware) control alias, multiple extremists using a single bank account, or all members of the same organization using the same e-mail address. The use of cyber-personas can make attributing responsibility for actions in cyberspace difficult. Because cyber-personas can be complex, with elements in many virtual locations not linked to a single physical location or form, their identification requires significant intelligence collection and analysis to provide enough insight and situational awareness to enable effective targeting or to create the JFC's desired effect.

See pp. 2-6 to 2-7 for expanded discussion of the these three interrelated layers from FM 3-12.

The Services' Cyberspace Doctrine

Ref: U.S. Army War College Strategic Cyberspace Operations Guide (Jun '16), pp. 99 to 106.

U.S. Army
Ref: FM 3-12, Cyberspace Operations and Electromagnetic Warfare (Aug '21).

FM 3-12 provides tactics and procedures for the coordination and integration of Army cyberspace and electronic warfare operations to support unified land operations and joint operations.

Army Cyber Electromagnetic Activities doctrine is Army cyberspace operations range from defensive to offensive. These operations establish and maintain secure communications, detect and deter threats in cyberspace to the DODIN, analyze incidents when they occur, react to incidents, and then recover and adapt while supporting Army and joint forces from strategic to tactical levels while simultaneously denying adversaries effective use of cyberspace and the EMS.

The Army plans, integrates, and synchronizes cyberspace operations through CEMA as a continual and unified effort. The continuous planning, integration, and synchronization of cyberspace and EW operations, enabled by SMO, can produce singular, reinforcing, and complementary effects. Though the employment of cyberspace operations and EW differ because cyberspace operates on wired networks, both operate using the EMS.

See chap. 2, Cyberspace Operations, and Chap. 3, Electromagnetic Warfare, for further discussion from FM 3-12, Cyberspace Operations and Electromagnetic Warfare (Aug '21)

U.S. Air Force
Ref: Annex 3-12, Cyberspace Operations (30 Nov '11).

Air Force Cyberspace Operations doctrine is set down in Annex 3-12 last updated 20 November 2011.

Cyberspace superiority is the operational advantage in, through, and from cyberspace to conduct operations at a given time and in a given domain without prohibitive interference. Cyberspace superiority may be localized in time and space, or it may be broad and enduring. The concept of cyberspace superiority hinges on the idea of preventing prohibitive interference to joint forces from opposing forces, which would prevent joint forces from creating their desired effects. "Supremacy" prevents effective interference, which does not mean that no interference exists, but that any attempted interference can be countered or should be so negligible as to have little or no effect on operations. While "supremacy" is most desirable, it may not be operationally feasible. Cyberspace superiority, even local or mission-specific cyberspace superiority, may provide sufficient freedom of action to create desired effects. Therefore, commanders should determine the minimum level of control required to accomplish their mission and assign the appropriate level of effort.

Cyberspace operations seek to ensure freedom of action across all domains for US forces and allies, and deny that same freedom to adversaries. Specifically, cyberspace operations overcome the limitations of distance, time, and physical barriers present in other domains. Exploiting improved technologies makes it possible to enhance the Air Force's global operations by delivering larger information payloads and increasingly sophisticated effects. Cyberspace links operations in other domains thus facilitating interdependent defensive, exploitative, and offensive operations to achieve situational advantage.

Refer to AFOPS2: The Air Force Operations & Planning SMARTbook, 2nd Ed. (Guide to Curtis E. LeMay Center & Joint Air Operations Doctrine), chap. 4, section XI Cyberspace Operations, pp. 4-89 to 4-94 for further discussion.

U.S. Navy
Ref: NWP 3-12, Cyberspace Operations.

NWP 3-12, Cyberspace Operations is classified. The following is an excerpt of the Executive Summary of Navy Cyber Power 2020, dated November 2012. This Strategic Plan provides the framework and vision necessary to ensure the U.S. Navy remains a critical insurer of our national security and economic prosperity well into the future.

U.S. maritime power is comprised of six core capabilities: forward presence, deterrence, sea control, power projection, maritime security, and humanitarian assistance/disaster response (HA/DR). In today's highly networked world each one of these core capabilities is enhanced by effective Navy cyberspace operations.

Navy Cyber Power 2020 (NCP 2020) is a strategy for achieving the Navy's vision for cyberspace operations (Figure 1). This document describes the key end-state characteristics that the Navy must create and the major strategic initiatives we will pursue to achieve success. It serves as a guidepost to inform our enterprise architecture, investment decisions, and future roadmaps.

Navy cyberspace operations provide Navy and Joint commanders with an operational advantage by:

- Assuring access to cyberspace and confident Command and Control
- Preventing strategic surprise in cyberspace
- Delivering cyber effects

Source: http://www.public.navy.mil/fcc-c10f/Strategies/Navy_Cyber_Power_2020.pdf, accessed 17 May 2016.

U.S. Marine Corps
Ref: MCIP 3-40.02, Marine Corps Cyberspace Operations (6 Oct '14).

MCIP 3-40.02, Marine Corps Cyberspace Operations, dated 6 October 2014, is not available to the public. The following is an excerpt of the Cyberspace Operations section from MCDP 1-0, Marine Corps Operations.

Cyberspace may be described as a global domain that leverages information and telecommunication technologies to create an environment of interdependent computer and telecommunication networks, including command and control systems, which can be used to produce outcomes in virtual and physical realms.

Cyberspace operations involve the employment of cyber capabilities where the primary purpose is to create military objectives or effects in or through cyberspace. Cyberspace operations comprise five broad categories—Department of Defense network operations, defensive cyber operations, offensive cyber operations, computer network exploitation, and information assurance.

Refer to our series of related Service-specific SMARTbooks.

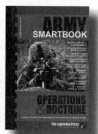

Permanent global cyberspace superiority is not possible due to the complexity of cyberspace. Even local superiority may be impractical due to the way IT is implemented; the fact US and other national governments do not directly control large, privately owned portions of cyberspace; the broad array of state and non-state actors; the low cost of entry; and the rapid and unpredictable proliferation of technology. Therefore, commanders should be prepared to conduct operations under degraded conditions in cyberspace. Commanders can manage resulting risks using threat mitigation actions; post-impact recovery measures; clear, defensive priorities; primary/secondary/tertiary communication means; and other measures to accomplish their mission and ensure critical data reliability. Once one segment of a network has been exploited or denied, the perception of data unreliability may inappropriately extend beyond the compromised segment due to uncertainty about how networks interact. Therefore, it is imperative commanders be well informed of the status of the portions of cyberspace upon which they depend and understand the impact to planned and ongoing operations.

B. Viewing Cyberspace Based on Location and Ownership

Maneuver in cyberspace is complex and generally not observable. Therefore, staffs that plan, execute, and assess CO benefit from language that describes cyberspace based on location or ownership in a way that aids rapid understanding of planned operations.

Blue Cyberspace

The term "blue cyberspace" denotes areas in cyberspace protected by the US, its mission partners, and other areas DOD may be ordered to protect. Although DOD has standing orders to protect only the Department of Defense information network (DODIN), cyberspace forces prepare, on order, and when requested by other authorities, to defend or secure other United States Government (USG) or other cyberspace, as well as cyberspace related to critical infrastructure and key resources (CI/KR) of the US and PNs.

Red Cyberspace

The term "red cyberspace" refers to those portions of cyberspace owned or controlled by an adversary or enemy. In this case, "controlled" means more than simply "having a presence on," since threats may have clandestine access to elements of global cyberspace where their presence is undetected and without apparent impact to the operation of the system. Here, controlled means the ability to direct the operations of a link or node of cyberspace, to the exclusion of others. All cyberspace that does not meet the description of either "blue" or "red" is referred to as "gray" cyberspace.

C. DOD Cyberspace (DODIN)

The DODIN is the set of information capabilities and associated processes for collecting, processing, storing, disseminating, and managing information on-demand to warfighters, policy makers, and support personnel, whether interconnected or stand-alone, including owned and leased communications and computing systems and services, software (including applications), data, security services, other associated services, and national security systems. The DODIN comprises all of DOD cyberspace, including the classified and unclassified global networks (e.g., NIPRNET, SIPRNET, Joint Worldwide Intelligence Communications System) and many other components, including DOD-owned smartphones, radio frequency identification tags, industrial control systems, isolated laboratory networks, and platform information technology (PIT). PIT is the hardware and software that is physically part of, dedicated to, or essential in real time to the mission performance of special purpose systems, including weapon systems. Nearly every military and civilian employee of DOD uses the DODIN to accomplish some portion of their mission or duties.

See chap. 6, DODIN Operations.

D. Connectivity and Access

Cyberspace consists of myriad different and often overlapping elements to include networks, nodes, links, interrelated applications, user data, and system data. Even though cyberspace continues to become increasingly interconnected, some elements are intentionally isolated or subdivided into enclaves using access controls, encryption, unique protocols, or physical separation. With the exception of actual physical isolation, none of these approaches eliminate the underlying physical connectivity; instead, they limit access to the logical network. Access, whether authorized or unauthorized, can be gained through a variety of means. Although CO require timely and effective connectivity and access, the USG may not own, control, or have access to the infrastructure needed to support US military operations. For CO, access means a sufficient level of exposure to, connectivity to, or entry into a device, system, or network to enable further operations. While some accesses can be created remotely with or without permission of the network owner, access to closed networks and other systems that are virtually isolated may require physical proximity or more complex, time-consuming processes. In addition, gaining access to operationally useful areas of cyberspace, including targets within them, is affected by legal, policy, or operational limitations. For all of these reasons, access is not guaranteed. Additionally, achieving a commander's objectives can be significantly complicated by specific elements of cyberspace being used by enemies, adversaries, allies, neutral parties, and other USG departments and agencies, all at the same time. Therefore, synchronization and deconfliction of CO access is critical to successful operations of all types.

III. The Operational Environment (OE)

The OE is a composite of the conditions, circumstances, and influences that affect the employment of capabilities and impact the decisions of the commander assigned responsibility for it. The information environment permeates the physical domains and therefore exists in any OE. The continuing advancement of IT has significantly reduced its cost of acquisition and cost of use, leading to the rapid proliferation of cyberspace capabilities, considerably complicating an already challenging OE. For instance, CO from moving platforms requires transmission through the EMS, which can be significantly affected by congestion (i.e., interference from commercial and military use), atmospheric conditions, and enemy electronic attack (EA). The decision to use CO to create effects may be affected by the political climate or even a single individual's use of cyberspace. Understanding the relationship of cyberspace to the physical domains and the information environment is essential for planning military operations in cyberspace.

The pervasiveness of mobile IT is forcing governments and militaries to reevaluate the impact of the information environment on operations. The nature of global social interaction has been changed by the rapid flow of information from around-the-clock news, including from nontraditional and unverifiable sources such as social networking, media sharing and broadcast sites, online gaming networks, topical forums, and text messaging. The popularity of these information sources enables unprecedented interaction among global populations, much of which is increasingly relevant to military operations. The ability of social networks in cyberspace to incite popular support (whether factually based or not) and to spread ideology is not geographically limited, and the continued proliferation of IT has profound implications for the joint force and US national security.

See pp. 0-6 to 0-9 for discussion of the contemporary operational environment (COE).

State and non-state threats use a wide range of advanced technologies, which represent an inexpensive way for a small and/or materially disadvantaged adversary to pose a significant threat to the US. The application of low-cost cyberspace capabilities can provide an advantage against a technology-dependent nation or organization. This can provide an asymmetric advantage to those who could not otherwise effectively oppose US military forces. Additionally, organized crime or other non-state, extralegal organizations often make sophisticated malware available for purchase or free, allowing even non-sophisticated threats to acquire advanced capabilities at little to no cost. Because of the low barriers to entry and the potentially high payoff, the US can expect an increasing number of adversaries to use cyberspace threats to attempt to negate US advantages in military capability.

A. Key Terrain

Key terrain in cyberspace is analogous to key terrain in the physical domains in that holding it affords any combatant a position of marked advantage. In cyberspace, It may only be necessary to maintain a secure presence on a particular location or in a particular process as opposed to seizing and retaining it to the exclusion of all others. Note that it is possible for the US and an adversary to occupy the same terrain or use the same process in cyberspace, potentially without knowing of the other's presence. An additional characteristic of terrain in cyberspace is that these localities have a virtual component, identified in the logical network layer or even the cyber-persona layer. Key terrain identification is an essential component of planning. The military aspects of terrain (obstacles, avenues of approach, cover and concealment, observation and fields of fire, and key terrain) provide a way to visualize and describe a network map. Obstacles in cyberspace may include firewalls and port blocks. Avenues of approach can be analyzed by identifying nodes and links, which connect endpoints to specific sites. Cover and concealment may refer to hidden IP addresses or password protected access. Cyberspace observation and fields of fire refer to areas where network traffic can be monitored, intercepted, or recorded. Examples of potential key terrain in cyberspace include access points to major lines of communications (LOCs), key waypoints for observing incoming threats, launch points for cyberspace attacks, and mission-relevant cyberspace terrain related to critical assets connected to the DODIN. Operators, planners, and intelligence staff work together to match plans' objectives with terrain analysis to determine key terrain in blue, gray, and red cyberspace for each plan. Correlating plan or mission objectives with key terrain ensures mission dependencies in cyberspace are identified and prioritized for protection in a standard manner across DOD. In many cases, the systems, networks, and infrastructure that support a mission objective will be interdependent. These complex interdependencies may require in-depth analysis to develop customized risk mitigation methodologies.

B. The Information Environment

The information environment is the aggregate of individuals, organizations, and systems that collect, process, disseminate, or act on information. Since all CO require the creation, processing, storage, and/or transmission of information, cyberspace is wholly contained within the information environment. The information environment is broken down into the physical, informational, and cognitive dimensions and includes many types of information not in cyberspace. Although the types of information excluded from cyberspace continue to dwindle, there remain individuals and organizations that handle their information requirements outside of cyberspace, particularly when security, durability, cost, and scope factors are significant.

See pp. 0-10 to 0-11 for further discussion of the information environment, information as a joint function, and information operations.

> ### The Relationship of Cyberspace Operations to Operations in the Information Environment
> Cyberspace is wholly contained within the information environment. CO and other information activities and capabilities create effects in the information environment in support of joint operations. Their relationship is both an interdependency and a hierarchy; cyberspace is a medium through which other information activities and capabilities may operate. These activities and capabilities include, but are not limited to, understanding information, leveraging information to affect friendly action, supporting human and automated decision making, and leveraging information (e.g., military information support operations [MISO] or military deception [MILDEC]) to change enemy behavior. CO can be conducted independently or synchronized, integrated, and deconflicted with other activities and operations.
>
> While commanders may conduct CO specifically to support information-specific operations, some CO support other types of military objectives and are integrated through appropriate cells and working groups. The lack of synchronized CO with other military operations planning and execution can result in friendly force interference and may counter the simplicity, agility, and economy of force principles of joint operations.
>
> *See p. 0-9 for discussion of information environment operations (IEO), p. 0-11 for information operations (IO), and pp. 0-12 to 0-15 for supporting activities.*

IV. Integrating Cyberspace Operations with Other Operations

During joint planning, cyberspace capabilities are integrated into the JFC's plans and synchronized with other operations across the range of military operations. While not the norm, some military objectives can be achieved by CO alone. Commanders conduct CO to obtain or retain freedom of maneuver in cyberspace, accomplish JFC objectives, deny freedom of action to the threat, and enable other operational activities.

The importance of CO support to military operations grows in direct proportion to the joint force's increasing reliance on cyberspace. Issues that may need to be addressed to fully integrate CO into joint planning and execution include centralized CO planning for DODIN operations and defense and other global operations; the JFC's need to integrate and synchronize all operations and fires across the entire OE, including the cyberspace aspects of joint targeting; deconfliction requirements between government entities; PN relationships; and the wide variety of authorities and legal issues related to the use of cyberspace capabilities. This requires all members of the commander's staff who conduct planning, execution, and assessment of operations to understand the fundamental processes and procedures for CO, including the organization and functions of assigned or supporting cyberspace forces.

Effective integration of CO with operations in the physical domains requires the active participation of CO planners and operators in each phase of joint operations on every staff supported by cyberspace forces. The physical and logical boundaries within which joint forces execute CO, and the priorities and restrictions on its use, should also be identified by the JFC, in coordination with other USG departments and agencies and national leadership. In particular, creation of effects in foreign cyberspace may have the potential to impact other efforts of the USG. Where the potential for such impact exists, national policy requires DOD coordination with interagency partners.

See pp. 1-39 to 1-56, CO Planning, Coordination, Execution, and Assessment.

(Joint Cyberpace Operations) I. Overview 1-9

IV. Cyberspace Operations Forces

A. United States Cyber Command (USCYBERCOM)

Commander, United States Cyber Command (CDRUSCYBERCOM), commands a preponderance of the cyberspace forces that are not retained by the Services. USCYBERCOM accomplishes its missions within three primary lines of operation: secure, operate, and defend the DODIN; defend the nation from attack in cyberspace; and provide cyberspace support as required to combatant commanders (CCDRs). The Services man, train, and equip cyberspace units and provide them to USCYBERCOM through the SCCs. Per the Memorandum of Agreement Between The Department of Defense and The Department of Homeland Security Regarding Department of Defense and US Coast Guard Cooperation on Cyberspace Security and Cyberspace Operations, the Commandant of the Coast Guard retains operational control (OPCON) of US Coast Guard Cyberspace forces when employed in support of DOD.

B. Cyber Mission Force (CMF)

The Secretary of Defense (SecDef) and Chairman of the Joint Chiefs of Staff (CJCS) established the CMF to organize and resource the force structure required to conduct key cyberspace missions. CDRUSCYBERCOM exercises combatant command (command authority) (COCOM) of the CMF, which is a subset of the DOD's total force for CO. Various Service tactical cyberspace units, assigned to CDRUSCYBERCOM, comprise the three elements of the CMF:

Cyber National Mission Force (CNMF)

The CNMF conducts CO to defeat significant cyberspace threats to the DODIN and, when ordered, to the nation. The CNMF comprises various numbered national mission teams (NMTs), associated national support teams (NSTs), and national-level CPTs for protection of non-DODIN blue cyberspace.

Cyber Protection Force (CPF)

The CPF conducts CO for internal protection of the DODIN or other blue cyberspace when ordered. The CPF consists of cyberspace protection teams (CPTs) organized, trained, and equipped to defend assigned cyberspace in coordination with and in support of segment owners, cybersecurity service providers (CSSPs), and users.

Cyber Combat Mission Force (CCMF)

The CCMF conducts CO to support the missions, plans, and priorities of the geographic and functional CCDRs. The CCMF comprises various numbered combat mission teams (CMTs) and associated combat support teams (CSTs).

See pp. 1-15 to 1-28, Cyberspace Operations Core Activities, for more information about the operations of CMF units.

C. USCYBERCOM Subordinate Command Elements

Subordinate headquarters (HQ) of USCYBERCOM execute C2 of the CMF and other cyberspace forces. These include the Cyber National Mission Force-Headquarters (CNMF-HQ), the Joint Force Headquarters-Department of Defense Information Network (JFHQ-DODIN), the joint force headquarters-cyberspace (JFHQ-C), and the SCC HQs. Each of the SCC commanders is dual-hatted by CDRUSCYBERCOM as a commander of one of the four JFHQs-C to enable synchronization of CO C2. In addition, there are other centers and staff elements that further enable unity of command for CO.

Figure I-2 (facing page) describes the organizational and subordination relationships of these command elements and the units of the CMF.

DoD Cyber Mission Force Relationships

Ref: JP 3-12, Cyberspace Operations (Jun '18), fig. I-2, p. I-10.

The Secretary of Defense (SecDef) and Chairman of the Joint Chiefs of Staff (CJCS) established the CMF to organize and resource the force structure required to conduct key cyberspace missions. CDRUSCYBERCOM exercises combatant command (command authority) (COCOM) of the CMF, which is a subset of the DOD's total force for CO. Various Service tactical cyberspace units, assigned to CDRUSCYBERCOM, comprise the three elements of the CMF:

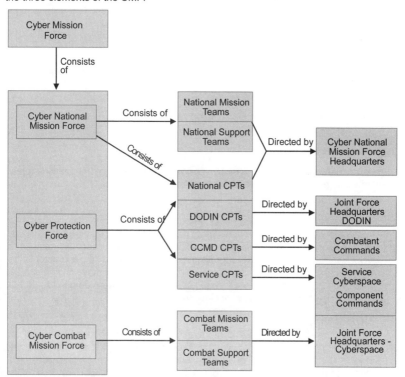

Cyber National Mission Force (CNMF)
The CNMF conducts CO to defeat significant cyberspace threats to the DODIN and, when ordered, to the nation. The CNMF comprises various numbered national mission teams (NMTs), associated national support teams (NSTs), and national-level CPTs for protection of non-DODIN blue cyberspace.

Cyber Protection Force (CPF)
The CPF conducts CO for internal protection of the DODIN or other blue cyberspace when ordered. The CPF consists of cyberspace protection teams (CPTs) organized, trained, and equipped to defend assigned cyberspace in coordination with and in support of segment owners, cybersecurity service providers (CSSPs), and users.

Cyber Combat Mission Force (CCMF)
The CCMF conducts CO to support the missions, plans, and priorities of the geographic and functional CCDRs. The CCMF comprises various numbered combat mission teams (CMTs) and associated combat support teams (CSTs).

D. Other Cyberspace Forces and Staff

Most cyberspace forces that protect the DODIN are Service-retained and some are employed in support of a specific CCDR. They may be used by the Service or SCCs to operationalize networks (i.e., design, build, configure and otherwise prepare to place into operation) and then secure, operate, and defend their Service enterprise portions of the DODIN. The Services may retain, or other CCDRs may organize, other scarce cyberspace forces that support CCMD missions as required, including CSSPs. Some of these Service-retained cyberspace forces that operate CCMD networks and systems are assigned directly to various CCDR staffs. In addition, the Defense Information Systems Agency (DISA) and various DOD agencies and activities employ civilian staff and contractors to do these same operationalizing and DODIN operations functions.

See pp. 1-48 to 1-50 for discussion of command and control (C2) of cyberspace operations.

VI. Challenges to the Joint Force's Use of Cyberspace

The JFC faces a unique set of persistent challenges executing CO in a complex global security environment.

A. Threats

Cyberspace presents the JFC's operations with many threats, from nation-states to individual actors to accidents and natural hazards.

Nation-State Threat

This threat is potentially the most dangerous because of nation-state access to resources, personnel, and time that may not be available to other actors. Some nations may employ cyberspace capabilities to attack or conduct espionage against the US. Nation-state threats involve traditional adversaries; enemies; and potentially, in the case of espionage, even traditional allies. Nation-states may conduct operations directly or may outsource them to third parties, including front companies, patriotic hackers, or other surrogates, to achieve their objectives.

Non-State Threats

Non-state threats are formal and informal organizations not bound by national borders, including legitimate nongovernmental organizations (NGOs), and illegitimate organizations such as criminal organizations, violent extremist organizations, or other enemies and adversaries. Non-state threats use cyberspace to raise funds, communicate with target audiences and each other, recruit, plan operations, undermine confidence in governments, conduct espionage, and conduct direct terrorist actions within cyberspace. Criminal organizations may be national or transnational in nature and steal information for their own use, including selling it to raise capital and target financial institutions for fraud and theft of funds. They may also be used as surrogates by nation-states or non-state threats to conduct attacks or espionage through cyberspace.

Individuals or Small Group Threat

Even individuals or small groups of people can attack or exploit US cyberspace, enabled by affordable and readily available techniques and malware. Their intentions are as varied as the number of groups and individuals. These threats exploit vulnerabilities to gain access to discover additional vulnerabilities or sensitive data or maneuver to achieve other objectives. Ethical hackers may share the vulnerability information with the network owners, but, more frequently, these accesses are used for malicious intent. Some threats are politically motivated and use cyberspace to spread their message. The activities of these small-scale threats can be co-opted by more sophisticated threats, such as criminal organizations or nation-states, often without their knowledge, to execute operations against targets while concealing the identity of the threat/sponsor and also creating plausible deniability.

Accidents and Natural Hazards

The physical infrastructure of cyberspace is routinely disrupted by operator errors, industrial accidents, and natural disasters. These unpredictable events can have greater impact on joint operations than the actions of enemies. Recovery from accidents and hazardous incidents can be complicated by the requirement for significant coordination external to DOD and/or the temporary reliance on back-up systems with which operators may not be proficient.

B. Anonymity and Difficulties with Attribution

To initiate an appropriate defensive response, attribution of threats in cyberspace is crucial for any actions external to the defended cyberspace beyond that authorized as authorized self-defense. The most challenging aspect of attributing actions in cyberspace is connecting a particular cyber-persona or action to a named individual, group, or nation-state, with sufficient confidence and verifiability to hold them accountable. This effort requires significant analysis and, often, collaboration with non-cyberspace agencies or organizations. The nature of cyberspace, government policies, and laws, both domestic and international, presents challenges to determining the exact origin of cyberspace threats. The ability to hide the sponsor and/ or the threat behind a particular malicious effect in cyberspace makes it difficult to determine how, when, and where to respond. The design of the Internet lends itself to anonymity and, combined with applications intended to hide the identity of users, attribution will continue to be a challenge for the foreseeable future.

C. Geography Challenges

In cyberspace, there is no stateless maneuver space. Therefore, when US military forces maneuver in foreign cyberspace, mission and policy requirements may require they maneuver clandestinely without the knowledge of the state where the infrastructure is located. Because CO can often be executed remotely, through a virtual presence enabled by wired or wireless access, many CO do not require physical proximity to the target but use remote actions to create effects, which represents an increase in operational reach not available in the physical domains. This use of global reach applies equally to both external operations in red and gray cyberspace, as well as internal protection effects in blue cyberspace. The cumulative effects of some CO may extend beyond the initial target, a joint operations area (JOA), or outside of a single area of responsibility (AOR). Because of transregional considerations and the requirement for high-demand forces and capabilities, some CO are coordinated, integrated, and synchronized using centralized execution from a location remote from the supported commander.

D. Technology Challenges

Using a cyberspace capability that relies on exploitation of technical vulnerabilities in the target may reveal its functionality and compromise the capability's effectiveness for future missions. This has implications for both offensive cyberspace operations (OCO) and defensive cyberspace operations (DCO) missions. Cyberspace capabilities without hardware components can be replicated for little or no cost. This means that once discovered, these capabilities will be widely available to adversaries, in some cases before security measures in the DODIN can be updated to account for the new threat. In addition, since similar technologies around the world share similar vulnerabilities, a single adversary may be able to exploit multiple targets at once using the same malware or exploitation tactic. Malware can be modified (or be designed to automatically modify itself), complicating efforts to detect and eradicate it.

(Joint Cyberpace Operations) I. Overview 1-13

E. Private Industry and Public Infrastructure

Many of DOD's critical functions and operations rely on contracted commercial assets, including Internet service providers (ISPs) and global supply chains, over which DOD and its forces have no direct authority. This includes both data storage services and applications provided from a cloud computing architecture. Cloud computing enables DOD to consolidate infrastructure, leverage commodity IT functions, and eliminate functional redundancies while improving continuity of operations. But, the overall success of these initiatives depends upon well-executed risk mitigation and protection measures, defined and understood by both DOD components and industry. Dependency on commercial Internet providers means DOD coordination with the Department of Homeland Security (DHS), other interagency partners, and the private sector is essential to establish and maintain security of DOD's information. DOD supports DHS, which leads interagency efforts to identify and mitigate cyberspace vulnerabilities in the nation's critical infrastructure. DOD has the lead for improving security of the defense industrial base (DIB) sector, which includes major sector contractors and major contractor support to operations regardless of corporate country of domicile and continues to support the development of whole-of¬government approaches for its risk management. The global technology supply chain affects mission-critical aspects of the DOD enterprise, and the resulting IT risks can only be effectively mitigated through public-private sector cooperation.

Globalization

The combination of DOD's global operations with its reliance on cyberspace and associated technologies means DOD often procures mission-essential IT products and services from foreign vendors. A prime example is our reliance on network backbones and transmission equipment in other countries, such as undersea cables, fiber optic networks and telecommunications services, satellite and microwave antennas, and leased channels on foreign satellites. These systems may normally be reliable and trustworthy, but they can also leave US forces vulnerable to access denial by service interruption, communications interception and monitoring, or infiltration and data compromise. Another example is DOD's use of commercial, globally interconnected, globally sourced IT components in mission-critical systems and networks. Leveraging rapid technology development of the commercial marketplace remains a key DOD advantage. While globally sourced technology provides innumerable benefits to DOD, it also provides adversaries the opportunity to compromise the supply chain to access or alter data and hardware, corrupt products, and to intercept or deny communications and other mission-critical functions. Supply chain risks threaten all users and our collective security; therefore, DOD cannot ignore these risks to its missions. Globalization, including by US companies, introduces risks across the entire system lifecycle, to include design, manufacturing, production, distribution, operation and maintenance, and disposal of a system or component. Each of these lifecycle stages presents the opportunity to manipulate, deny, or collect information on such systems. It is not feasible to eliminate our reliance on foreign-owned services and products, but our reliance on them makes it essential every reasonable avenue for risk mitigation be pursued, to include user and commander education at all levels, encryption, C2 system redundancy, OPSEC, and careful inspection of vendor-provided equipment in accordance with DOD IT procurement policy.

Mitigations

DOD partners with the DIB to increase the security of information about DOD programs residing on or transiting DIB unclassified networks. The Department of Defense Cyber Crime Center (DC3) serves as DOD's operational focal point for voluntary cyberspace information sharing and incident reporting program. In addition, DOD is strengthening its acquisition regulations to require consideration of applicable cybersecurity policies during procurement of all DODIN components to reduce risks to joint operations

Chap 1
II. Cyberspace Operations Core Activities

Ref: JP 3-12, Cyberspace Operations (Jun '18), chap. II.

Cyberspace Operations (CO)
CO are the employment of cyberspace capabilities where the primary purpose is to achieve objectives in or through cyberspace. CO comprise the military, national intelligence, and ordinary business operations of DOD in and through cyberspace. Although commanders need awareness of the potential impact of the other types of DOD CO on their operations, the military component of CO is the only one guided by joint doctrine and is the focus of this publication. CCDRs and Services use CO to create effects in and through cyberspace in support of military objectives. Military operations in cyberspace are organized into missions executed through a combination of specific actions that contribute to achieving a commander's objective. Various DOD agencies and components conduct national intelligence, ordinary business, and other activities in cyberspace. Although discussed briefly here for context, these activities are guided by DOD policies concerning CO. While joint doctrine does apply to CSAs where it directly relates to their mission to support military forces, CSAs and other DOD agencies and activities also conduct various CO activities that are considered cyberspace-enabled activities.

Cyberspace-Enabled Activities
Most DOD cyberspace actions use cyberspace to enable other types of activities, which employ cyberspace capabilities to complete tasks but are not undertaken as part of one of the three CO missions: OCO, DCO, or DODIN operations. These uses include actions like operating a C2 or logistics system, sending an e-mail to support an information objective, using the Internet to complete an online training course, or developing a briefing. Other than being an authorized user of the network, DOD personnel need no special authorities to use cyberspace capabilities in this way. It is through these uses of cyberspace that the majority of DODIN vulnerabilities are exposed to, and exploited by, our adversaries. The challenge is to train all DODIN users to understand the significance of cyberspace threats and to recognize threat tactics so these uses of cyberspace do not create unnecessary risk to the mission. Protecting the DODIN by establishing a culture of vulnerability awareness, particularly through DOD and interagency policies, practices, and training, is critical to the success of all types of cyberspace-enabled DOD missions.

I. Cyberspace Missions
All actions in cyberspace that are not cyberspace-enabled activities are taken as part of one of three cyberspace missions: OCO, DCO, or DODIN operations. These three mission types comprehensively cover the activities of the cyberspace forces. The successful execution of CO requires integration and synchronization of these missions. Military cyberspace missions and their included actions are normally authorized by a military order (e.g., execute order [EXORD], operation order [OPORD], tasking order, verbal order), referred to hereafter as mission order, and by authority derived from DOD policy memorandum, directive, or instruction. Cyberspace missions are categorized as OCO, DCO, or DODIN operations based only on the intent or objective of the issuing authority, not based on the cyberspace actions executed, the type of military authority used, the forces assigned to the mission, or the cyberspace capabilities used. Some orders may cover multiple types of missions. For example, a standing order to protect the DODIN may include both DODIN operations

(Joint Cyberspace Operations) II. Core Activities 1-15

Military Operations In and Through Cyberspace

Ref: JP 3-12, Cyberspace Operations (Jun '18), pp. II-2 to II-9.

Referring to Adversary Activities in Cyberspace

DOD CO planning terms may not accurately describe the actions of our adversaries and enemies in cyberspace because their mission objectives and commander's intent may not be known with certainty. Therefore, the term "malicious cyberspace activity" refers to all such activities. If the context of the discussion requires more specific descriptions of this activity, use generic terms (e.g., attack, exploitation, sabotage, maneuver), depending upon the specific effects of the malicious actions.

Fig. II-1 below depicts the primary relationships between the cyberspace missions and actions. The depiction in Figure II-1 of the types of forces that normally conduct each type of CO mission is not intended to limit a JFC's ability to employ the best-qualified unit on any particular mission.

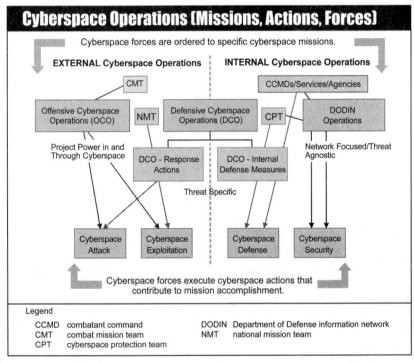

Ref: JP 3-12, Cyberspace Operations (Jun '18), fig. II-1. Cyberspace Operations Missions, Actions, and Forces.

Cyberspace MISSIONS

All actions in cyberspace that are not cyberspace-enabled activities are taken as part of one of three cyberspace missions: OCO, DCO, or DODIN operations. These three mission types comprehensively cover the activities of the cyberspace forces. The successful execution of CO requires integration and synchronization of these missions. Military cyberspace missions and their included actions are normally authorized by a military order (e.g., execute order [EXORD], operation order [OPORD], tasking order, verbal order), referred to hereafter as mission order, and by authority derived from DOD policy memorandum, directive, or instruction. Cyberspace missions are categorized as OCO, DCO, or DODIN operations based only on the intent or objective of the issuing authority, not based on the cyberspace actions executed, the type of military authority used, the forces assigned to the mission, or the cyberspace capabilities used. Some orders may cover multiple types of missions.

See pp. 1-15 to 1-19 for discussion of cyberspace MISSIONS.

Cyberspace ACTIONS

Execution of any OCO, DCO, or DODIN operations mission requires completion of specific tactical-level actions or tasks that employ cyberspace capabilities to create effects in cyberspace. All cyberspace mission objectives are achieved by the combination of one or more of these actions, which are defined exclusively by the types of effects they create. To plan for, authorize, and assess these actions, it is important the commander and staff clearly understand which actions have been authorized under their current mission order. For example, the transition from DODIN operations to DCO-IDM missions may need to occur quickly whenever the DODIN is threatened and cyberspace operators begin to take cyberspace defense actions. To enable and synchronize this transition and subsequent cyberspace defense actions, clear orders are required that communicate to cyberspace operators the applicable constraints, restraints, and authorities. Since they will always be necessary, standing orders for DODIN operations and DCO-IDM missions cover most cyberspace security and initial cyberspace defense actions. However, OCO and DCO-RA missions are episodic. They may require clandestine maneuver and collection actions or may require overt actions, including fires. Therefore, the approval for CO actions in foreign cyberspace requires separate OCO or DCO-RA mission authorities.

See pp. 1-20 to 1-22 for discussion of cyberspace ACTIONS.

National Intelligence Operations In and Through Cyberspace

National-level intelligence organizations conduct intelligence activities in, through, and about cyberspace in response to national intelligence priorities. This intelligence can support a military commander's planning and preparation. Although DOD's cyberspace forces may collect tactically and operationally useful information while maneuvering to and through foreign cyberspace, like all joint forces, they also depend on intelligence support from traditional military and national intelligence sources.

Department of Defense Ordinary Business Operations In and Through Cyberspace

Ordinary business operations in and through cyberspace are "cyberspace-enabled activities" that comprise those non-intelligence and non-warfighting capabilities, functions, and actions used to support and sustain DOD forces and components. This includes the cyberspace-enabled functions of the civilian-run DOD agencies and activities, such as the Defense Finance and Accounting Service and the Defense Contract Audit Agency. Since the conduct of DOD ordinary business operations in cyberspace is guided by DOD policy and not generally by joint doctrine, it is not discussed here in detail. However, vulnerabilities that may exist in the applications and devices used for DOD ordinary business operations might be exploited in a manner that directly impacts a military commander's mission. Since DOD agencies and activities use many of the same networks as military commanders, a compromise in any area of the DODIN used for business operations might result in a loss of mission assurance in cyberspace for military operations.

(Joint Cyberspace Operations) II. Core Activities 1-17

and DCO mission components, and an order for an external mission could support both offensive and defensive objectives. "Cyberspace Actions," discusses the specific actions used in the execution of these missions. Effective execution of all cyberspace missions requires timely intelligence and threat indicators from traditional and cyberspace sensors, vulnerability information from DOD and non-DOD sources, and accurate assessment of previous missions. IAW current USG policy, DOD deconflicts missions in foreign cyberspace with the other USG department and agency mission partners who share this responsibility.

Cyberspace Missions

 DODIN Operations

 Offensive Cyberspace Operations (OCO)

 Defensive Cyberspace Operations (DCO)

A. DODIN Operations

The DODIN operations mission includes operational actions taken to secure, configure, operate, extend, maintain, and sustain DOD cyberspace and to create and preserve the confidentiality, availability, and integrity of the DODIN. These include proactive cyberspace security actions which address vulnerabilities of the DODIN or specific segments of the DODIN. It also includes the set-up of tactical networks by deployed forces to extend existing networks, maintenance actions and other non-security actions necessary for the sustainment of the DODIN, and the operation of red teams and other forms of security evaluation and testing. DODIN operations are network-focused and threat-agnostic: the cyberspace forces and workforce undertaking this mission endeavor to prevent all threats from negatively impacting a particular network or system they are assigned to protect. They are threat-informed and use all available intelligence about specific threats to improve the security posture of the network. DODIN operations does not include actions taken under statutory authority of a chief information officer (CIO) to provision cyberspace for operations, including IT architecture development; establishing standards; or designing, building, or otherwise operationalizing DODIN IT for use by a commander.

See chap. 6, DODIN Operations, for a detailed discussion of DODIN operations and the management of networked communication systems.

B. Offensive Cyberspace Operations (OCO)

OCO are CO missions intended to project power in and through foreign cyberspace through actions taken in support of CCDR or national objectives. OCO may exclusively target adversary cyberspace functions or create first-order effects in cyberspace to initiate carefully controlled cascading effects into the physical domains to affect weapon systems, C2 processes, logistics nodes, high-value targets, etc. All CO missions conducted outside of blue cyberspace with a commander's intent other than to defend blue cyberspace from an ongoing or imminent cyberspace threat are OCO missions. Like DCO-RA missions, some OCO missions may include actions that rise to the level of use of force, with physical damage or destruction of enemy systems. Specific effects created depend on the broader operational context, such as the existence or imminence of open hostilities and national policy considerations. OCO missions require a properly coordinated military order and careful consideration of scope, ROE, and measurable objectives.

C. Defensive Cyberspace Operations (DCO)

Ref: JP 3-12, Cyberspace Operations (Jun '18), pp. II-3 to II-5.

DCO missions are executed to defend the DODIN, or other cyberspace DOD cyberspace forces have been ordered to defend, from active threats in cyberspace. Specifically, they are missions intended to preserve the ability to utilize blue cyberspace capabilities and protect data, networks, cyberspace-enabled devices, and other designated systems by defeating on-going or imminent malicious cyberspace activity. This distinguishes DCO missions, which defeat specific threats that have bypassed, breached, or are threatening to breach security measures, from DODIN operations, which endeavor to secure DOD cyberspace from all threats in advance of any specific threat activity. DCO are threat-specific and frequently support mission assurance objectives. DCO missions are conducted in response to specific threats of attack, exploitation, or other effects of malicious cyberspace activity and leverage information from maneuver, intelligence collection, counterintelligence (CI), law enforcement (LE), and other sources as required. DCO include outmaneuvering or interdicting adversaries taking or about to take actions against defended cyberspace elements, or otherwise responding to imminent internal and external cyberspace threats. The goal of DCO is to defeat the threat of a specific adversary and/or to return a compromised network to a secure and functional state. The components of DCO are:

Defensive Cyberspace Operations-Internal Defensive Measures (DCO-IDM)

DCO-IDM are the form of DCO mission where authorized defense actions occur within the defended network or portion of cyberspace. DCO-IDM of the DODIN is authorized by standing order and includes cyberspace defense actions to dynamically reconfirm or reestablish the security of degraded, compromised, or otherwise threatened DOD cyberspace to ensure sufficient access to enable military missions. For compromised DODIN elements, specific tactics include rerouting, reconstituting, restoring, or isolation.

Defensive Cyberspace Operations-Response Actions (DCO-RA)

DCO-RA are the form of DCO mission where actions are taken external to the defended network or portion of cyberspace without the permission of the owner of the affected system. DCO-RA actions are normally in foreign cyberspace. Some DCO-RA missions may include actions that rise to the level of use of force, with physical damage or destruction of enemy systems, depending on broader operational context, such as the existence or imminence of open hostilities, the degree of certainty in attribution of the threat, the damage the threat has caused or is expected to cause, and national policy considerations. DCO-RA missions require a properly coordinated military order and careful consideration of scope, rules of engagement (ROE), and measurable objectives.

Defense of Non-DOD Cyberspace

While DCO generally focus on the DODIN, which includes all of DOD cyberspace, military cyberspace forces prepare to defend any US or other blue cyberspace when ordered. DOD operations rely on many non-DOD segments of cyberspace, including private sector and mission partner networks. Security of this cyberspace is the responsibility of the resource owners, which include other USG departments and agencies, private sector entities, and other partners. Since DOD-associated cyberspace are known targets for malicious cyberspace activity, protection of these non-DOD networks and systems can be a vital component of mission assurance. However, DOD cannot guarantee the robustness of the security standards applied to such networks. The commander's mission risk analysis should account for this uncertainty in the security of non-DOD cyberspace.

II. Cyberspace Actions

Execution of any OCO, DCO, or DODIN operations mission requires completion of specific tactical-level actions or tasks that employ cyberspace capabilities to create effects in cyberspace. All cyberspace mission objectives are achieved by the combination of one or more of these actions, which are defined exclusively by the types of effects they create. To plan for, authorize, and assess these actions, it is important the commander and staff clearly understand which actions have been authorized under their current mission order. For example, the transition from DODIN operations to DCO-IDM missions may need to occur quickly whenever the DODIN is threatened and cyberspace operators begin to take cyberspace defense actions. To enable and synchronize this transition and subsequent cyberspace defense actions, clear orders are required that communicate to cyberspace operators the applicable constraints, restraints, and authorities. Since they will always be necessary, standing orders for DODIN operations and DCO-IDM missions cover most cyberspace security and initial cyberspace defense actions. However, OCO and DCO-RA missions are episodic. They may require clandestine maneuver and collection actions or may require overt actions, including fires. Therefore, the approval for CO actions in foreign cyberspace requires separate OCO or DCO-RA mission authorities. The cyberspace actions are:

Cyberspace Actions

- A. Cyberspace Security
- B. Cyberspace Defense
- C. Cyberspace Exploitation
- D. Cyberspace Attack

A. Cyberspace Security

Cyberspace security actions are taken within protected cyberspace to prevent unauthorized access to, exploitation of, or damage to computers, electronic communications systems, and other IT, including PIT, as well as the information contained therein, to ensure its availability, integrity, authentication, confidentiality, and nonrepudiation.

Note: Joint doctrine uses the term "cyberspace security" to distinguish this tactical-level cyberspace action from the policy and programmatic term "cybersecurity" used in Department of Defense (DOD) and United States Government (USG) policy. To enable effective planning, execution, and assessment, doctrine distinguishes between cyberspace security and cyberspace defense actions, a distinction not made in DOD and USG cybersecurity policy, where the term cybersecurity includes the ideas of both security and defense. Doctrine uses both "cyberspace security" and "cybersecurity," depending upon the context.

See chap. 7, Cybersecurity.

Although they are threat-informed, cyberspace security actions occur in advance of a specific security compromise and are a primary component action of the DODIN operations mission. Cyberspace security actions protect from threats within cyberspace by reducing or eliminating vulnerabilities that may be exploited by an adversary and/or implementing measures to detect malicious cyberspace activities. Examples of cyberspace security actions include increasing password strength, installing a software patch to remove vulnerabilities, encrypting stored data, training users on cyberspace security best practices, restricting access to suspicious Web sites, or blocking traffic on unused router ports.

B. Cyberspace Defense

Cyberspace defense actions are taken within protected cyberspace to defeat specific threats that have breached or are threatening to breach the cyberspace security measures and include actions to detect, characterize, counter, and mitigate threats, including malware or the unauthorized activities of users, and to restore the system to a secure configuration. The CCMD, Service, or DOD agency that owns or operates the network is generally authorized to take these defensive actions except in cases when they would compromise the operations of elements of cyberspace outside the responsibility of the respective CCMD, Service, or agency. In some cases, a CPT will be assigned to assist with re-securing and mitigation actions. JFHQ-DODIN coordinates all defensive actions that impact more than one CCMD or have impacts outside the realm of the network owner. Cyberspace defense actions are the component actions of a DCO-IDM mission. Since the same personnel often perform both cyberspace security and cyberspace defense actions, these actions are collectively referred to as protection.

C. Cyberspace Exploitation

Cyberspace exploitation actions include military intelligence activities, maneuver, information collection, and other enabling actions required to prepare for future military operations. Cyberspace exploitation actions are taken as part of an OCO or DCO-RA mission and include all actions in gray or red cyberspace that do not create cyberspace attack effects. Cyberspace exploitation includes activities to gain intelligence and support operational preparation of the environment for current and future operations through actions such as gaining and maintaining access to networks, systems, and nodes of military value; maneuvering to positions of advantage; and positioning cyberspace capabilities to facilitate follow-on actions. Cyberspace exploitation also supports current and future operations through collection of information, including mapping red and gray cyberspace to support situational awareness; discovering vulnerabilities; enabling target development; and supporting the planning, execution, and assessment of military operations. Cyberspace exploitation actions are deconflicted with other USG departments and agencies IAW national policy.

D. Cyberspace Attack

Cyberspace attack actions create noticeable denial effects (i.e., degradation, disruption, or destruction) in cyberspace or manipulation that leads to denial effects in the physical domains. Unlike cyberspace exploitation actions, which are often intended to remain clandestine to be effective, cyberspace attack actions will be apparent to system operators or users, either immediately or eventually, since they remove some user functionality. Cyberspace attack actions are a form of fires, are taken as part of an OCO or DCO-RA mission, are coordinated with other USG departments and agencies, and are carefully synchronized with planned fires in the physical domains.

Cyberspace attack includes actions to:

Deny

To prevent access to, operation of, or availability of a target function by a specified level for a specified time, by:

- **Degrade**. To deny access to, or operation of, a target to a level represented as a percentage of capacity. Level of degradation is specified. If a specific time is required, it can be specified.

- **Disrupt**. To completely but temporarily deny access to, or operation of, a target for a period of time. A desired start and stop time are normally specified. Disruption can be considered a special case of degradation where the degradation level is 100 percent.

- **Destroy**. To completely and irreparably deny access to, or operation of, a target. Destruction maximizes the time and amount of denial. However, destruction is scoped according to the span of a conflict, since many targets, given enough time and resources, can be reconstituted.

Manipulate

Manipulation, as a form of cyberspace attack, controls or changes information, information systems, and/or networks in gray or red cyberspace to create physical denial effects, using deception, decoying, conditioning, spoofing, falsification, and other similar techniques. It uses an adversary's information resources for friendly purposes, to create denial effects not immediately apparent in cyberspace. The targeted network may appear to operate normally until secondary or tertiary effects, including physical effects, reveal evidence of the logical first-order effect.

Countermeasures in Cyberspace

Countermeasures are that form of military science that, by the employment of devices and/or techniques, has as its objective the impairment of the operational effectiveness of enemy activity. In cyberspace, the term applies to any CO actions that fit the description of the term, regardless of where the countermeasure is taken. As in the physical domains, countermeasure actions can be taken either internal or external to the defended terrain and can be used preemptively or reactively. Internal countermeasures are cyberspace defense actions taken as part of a DCO-IDM mission; for example, closing router ports being used by an adversary for unauthorized access or blocking malware that is beaconing out of the DODIN. External countermeasures, which would be part of a DCO-RA or OCO mission, are employed beyond the DODIN boundary against a specific malicious cyberspace activity. In support of an OCO mission, they may be cyberspace attack actions that spoof or otherwise negate the effectiveness of adversary sensors or defenses. As part of a DCO-RA mission, they may be used to identify the source of a threat and/or use non-intrusive or minimally intrusive techniques to interdict or mitigate threats. External defensive countermeasures are normally nondestructive/nonlethal in nature, typically impact only malicious activity but not the associated threat systems and terminate when the threat stops. All external countermeasures are subject to the same synchronization, deconfliction, legal, and policy guidance as any other aspect of an OCO or DCO-RA mission.

III. Assignment of Cyberspace Forces to CO
Ref: JP 3-12, Cyberspace Operations (Jun '18), pp. II-8 to II-9.

Mission orders or other directives assign cyberspace forces to specific cyberspace missions, as depicted in figure II-1 *(see p. 1-16)*.

Forces and Workforce Conducting DODIN Operations and DCO-IDM

Service-retained cyberspace forces, CCMD cyberspace forces, RC forces, and DOD agency and activity staffs execute much of the DODIN operations required to secure and operate the various backbones, sub-nets, segments, enclaves, and private networks of the DODIN under the planning, direction, integration, and synchronization of the JFHQ¬DODIN. These staffs include CSSPs established by the Services and DOD agencies to provide DODIN protection services under support agreements with system owners. Although they are not military forces, contracted personnel protect some segments of the DODIN. Note also that other, non-cyberspace forces conduct DODIN operations as an integral part of assigned duties. For example, operators of PIT have an implied responsibility to protect their equipment from threats in cyberspace and require specialized training to detect and defeat cyberspace threats. Protecting PIT from malicious cyberspace activity is complicated by the design of these systems, which are often developed with little consideration of cyberspace threats. Regardless of which personnel and DODIN segments are involved, when personnel with DODIN security responsibilities detect compromise of cyberspace security measures, they transition, IAW standing authorities delegated by the commander, to the cyberspace defense actions of DCO-IDM to restore security to their assigned portion of the DODIN. Their effectiveness in making this transition depends upon their level of training and resources to detect and respond to threats. If discovery and mitigation of malicious cyberspace activity requires expertise beyond that available to the network operator and/or the ISP, CPTs may respond to provide support conducting cyberspace defense actions, either remotely or by deploying to the affected location. CPTs perform other tasks to support network operators, including penetration testing, security surveys, and assessment. National-level CPT support can be extended to defend non-DOD mission partner or critical infrastructure networks when ordered by SecDef.

Forces Conducting DCO-RA and OCO

DCO-RA missions are normally assigned to NMTs, which are tactical units of the CNMF that defend the DODIN, or other blue cyberspace when ordered. The NMTs are aligned under the CNMF-HQ against specific cyberspace threats. OCO missions are normally assigned to CMTs, tactical units of the CCMF that support CCDR plans and priorities to project power in support of national objectives. The CMTs are aligned, under the JFHQs-C, in support of CCMDs. In addition to NMTs and CMTs, there are NSTs and CSTs not depicted in Figure II-1 that provide specialized technical and analytic support for the units of the CMF. This support includes intelligence analysis, cyberspace capability development, linguist support, and planning.

See pp. 1-48 to 1-50 for discussion of command and control (C2) of cyberspace operations.

IV. The Joint Functions and Cyberspace Operations

JP 3-0, Joint Operations, delineates joint functions common to joint operations at all levels of warfare. These joint functions comprise related capabilities and activities grouped together to help commanders integrate, synchronize, and direct joint operations. This section presents an overview of how military operations leverage cyberspace capabilities to enable these functions in support of all DOD missions and how the functions themselves are accomplished in cyberspace during CO.

Joint Functions

- **A** Command and Control
- **B** Intelligence
- **C** Fires
- **D** Movement and Maneuver
- **E** Sustainment
- **F** Protection
- **G** Information

Joint Functions

Joint functions are related capabilities and activities grouped together to help JFCs integrate, synchronize, and direct joint operations. Functions common to joint operations at all levels of warfare fall into seven basic groups—C2, information, intelligence, fires, movement and maneuver, protection, and sustainment. Some functions, such as C2, information, and intelligence, apply to all operations. Others, such as fires, apply as the JFC's mission requires. A number of subordinate tasks, missions, and related capabilities help define each function, and some could apply to more than one joint function.

Refer to JFODS5-1: The Joint Forces Operations & Doctrine SMARTbook (Guide to Joint, Multinational & Interorganizational Operations). Updated for 2019, topics include joint doctrine fundamentals (JP 1), joint operations (JP 3-0 w/Chg 1), an expanded discussion of joint functions, joint planning (JP 5-0), joint logistics (JP 4-0), joint task forces (JP 3-33), joint force operations (JPs 3-30, 3-31, 3-32 & 3-05), multinational operations (JP 3-16), interorganizational cooperation (JP 3-08), & more!

A. Command and Control (C2)

Discussion of C2 and cyberspace requires a distinction between using cyberspace systems that implement the C2 of military operations and the C2 of forces that execute CO. C2 encompasses the exercise of authority and direction by commanders over assigned and attached forces in the accomplishment of their mission. Use of cyberspace as a means of exchanging communications is overwhelmingly the most common method at the strategic and operational levels of warfare and is increasingly important in tactical warfare. Digital communications methods have largely supplanted analog communications, except at the tactical level, where analog signaling methods remain. Analog communications will likely persist indefinitely in tactical operations for reasons of simplicity, reliability, and security. However, military C2 systems that function by the transmission of digital data are part of the DODIN. Cyberspace provides communications pathways, planning and decision-support aids, and cyberspace-related intelligence to enable timely decision making and execution of those decisions. This provides the commander the advantage of controlling the timing and tempo of operations. Cyberspace offers an exceptionally diverse array of circuits for issuance of commands and signals to forces and for those forces to relay operational information back up the chain of command. Military orders converted to digital form, including digital voice and video, can travel on circuits that transit all of the physical domains, significantly increasing the likelihood of timely delivery. However, a commander's confidence in the C2 system can be easily compromised when the security of the DODIN becomes suspect; therefore, the more the commander relies on cyberspace for C2, the more important protection of supporting cyberspace assets is to this joint function.

Refer to JP 3-30, Command and Control of Joint Air Operations; JP 3-31, Command and Control for Joint Land Operations; and JP 3-32, Command and Control of Joint Maritime Operations, for more information on how cyberspace is used to enable operations in the physical domains.

B. Intelligence

Understanding the OE is fundamental to all joint operations, including CO. Intelligence may be derived from information gained during military operations in cyberspace or from other sources. The process includes:

- Planning and direction, to include identification of target vulnerabilities to enable continuous planning and direction of CI activities to protect against espionage, sabotage, and attacks against US citizens/facilities and continuously examining mission success criteria and associated metrics to assess the impact of CO and inform the commander's decisions.
- Collection sensors with access to information about cyberspace.
- Processing and exploitation of collected data, including identification of useful information from collected data, either real-time or after-the-fact.
- Analysis of information and production of intelligence products.
- Dissemination and integration of intelligence related to cyberspace with operations.
- Evaluation and feedback regarding intelligence effectiveness and quality.

Intelligence operations in cyberspace not conducted by a military commander are covered on p. 1-17, National Intelligence Operations In and Through Cyberspace. All-source intelligence support to CO utilizes the same intelligence process used by all other military operations, with unique attributes necessary for support of CO planning. See pp. 1-43 to 1-45, Intelligence and Operational Analytic Support to Cyberspace Operations Planning, for further discussion.

C. Fires

Cyberspace attack capabilities create fires in and through cyberspace and are often employed with little or no associated physical destruction. However, modification or destruction of computers that control physical processes can lead to cascading effects (including collateral effects) in the physical domains. Depending upon the commander's objective, fires in cyberspace can be offensive or defensive, supporting or supported. Like all forms of fires, fires in and through cyberspace should be included in the joint planning and execution processes to facilitate synchronization and unity of effort and must comply with the law of war and ROE. Fires in and through cyberspace encompass a number of tasks, actions, and processes, including targeting, coordination, and deconfliction. If multiple USG or allied entities have requirements to create effects or collect intelligence on the same target in cyberspace, synchronization and deconfliction across all USG entities will be required, otherwise their uncoordinated actions could expose or interfere with each other. Even if effects can be created independently and are sufficiently justified, a technical analysis is still required to determine if the capabilities can operate as planned in the same environment without interference or increasing the chances of unwanted detection.

See pp. 1-46 to 1-48 for discussion of targeting during CO. See pp. 4-29 to 4-34, Targeting (D3A), for related discussion from FM 3-12. Refer to JP 3-60, Joint Targeting, for more information on joint targeting.

D. Movement and Maneuver

Movement and maneuver involves deploying forces and capabilities into an OA and positioning within that area to gain operational advantage in support of mission objectives, including accessing and, as necessary, controlling key terrain. Cyberspace operations enable force projection without the need to establish a physical presence in foreign territory. Maneuver in the DODIN or other blue cyberspace includes positioning of forces, sensors, and defenses to best secure areas of cyberspace or engage in defensive actions as required. Maneuver in gray and red cyberspace is a cyberspace exploitation action and includes such activities as gaining access to adversary, enemy, or intermediary links and nodes and shaping this cyberspace to support future actions. The ability to access or even control such terrain can change the outcome of an engagement. A significant factor in maneuverability in cyberspace is gaining and maintaining logical access to the environment. This capability to maneuver and provide operational reach may be lost at any time if the configuration of the relevant cyberspace nodes are modified. The ubiquitous nature of cyberspace creates another major consideration, because it enables an adversary or enemy to establish key points of presence outside the physical OA, in third-party countries, protected areas, or even inside the US. Additionally, adversaries or enemies may conduct CO from physical network connections within the US, PNs, or third-party nations, thereby limiting the JFC's maneuver space based on law and policy restriction and creating dependencies on our ability to coordinate with interagency and other mission partners.

Another component of maneuver in cyberspace is the ability to move data to a place or process where it has maximum military utility, including movement of data out of harm's way and into a secure location or process. Because of network latencies and performance differences between system messaging models, remote data stores are generally slower than local data stores. This could make the difference between success and failure in CO. In this context, having access to secure wired or wireless bandwidth is analogous to maintaining LOCs in the physical domains. The ability to divert the flow of data from one physical link to another in the face of threats, for example from terrestrial cables to satellite communications (SATCOM) links, is an example of freedom of maneuver in cyberspace. Therefore, managing the EMS within the battlespace is a key planning consideration for CO.

E. Sustainment

Sustainment is the provision of logistics and personnel services to maintain operations through mission accomplishment and redeployment of the force. From the perspective of cyberspace-enabled activities in support of global logistics, DOD relies on protected DODIN and commercial network segments to coordinate sustainment of forces.

Rapid advancements in IT require the development, fielding, and sustainment of cyberspace capabilities adaptable to the changing OE. For example, secure, wireless mobile devices provide anonymity for adversary Internet users; an adversary might update or change operating systems; or they may transition to using more secure virtual machines in their network architecture. Joint forces need the capability to adapt by rapidly incorporating new cyberspace capabilities into their arsenal. Additionally, the joint force may need the capability to quickly upgrade their own cyberspace to leverage these same new technologies. However, pressure to deploy new technology should be balanced against the potential for increased risk and the requirements of cybersecurity policy, and implementation should be carefully orchestrated to prevent divergence among Service-provisioned cyberspace that could create vulnerabilities in DODIN architecture.

Sustainment planning should identify and address legacy systems. Many legacy mission-critical systems were not designed and configured to be easily updated. As a result, many of the vulnerabilities incurred on the DODIN are introduced via unpatched (and effectively un-patchable) systems. These vulnerabilities can be mitigated through additional layers of protection, which must then be sustained. Additionally, hardware capabilities, including sensors and other forward-deployed cyberspace capabilities, can deteriorate over time due to wear and tear or adversary discovery, requiring component repair or replacement to remain operable. This can be particularly problematic when physically inaccessible systems (such as those deployed to remote sites) require replacement or upgrade. It is vital that commanders understand the mission risk created by leaving such cyberspace capabilities in place over long periods, not just to current operations but to the success of future DOD missions that rely on such capabilities. Finally, contingency software capabilities that are infrequently accessed may also require periodic refreshing and retesting to verify they are still secure and capable of creating the required effects, despite changes in the OE.

F. Protection

Protection of the DODIN and other critical US cyberspace includes the continuous and synchronized integration of cyberspace security and, when required, cyberspace defense actions. Protection of cyberspace assets is complicated by their logical connectivity that can enable enemies to create multiple, cascading effects that may not be restricted by physical geography and civil/military boundaries. Cyberspace capabilities requiring protection include not only the infrastructure (computers, cables, antennas, and switching and routing equipment) but also parts of the EMS (datalink frequencies to include satellite downlink, cellular, and wireless) and the content (both data and applications) on which military operations rely. Key to cyberspace protection is the positive control of all direct connections between the DODIN and the Internet and other public portions of cyberspace, as well as the ability to monitor, detect, and prevent the entrance of malicious network traffic and unauthorized exfiltration of information through these connections.

Protection of blue cyberspace uses a combination of security and defensive cyberspace capabilities. Due to the speed of effects and the number of elements in cyberspace, automated procedures to defend cyberspace, verify configurations, and discover network vulnerabilities often provide a better chance of initial success against an aggressor than the manual equivalents. Several factors work against achieving perfect security of a collection of networks and systems as complex as the

DODIN. Therefore, mission-critical parts of the DODIN which provide an advantage to either combatant are considered key terrain and given priority for protection. Even the strongest encryption and most secure protocols cannot protect the DODIN from poorly trained and/or unmotivated users who do not employ proper security practices. Therefore, the training of all DODIN users on appropriate behaviors and commander's strict enforcement of cyberspace security best practices is part of an overall risk management program. Commanders are accountable for the actions of their personnel in cyberspace and should ensure clear understanding at all levels of the command of cyberspace security standards, expectations, and best practices to protect cyberspace.

Protection of cyberspace capabilities requires strict adherence to unique OPSEC countermeasures, since these operations might be thwarted if discovered in advance of their effects. Concealment of movement within cyberspace uses different techniques than concealment in the physical domains. Skills such as avoiding detection are fundamental to most external missions and, therefore, essential to many joint military CO.

For more information on OPSEC, refer to JP 3-13.3, Operations Security.

G. Information

The information function encompasses the management and application of information and its deliberate integration with other joint functions to influence relevant actor perceptions, behavior, and/or action or inaction and support human and automated decision making. The information function helps commanders and staffs understand and leverage the pervasive nature of information, its military uses, and its application during all military operations. This function provides JFCs the ability to integrate the generation and preservation of friendly information while leveraging the inherent informational aspects of all military activities to achieve the commander's objectives and attain the end state. This joint force function supports actions that achieve objectives within the operational and information environments. Given the aim of CO is to achieve objectives within cyberspace and cyberspace is wholly contained within the information environment, it is important to understand its relationship with the information joint function.

The joint force conducts CO in concert with other capabilities, to gain and maintain an advantage. Cyberspace is a medium through which specific information capabilities, such as MISO or MILDEC may be employed. Note that while some operations in the information environment may be done using only CO, they are still synchronized, integrated, and deconflicted with other activities and operations that impact the commander's objectives.

It is important to understand, that although CO will enable certain primary activities within the information function, there are information activities that do not involve CO. Therefore, failure to synchronize CO with other military operations planning and execution can result in friendly forces conducting redundant or conflicting information activities, resulting in wasted time and resources and loss of operational advantage.

See pp. 0-12 to 0-15 for discussion of the information environment, information as a joint function, information operations, and the primary activities that support the information joint function. See also pp. 4-45 to 4-50 for related discussion of the integrating/coordinating functions of information operations (IO) and pp. 4-51 to 4-54 for related discussion of IO planning considerations.

III. Authorities, Roles, & Responsibilities

Ref: JP 3-12, Cyberspace Operations (Jun '18), chap. III.

> "The Defense Department (DOD) requires the commitment and coordination of multiple leaders and communities across DOD and the broader US [G] government to carry out its missions and execute this strategy. Defense Department law enforcement, intelligence, counterintelligence, and policy organizations all have an active role, as do the men and women that build and operate DOD's networks and information technology systems. Every organization needs to play its part."
>
> - Ashton B. Carter Secretary of Defense, The Department of Defense Cyber Strategy, April 17, 2015

I. Introduction

Under the authorities of SecDef, DOD uses cyberspace capabilities to shape cyberspace and provide integrated offensive and defensive options for the defense of the nation. USCYBERCOM coordinates with CCMDs, the JS, and the Office of the Secretary of Defense (OSD); liaises with other USG departments and agencies; and, in conjunction with DHS, DOD's DC3, and the Defense Security Service, liaises with members of the DIB. Similarly, as directed, DOD deploys necessary resources to support efforts of other USG departments and agencies, and allies.

The National Military Strategy and The Department of Defense Cyber Strategy provide high-level requirements for national defense in cyberspace and DOD's role in defending DOD and larger US national security interests through CO.

DOD's Roles and Initiatives in Cyberspace

DOD's roles in cyberspace are, for the most part, the same as they are for the physical domains. As a part of its role to defend the nation from threats in cyberspace, DOD prepares to support DHS and the Department of Justice (DOJ), the USG leads for incident response activities during a national cybersecurity incident of significant consequences. To fulfill this mission, DOD conducts military operations to defend DOD elements of CI/KR and, when ordered, defend CI/KR related to vital US interests. DOD's national defense missions, when authorized by Presidential orders or standing authorities, take primacy over the standing missions of other departments or agencies. The Department of Defense Cyber Strategy establishes strategic initiatives that offer a roadmap for DOD to operate effectively in cyberspace, defend national interests, and achieve national security objectives.

National Incident Response

When directed, DOD provides cyberspace defense support during major cyberspace threat events to the US. DOD coordinates with the requesting agency or department through the lead response department or agency, as described in the Presidential Policy Directive (PPD)-41, United States Cyber Incident Coordination. When DHS requests such support, the fundamental principles of DSCA used to respond to domestic emergencies in the physical domains also apply to CO support.

CI/KR Protection

CI/KR consist of the infrastructure and assets vital to the nation's security, governance, public health and safety, economy, and public confidence. IAW the National Infrastructure Protection Plan, DOD is designated as the sector-specific agency for the DIB. DOD provides cyberspace analysis and forensics support via the DIB Cybersecurity and Information Assurance Program and the DC3. Concurrent with its national defense and incident response missions, DOD may be directed to support DHS and other USG departments and agencies to help ensure all sectors of cyberspace CI/KR are available to support national objectives. CI/KR protection relies on analysis, warning, information sharing, risk management, vulnerability identification and mitigation, and aid to national recovery efforts. Defense critical infrastructure (DCI) is a subset of CI/KR that includes DOD and non-DOD assets essential to project, support, and sustain military forces and operations worldwide. Geographic combatant commanders (GCCs) have the responsibility to prevent the loss or degradation of DCI within their AORs and coordinate with the DOD asset owner, heads of DOD components, and defense infrastructure sector lead agents to fulfill this responsibility. CCDRs may act to prevent or mitigate the loss or degradation of non-DOD-owned DCI only in coordination with the CJCS and the Under Secretary of Defense for Policy (USD[P]) and at the direction of SecDef IAW Department of Defense Directive (DODD) 3020.40, Mission Assurance (MA). As the lead agent of the DODIN sector of the DCI, the Commander, JFHQ¬DODIN, is responsible for matters pertaining to the identification, prioritization, and remediation of critical DODIN infrastructure issues. Likewise, DOD coordinates and integrates when necessary with DHS for support of efforts to protect the DIB.

II. Authorities

Authority for CO actions undertaken by the US Armed Forces is derived from the US Constitution and federal law. Key laws that apply to DOD include Title 10, USC, Armed Forces; Title 50, USC, War and National Defense; and Title 32, USC, National Guard.

See fig. III-1 (facing page) for a summary of applicable titles of USC as they apply to CO.

Authorities for specific types of military CO are established within SecDef policies, including DOD instructions, directives, and memoranda, as well as in EXORDs and OPORDs authorized by the President or SecDef and subordinate orders issued by commanders approved to execute the subject missions. These include the directive authority for cyberspace operations (DACO), established by CJCS EXORD, that enables DOD-wide synchronized protection of the DODIN. The military missions and related actions of the cyberspace forces remain as described in Chapter II, "Cyberspace Operations Core Activities," regardless of the type of authority under which they are executed.

Refer to JP 3-12 Appendix A, "Classified Planning Considerations for Cyberspace Operations," for additional information on authorities for CO.

III. Roles and Responsibilities

A. SecDef

- Directs the military, intelligence, and ordinary business operations of DOD in cyberspace.
- Provides policy and guidance for employment of forces conducting cyberspace missions through the USD(P), the SecDef's Principal Cyber Advisor, and the Deputy Assistant Secretary of Defense for Cyber Policy.
- Develops and issues the DOD Information Resources Management Strategic Plan through the DOD CIO. The DOD CIO is the DODIN architect and, as

United States Code

Ref: JP 3-12, Cyberspace Operations (Jun '18), fig. III-1, p. III-3.

United States Code (USC)	Title	Key Focus	Principal Organization	Role in Cyberspace
Title 6	Domestic Security	Homeland security	Department of Homeland Security	Security of US cyberspace
Title 10	Armed Forces	National defense	Department of Defense	Man, train, and equip US forces for military operations in cyberspace
Title 18	Crimes and Criminal Procedure	Law enforcement	Department of Justice	Crime prevention, apprehension, and prosecution of criminals operating in cyberspace
Title 28	Judiciary and Judicial Procedure			
Title 32	National Guard	National defense and civil support training and operations, in the US	State Army National Guard, State Air National Guard	Domestic consequence management (if activated for federal service, the National Guard is integrated into the Title 10, USC), *Armed Forces*
Title 40	Public Buildings, Property, and Works	Chief Information Officer roles and responsibilities	All Federal departments and agencies	Establish and enforce standards for acquisition and security of information technologies
Title 44	Public Printing and Documents	Defines basic agency responsibilities and authorities for information security policy	All Federal departments and agencies	The foundation for what we now call cybersecurity activities, as outlined in Department of Defense Instruction, 8530.01, *Cybersecurity Activities Support to DOD Information Network Operations*.
Title 50	War and National Defense	A broad spectrum of military, foreign intelligence, and counterintelligence activities	Commands, Services, and agencies under the Department of Defense and intelligence community agencies aligned under the Office of the Director of National Intelligence	Secure US interests by conducting military and foreign intelligence operations in cyberspace

such, develops, maintains, and enforces compliance with DODIN architecture standards and cybersecurity policy. Inherent in the DOD CIO's architecture responsibility are the responsibilities for interoperability, data sharing, effective use of enterprise services, spectrum management, and DODIN program synchronization.

- Develops and oversees implementation of DOD policy, strategy, programs, and guidance regarding: intelligence; CI; security; sensitive activities; and other intelligence-related matters in cyberspace, to include all intelligence, surveillance, and reconnaissance (ISR) cyberspace activities and associated tasking, processing, exploitation, and dissemination through the Under Secretary of Defense for Intelligence IAW DODD 5143.01, Under Secretary of Defense for Intelligence (USD[I]).
- Coordinates with secretaries of other USG departments to establish appropriate representation and participation of personnel on joint interagency coordination groups (JIACGs), working groups, task forces, and collaboration and deconfliction bodies.

B. CJCS

- As the global integrator, advises the President and SecDef on operational policies, responsibilities, and programs.
- Assists SecDef in implementing operational responses to threats in cyberspace.
- Translates SecDef guidance into orders.
- Ensures cyberspace plans and operations are compatible with other military plans and operations.
- Assists CCDRs in meeting SecDef-approved operational requirements.

C. Service Chiefs

- Provide appropriate administration of and support to cyberspace forces, including Service-retained forces and forces assigned or attached to CCMDs.
- Train and equip cyberspace forces and develop cyberspace capabilities for deployment/support to CCMDs, as directed by SecDef.
- Comply with CDRUSCYBERCOM's direction for security, operation, and defense of their respective Service segments of the DODIN, including applicable direction issued under CDRUSCYBERCOM's DACO, either from USCYBERCOM directly or from JFHQ-DODIN or the SCCs, as delegated.
- Coordinate with CDRUSCYBERCOM to prioritize cyberspace mission requirements and force capabilities.
- Provide users of the EMS with regulatory and operational guidance in the use of frequencies through the authority of Army (Army Spectrum Management Office), Navy (Navy and Marine Corps Spectrum Center), and Air Force (Air Force Spectrum Management Office).

D. Chief, National Guard Bureau (NGB)

- Advises CDRUSCYBERCOM on NGB matters pertaining to CCMD CO missions, and supports planning and coordination for such activities as requested by the CJCS or the CCDRs.
- Serves as the channel of communications on all CO matters pertaining to the NG between USCYBERCOM and the 50 states, the Commonwealth of Puerto Rico, the District of Columbia, Guam, and the US Virgin Islands.
- Responds to direction from USCYBERCOM and JFHQ-DODIN, issued under DACO, to secure, operate, and defend the NGB segments of the DODIN.

E. CDRUSCYBERCOM

Ref: JP 3-12, Cyberspace Operations (Jun '18), fig. III-1, p. III-5 to III-6.

As the coordinating authority for CO, plans, coordinates, integrates, synchronizes, and conducts activities to:
- Direct the security, operations, and defense of the DODIN.
- Prepare to, and when directed, conduct military CO external to the DODIN, including in gray and red cyberspace, in support of national objectives.
- Deconflicts cyberspace exploitation and attack actions IAW national and DOD policy.
- For CO events requiring actions and effects across multiple geographic AORs, CDRUSCYBERCOM is the supported commander. For theater-specific events, CDRUSCYBERCOM may be designated a supporting or supported commander, depending upon the order issued.
- Leverages intelligence community (IC) sensors and directs DODIN sensors, as appropriate, to establish and share comprehensive situational awareness of red and gray cyberspace in support of assigned mission.
- Coordinates with the IC, CCMDs, Services, DOD agencies and activities, and multinational partners to facilitate development of improved cyberspace accesses to support planning and operations.
- As directed, provides military representation to USG departments and agencies, US commercial entities, and international organizations for cyberspace matters.
- Notifies the CCMDs of ongoing or developing cyberspace threats and anomalies to reduce potential risks and effectively integrate systems, networks, services, and EMS usage and to ensure compliance with DOD-mandated DODIN configuration standards.
- Performs analysis of threats to the DODIN, including threat analysis of foreign malicious cyberspace activity. In coordination with CCMDs, changes the global protection posture of the DODIN, as warranted by threat assessments.
- Plans for and, as directed, coordinates or executes DCO of US CI/KR.
- **Commander, JFHQ-DODIN.** In coordination with all CCDRs and other DOD components, conducts the operational-level planning, direction, coordination, execution, and oversight of global DODIN operations and DCO-IDM missions. Maintains support relationships, as established by CDRUSCYBERCOM, with all CCDRs for theater/functional DODIN operations and DCO-IDM. Commander, JFHQ-DODIN, is supported for global DODIN operations and DCO-IDM, and CCDRs are supported for DODIN operations and DCO-IDM with effects contained within their AOR or functional mission area. Exercises DACO over all DOD components as delegated by CDRUSCYBERCOM.
- **Commander, CNMF-HQ.** When directed, conducts the defense of the nation's cyberspace through operational-level planning, coordination, execution, and oversight of DCO-RA missions and, when directed, employment of national CPTs on DCO-IDM missions focused on internal threats to critical blue cyberspace outside the DODIN.
- **Commanders, SCCs.** In coordination with Commander, JFHQ-DODIN, conduct the operational-level planning, direction, coordination, execution, and oversight of DODIN operations and DCO-IDM within their Service portion of the DODIN. To achieve unity of action for protection of the DODIN, as directed, exercise DACO over organizations within their Service that take cyberspace security and cyberspace defense actions. Exercise administrative control of Service cyberspace forces, to include those that are GFMIG-assigned to USCYBERCOM.
- **Commanders, JFHQ-C.** Analyze, plan, and execute CO missions in support of the CCDRs. Focus on refining intelligence requirements (IRs), providing tactical expertise regarding feasibility of courses of action, and integrating CO into CCDR plans and orders.
- **USCYBERCOM Cyberspace Operations-Integrated Planning Element (CO-IPE).** Integrates within a CCDR's CO support staff to provide CO expertise and reachback capability to USCYBERCOM. CO-IPEs are organized from USCYBERCOM, JFHQ-DODIN, and JFHQ-C personnel and are co-located with each CCMD for full integration into their staffs. CO-IPEs provide a CCDR with CO planners and other subject matter experts required to support development of CCMD requirements for CO and to assist CCMD planners with coordinating, integrating, and deconflicting CO.

F. Other CCDRs

- Secure, operate, and defend tactical and constructed DODIN segments within their commands and AORs.
- Integrate CO into plans (e.g., theater and functional campaign plans, concept plans [CONPLANs], and operation plans [OPLANs]); integrate cyberspace capabilities into military operations as required; and work closely with the joint force, USCYBERCOM, SCCs, and DOD agencies to create fully integrated capabilities.
- In coordination with USCYBERCOM, CCDRs orchestrate planning efforts for CO, designate the desired effects of CO, and determine the timing and tempo for CO conducted in support of their missions. Functional CCDRs direct DODIN operations and DCO-IDM over DODIN segments under their control, consistent with their functional responsibilities.
- GCCs lead, prioritize, and direct theater-specific DCO-IDM in response to compromises of DODIN security through the unified command theater network control center or equivalent organization. For cybersecurity events that have been categorized as a global event by USCYBERCOM, CDRUSCYBERCOM is the supported commander for the DCO-IDM, and other CCDRs support response efforts and tasking from JFHQ-DODIN.
- Serve as a focal point for in-theater DODIN operations that integrate multinational partners.
- Plan for communications system support of operations that may be directed by SecDef and ensure the interoperability of DOD forces with non-DOD mission partners in terms of equipment, procedures, and standards.
- Retain authority to approve or deny DOD component-initiated modifications to the DODIN that will impact in-theater operations only.
- In coordination with the DOD asset owner, heads of DOD components, and DOD infrastructure sector lead agents, GCCs act to prevent the loss, degradation, or other denial of DOD-owned DCI within their AORs. Act only in coordination with the CJCS and USD(P) to prevent or mitigate the loss or degradation for non-DOD-owned DCI.
- In coordination with CDRUSCYBERCOM, advocate for cyberspace capabilities and resources needed to support the CCDR's missions.
- Provide users of the EMS with regulatory and operational guidance in the use of required frequencies for CO IAW coordinated agreements between US forces and PNs.

G. Commanders, US Pacific Command and US Northern Command

In addition to responsibilities "Other CCDRs," these CCDRs fulfill specific CO responsibilities related to DSCA and homeland defense with CDRUSCYBERCOM and others, as required.

H. Commander, United States Strategic Command (CDRUSSTRATCOM)

In addition to responsibilities "Other CCDRs," CDRUSSTRATCOM fulfills specific CO-related SATCOM responsibilities.

- Represents the DOD SATCOM community by coordinating and orchestrating consolidated user positions with CCMDs, Services, DOD agencies, and international partners. CDRUSSTRATCOM has operational and configuration management authority for the SATCOM component of the DODIN, including

on-orbit assets, control systems, and DOD ground terminal and gateway infrastructure. Directs day-to-day operations of DOD-owned and leased SATCOM resources, as well as international partner and non-DOD SATCOM resources used by DOD to support mission requirements.
- Develops, coordinates, and executes SATCOM operations policies and procedures; constellation deployment plans; and satellite positioning, repositioning, and disposal plans. Assesses, in collaboration with DISA and JFHQ-DODIN, how these various plans impact communications support to current and future operations, OPLANs, and CONPLANs. Except in the case of emergencies, CDRUSSTRATCOM coordinates SATCOM actions with users prior to execution.

I. Director, Defense Information Systems Agency (DISA)

- Complies with CDRUSCYBERCOM direction, through the commander of JFHQ-DODIN, to execute DODIN operations and DCO-IDM missions at the global and enterprise level, within DISA-operated portions of the DODIN.
- Provides engineering, architecture, and provisioning support for integrated DODIN operations, including enterprise management, content management, and mission assurance.
- Provides shared situational awareness of DISA-operated portions of the DODIN.
- Supports compliance inspections IAW Department of Defense Instruction (DODI) 8530.01, Cybersecurity Activities Support to DOD Information Network Operations.
- Acquires all commercial SATCOM resources (unless the DOD CIO has granted a waiver to the requesting organization). Supports CDRUSSTRATCOM as the Consolidated SATCOM System Expert for commercial SATCOM and DOD gateways.
- Plans, mitigates, and executes service restoration at the global and enterprise level, as directed by commander of JFHQ-DODIN.
- Provides and maintains a critical nodes defense plan for long-haul communications.

J. Director, National Security Agency/Chief, Central Security Service

- Provides signals intelligence (SIGINT) support and cybersecurity guidance and assistance to DOD components and national customers, pursuant to DOD policy (DODI 8500.01, Cybersecurity; DODI, 8530.01, Cybersecurity Activities Support to DOD Information Network Operations; and DODI 8560.01, Communications Security [COMSEC] Monitoring and Information Assurance [IA] Readiness Testing); Executive Order 12333, US Intelligence Activities; and National Security Directive 42, National Policy for the Security of National Security Telecommunications and Information Systems.
- Provides DOD with capacity/capability in both cyberspace security and cyberspace defense products and expertise and intelligence support required to execute CO, including operation of cyberspace perimeter defenses under direction of USCYBERCOM; target development assistance; situational awareness and attack sensing and warning; threat analysis; internal threat hunting; red-teaming and security assist visits; communications monitoring; forensics; linguist support; and other specialized support, as authorized.

K. Director, Defense Intelligence Agency (DIA)

- Provides timely, objective, and cogent military intelligence to warfighters, defense planners, and defense and national security policy makers.
- Conducts all-source analysis in support of CO, to include contributing to the development of CO-related joint intelligence preparation of the OE products.
- Serves as the DOD focal point for all CI cyberspace investigations and operations. In conjunction with the Military Departments and DOD agencies, DIA strives to identify and neutralize all CI-related cyberspace threats to DOD. DIA supports CI operations in cyberspace to promote cyberspace superiority and provides worldwide cyberspace CI situational awareness and coordination.
- In coordination with JS, Services, other DOD agencies and activities, and OSD, engineers, develops, implements, and manages the sensitive compartmented information portion of the DODIN, including the configuration of information, data, and communications standards for intelligence systems. Included within this is the overall responsibility for the operation of Joint Worldwide Intelligence Communications System, a strategic, secure, high-capacity telecommunications network serving the IC with voice, data, and video services. DIA establishes defense-wide intelligence priorities for achieving interoperability between tactical, theater, and national intelligence-related systems and between intelligence-related systems and the tactical, theater, and national elements of the DODIN.
- Sets policies, standards, and requirements for targets, including the virtual elements of facility, individual, organization, and equipment targets. All target development, to include targets in support of CO, adheres to the standards put forth in Chairman of the Joint Chiefs of Staff Instruction (CJCSI) 3370.01, Target Development Standards.

L. Director, DC3

Administratively assigned to the Department of the Air Force but supporting the entire DOD, the DC3:

- Provides digital and multimedia forensics; cyberspace investigative training; research, development, test and evaluation; and cyberspace vulnerability analysis for DODIN protection, LE, IC, CI, and counterterrorism organizations.
- Serves as the DOD center of excellence and establishes DOD standards for digital and multimedia forensics.
- Serves as the operational focal point for the DIB cyberspace security information sharing activities performed to protect unclassified DOD information that transits or resides on unclassified DIB information systems and networks.

M. Other DOD Agencies and Activities

All DOD agencies and activities are responsible for developing and maintaining their IT in a manner consistent with and reflective of applicable DODIN architecture and cybersecurity standards, and they plan, resource, acquire, implement, and maintain agency-specific IT IAW the DOD policy and resource priorities. Those DOD agencies, which are also part of the IC, are additionally subject to the policies and guidance of the IC CIO. All DOD agencies and activities respond to direction from USCYBERCOM and JFHQ-DODIN, issued under DACO, to secure, operate, and defend their segments of the DODIN.

N. Department of Homeland Security (DHS)

- DHS has the responsibility to secure US cyberspace, at the national level, by protecting non-DOD USG networks against cyberspace intrusions and attacks, including actions to reduce and consolidate external access points, deploy passive network defenses and sensors, and define public and private partnerships in support of national cybersecurity policy.

- DHS protects USG network systems from cyberspace threats and partners with government, industry, and academia, as well as the international community, to make cybersecurity a national priority and a shared responsibility.

- Pursuant to the Homeland Security Act of 2002 and Homeland Security Presidential Directive-5, Management of Domestic Incidents, the Secretary of Homeland Security is the principal federal official for domestic incident management. Pursuant to PPD-41, United States Cyber Incident Coordination, DHS is the lead federal agency for cyberspace incident asset response. For significant cybersecurity incidents external to the DODIN and IC networks, DHS's National Cybersecurity and Communications Integration Center is the lead federal agency for technical assistance and vulnerability mitigation.

O. Department of Justice (DOJ)

- DOJ, including the Federal Bureau of Investigation (FBI), leads counterterrorism and CI investigations and related LE activities associated with government and commercial CI/KR. DOJ investigates, defeats, prosecutes, and otherwise reduces foreign intelligence, terrorist, and other cyberspace threats to the nation's CI/KR. The FBI is the lead agency for significant cybersecurity incident threat response activities, except those that affect the DODIN or the IC. Given the ability of malicious cyberspace activity to spread, investigation of threats to the DODIN will need to be coordinated with the FBI.

- The FBI also conducts domestic collection, analysis, and dissemination of cybersecurity threat information and operates the National Cyber Investigative Joint Task Force, a multi-agency focal point for coordinating, integrating, and sharing pertinent information related to cybersecurity threat investigations, with representation from DHS, the IC, DOD, and other agencies as appropriate.

IV. Legal Considerations

Ref: JP 3-12, Cyberspace Operations (Jun '18), fig. III-1, p. III-11 to III-12.

DOD conducts CO consistent with US domestic law, applicable international law, and relevant USG and DOD policies. The laws that restrict military actions in US territory also apply to cyberspace. Therefore, DOD cyberspace forces that operate outside the DODIN, when properly authorized, are generally limited to operating in gray and red cyberspace only, unless they are issued different ROE or conducting DSCA under appropriate authority. Since each CO mission has unique legal considerations, the applicable legal framework depends on the nature of the activities to be conducted, such as OCO or DCO, DSCA, ISP actions, LE and CI activities, intelligence activities, and defense of the homeland. Before conducting CO, commanders, planners, and operators require clear understanding of the relevant legal framework to comply with laws and policies, the application of which may be challenging given the global nature of cyberspace and the geographic orientation of domestic and international law. It is essential commanders, planners, and operators consult with legal counsel during planning and execution of CO.

Application of the Law of War

Members of DOD comply with the law of war during all armed conflicts and in all other military operations. The law of war encompasses all international law for the conduct of armed hostilities binding on the US or its individual citizens, including treaties and international agreements to which the US is a party and applicable customary international law. The law of war rests on fundamental principles of military necessity, proportionality, distinction (discrimination), and avoidance of unnecessary suffering, all of which may apply to certain CO.

Refer to JP 1-04, Legal Support to Military Operations; DODD 2311.01E, DOD Law of War Program; CJCSI 5810.01, Implementation of the DOD Law of War Program; and the Department of Defense Law of War Manual for more information on the law of war.

IV. Planning, Coordination, Execution & Assessment

Ref: JP 3-12, Cyberspace Operations (Jun '18), chap. IV.

I. Joint Planning Process (JPP) and Cyberspace Operations

Commanders integrate CO into their operations at all levels. Their plans should address how to effectively integrate cyberspace capabilities, counter adversaries' use of cyberspace, identify and secure mission-critical cyberspace, access key terrain in cyberspace, operate in a degraded environment, efficiently use limited cyberspace assets, and pair operational requirements with cyberspace capabilities. The commander provides initial planning guidance, which may specify time constraints, outline initial coordination requirements, authorize the movement of forces within the commander's authority, and direct other actions as necessary. Supporting CO plans and concepts describe the role and scope of CO in the commander's effort and address how CO support the execution of the supported plan. If requested by a commander, CDRUSCYBERCOM provides assistance in integrating cyberspace forces and capabilities into the commander's plans and orders.

See pp. 4-41 to 4-44 for discussion of integrating cyberspace planning into the joint planning process.

II. Cyberspace Operations Planning Considerations

Although CO planners are presented the same operational design considerations and challenges as planners for operations in the physical domains, there are some unique considerations for planning CO. For instance, because of unforeseen linkages in cyberspace, higher-order effects of some CO may be more difficult to predict. This may require more branch and sequel planning. Further, while many elements of cyberspace can be mapped geographically, a full understanding of an adversary's disposition and capabilities in cyberspace involves understanding the target, not only at the underlying physical network layer but also at the logical network layer and cyber-persona layer, including profiles of system users and administrators and their relationship to adversary critical factors. For planning internal operations within DOD cyberspace, DODIN operations and DCO-IDM planners require a clear understanding of which friendly forces or capabilities might targeted by an adversary; what DODIN vulnerabilities are most likely to be targeted and the potential effects of the adversary's action; the mission assurance risks involved; and an understanding of applicable domestic, foreign, and international laws and USG policy. Threats in cyberspace may be nation-states, non-state groups, or individuals, and the parts of cyberspace they control are not necessarily within the geographic borders associated with the threat's nationality or proportional to their geopolitical influence. A criminal element, a politically motivated group, or even a well-resourced individual may have a greater presence and capability in cyberspace than do many nations. Moreover, many adversaries operate cyberspace capabilities from portions of cyberspace geographically associated with the US or owned by a US entity. Each of these factors complicates the planning of CO.

A. Planning Timelines

For external missions, it is essential OCO and DCO-RA planners understand the authorities required to execute the specific CO actions proposed. The applicable authorities may vary depending upon the phase of the operation. This includes accounting for the lead time required to obtain the necessary intelligence to define the correct target; develop target access; confirm the appropriate authorities; complete necessary coordination, including interagency coordination and/or synchronization; and to verify the cyberspace capability matches the intended target using the results of technical assurance evaluations. For internal missions, the timelines for DCO-IDM and DODIN operations planners are impacted by other factors, including levels of automation available to manage network posture, availability of security solutions from commercial providers and their licensing requirements, and operational considerations that may impact a defender's abilities to maneuver or take systems off-line to better manage their protection. However, the planning fundamentals remain the same, and despite the additional considerations and challenges of integrating CO, planners use most elements of the traditional processes to implement the commander's intent and guidance.

B. Planning Considerations for Operating in Red and Gray Cyberspace

Characteristics of Cyberspace Capabilities

While cyberspace is complex and ever changing, cyberspace capabilities, whether devices or computer programs, must reliably create the intended effects. However, cyberspace capabilities are developed based on environmental assumptions and expectations about the operating conditions that will be found in the OE. These conditions may be as simple as the type of computer operating system being used by an adversary or as complex as the exact serial number of the hardware or version of the software installed, what system resources are available, and what other applications are expected to be running (or not running) when the cyberspace capability activates on target. These expected conditions should be well documented by the capability developer and are important for planners and targeting personnel to understand as capability limitations. The extent to which the expected environmental conditions of a target cannot be confirmed through ISR sources represents an increased level of risk associated with using the capability. All other factors being equal, cyberspace capabilities that have the fewest environmental dependencies and/or allow the operator to reconfigure the capability are preferred. DODI O-3600.03, Technical Assurance Standard (TAS) for Computer Network Attack (CNA) Capabilities, provides detailed requirements for technical assurance evaluations that document these characteristics.

Cascading, Compounding, and Collateral Effects

Overlaps among military, other government, corporate, and private activities on shared networks in cyberspace make the evaluation of probable cascading, compounding, and collateral effects particularly important when targeting for CO. The effects can ripple through a targeted system, sometimes cascading through links with related systems that were not evident to the planner. Cascading effects sometimes travel through systems subordinate to the one targeted but can also move laterally to peer systems or up to higher-level systems. Compounding effects are an aggregation of various levels of effects that have interacted in ways that may be intended or may have been unforeseen. Collateral effects, including collateral damage, are the incidental effects of military operations on non-combatants and civilian property that were not the intended targets of the strike. Depending upon the strategic and operational situation, an order or applicable ROE may limit CO to only those actions likely to result in no or low levels of collateral effects. A collateral effects estimate to meet policy restrictions is separate from the proportionality analysis required by the law of war. This estimate is a tool for the commander to understand risk when considering approval of operations. Therefore, even if a proposed CO is permissible after a collateral effects analysis, the likely effects of the proposed CO must also be permissible under a law of war proportionality analysis, as applicable.

Reversibility of Effects
An important consideration for planning cyberspace attack and cyberspace exploitation effects is the level of control over the duration of the effect that can be exercised by friendly forces. There are two basic ways to categorize effects by this standard:

- **Operator Reversible Effects**. Effects that can be recalled, recovered, or terminated by friendly forces. These effects may represent a lower risk of undesired consequences, including discovery or retaliation.

- **Non-Operator Reversible Effects**. Effects that cannot be recalled, recovered, or terminated by friendly forces after execution. These effects may represent a higher risk of response from the threat or other undesired consequences and may require more coordination.

Refer to JP 3-12 Appendix A, "Classified Planning Considerations for Cyberspace Operations," for additional planning considerations for external missions. Refer to JP 3-60, Joint Targeting, for additional information on creation of effects. Refer to CJCSI 3160.01, No-Strike and the Collateral Damage Estimation Methodology, for additional information on collateral damage.

C. Planning Considerations for Protecting the DODIN

For Specific Plans and Operations
DODIN operations underpin nearly every aspect of military operations, and this reliance on cyberspace is well understood by our adversaries. However, a commander's reliance on specific segments of the DODIN is often not considered during plans development, but planning for DODIN resiliency is essential. JFC planning staffs should incorporate DCO-IDM branches and sequels for any operations that pose an increased threat to the DODIN. The CCDR's CO staff coordinates and deconflicts DCO-IDM mission activities with the USCYBERCOM CO-IPEs. If the planned defensive actions will create effects in cyberspace outside of the GCC's AOR, JFHQ-DODIN will ensure the cyberspace defense actions are coordinated and synchronized globally.

Prioritizing DODIN Protection
Cybersecurity policies generally apply to all of the DODIN, unless specific exceptions or waivers are granted. Each segment of the DODIN has an organization responsible for its security and first-line defensive actions, including administrative and non-mission-critical networks, which are protected primarily by their operators and their CSSP. Some of these protection services may be contracted, particularly when the creation and operation of the network itself has been contracted. The determination of whether or not a specific piece of contractor hardware or a specific contractor network segment is considered part of the DODIN is determined by the exact language of the contract. Given the limited number of CPTs and other cyberspace forces, the significant scope of the DODIN means not every segment can be defended in the same depth. Primarily, these specialized cyberspace forces focus on protecting the highest priority segments of the DODIN, including mission-critical, classified, and those directly supporting operations. As resources allow, CPTs may assist service providers and network segment operators with defense of lower priority networks.

Coordinating DODIN Defense
Effective response to intrusions or other malicious activity on the DODIN requires coordinated action. Although the ultimate goal of DCO is to defeat the threat and

reestablish secure cyberspace, the nature of the threat determines the specific response to each incident. All cybersecurity incidents are reported IAW DOD policy, but some threat adversary activity may be effectively remediated by well-trained, local cyberspace forces without external support. Sophisticated nation-state threats that penetrate our security measures require a different type of response. Each encounter with a peer or near-peer adversary in cyberspace warrants careful consideration of the response. Choosing when, where, and how to engage the threat is as important in DCO as it is to defense in the physical domains. If circumstances allow, including a consideration of threat to the supported mission, intelligence gain/loss (IGL) considerations may suggest careful observation of the threat while limiting its maneuver. When a command is engaged with a threat in cyberspace, the global enterprise adapts to support that command IAW defensive priorities. Reachback support for analytics, intelligence, and even fires is provided to maintain continuity of operations at the supported command. Local and Service commanders consult with USCYBERCOM and its subordinate HQ staffs to create tailored responses to specific threats. Some incidents require remote or on-site response by CPTs to assist network operators and the assigned CSSP with remediation and restoration of the affected network segment.

Situational Awareness

Cyberspace situational awareness is the requisite current and predictive knowledge of cyberspace and the OE upon which CO depend, including all factors affecting friendly and adversary cyberspace forces. A commander continually assesses the OE through a combination of staff element and other reporting; personal observation; intelligence, to include threat warning; and representations of various activities occurring in the OE using a common operational picture (COP). The DODIN is a primary source of information used to support the commander's situational awareness of the OE, including the status of the DODIN itself. Sustainment of DODIN sensors, communication channels, data feeds, and user interfaces is a key outcome of DODIN operations. Accurate and comprehensive situational awareness is critical for rapid decision making in a constantly changing OE and while engaging an elusive, adaptive adversary. Situational awareness of adversary activity in gray and red cyberspace relies heavily on cyberspace exploitation and SIGINT, but contributions can come from all sources of intelligence. Situational awareness within the DODIN is provided by the Services and agencies operating their portions of the DODIN, by DISA and JFHQ-DODIN through the network operations and security centers, by USCYBERCOM's Joint Operations Center, and by the Joint Functional Component Command for Space's Joint Space Operations Center for SATCOM. They coordinate with each other as required for operational effectiveness and shared situational awareness. The ever-increasing complexity and scope of cyberspace means a commander never has perfect or even optimal situational awareness of cyberspace factors that could impact operations and should consider the risks represented by this lack of information when making decisions.

D. Preparing for Assessment

Assessment is used to measure progress of the joint force toward mission accomplishment. Commanders continuously assess the OE and the progress of operations and compare them to their initial vision and intent. The assessment process begins during the planning process and helps the commander and staff decide what to measure and how to measure it, in order to determine progress toward accomplishing a task, creating an effect, or achieving an objective. The data collected to support these measures can range from simply noting an inability to reach the target network after a cyberspace attack to complex network monitoring and statistical analysis. Data gathered about the target's state prior to the operation, through access, execution, and possibly its long-term post-attack state, may facilitate later assessment of higher-order effects. Assessment of internal missions to protect the DO-

DIN requires similar preparation. It is difficult to determine the degree that protection measures reduce risk to mission without accurate knowledge of the initial conditions of the network. Assessment of CO is not limited to analysis of data from within cyberspace. For example, if the desired effect of an OCO mission was to cause a power outage, the assessment might be made using visual sensors to observe indications of an outage. Planners submit assessment requests, with sufficient justification, as early as is necessary for the appropriate allocation of resources.

See p. 1-55 to 1-56 for discussion of assessment of cyberspace perations. Refer to JP 3-12 Appendix A, Classified Planning Considerations for Cyberspace Operations, for additional information on planning CO.

III. Intelligence and Operational Analytic Support to Cyberspace Operations Planning

A. Intelligence Requirements (IRs)

During mission analysis, the joint force staff identifies significant information gaps about the adversary and other relevant aspects of the OE. After gap analysis, the staff formulates IRs, which are general or specific subjects upon which there is a need for the collection of information or the production of intelligence. Based upon identified IRs, the staff develops more specific questions known as information requirements (those items of information that must be collected and processed to develop the intelligence required by the commander). Information requirements related to cyberspace can include such things as network infrastructures and status, readiness of adversary's equipment and personnel, and unique cyberspace signature identifiers such as hardware/software/firmware versions and configuration files. These IRs are met through a combination of military intelligence and national intelligence sources.

Refer to JP 2-0, Joint Intelligence, for additional information on IRs.

Requests for Information (RFIs)

CO planners can submit an RFI to generate intelligence collection efforts in any part of the OE or discipline in support of the JPP. RFIs are specific, time-sensitive, ad hoc requirements for intelligence information to support an ongoing crisis or operation and not necessarily related to standing requirements or scheduled intelligence production. RFIs fulfill customer requirements and range from disseminating existing products through integrating or tailoring on-hand information to scheduling new collection and production. The RFI manager translating the customer's requirement and the primary intelligence producer determine how best to meet the customer's needs. In addition to information collected during military operations, information required to support CO planning can come from SIGINT, human intelligence, CI, measurement and signature intelligence, geospatial intelligence, or open-source intelligence (OSINT). Regardless of source, the information should be timely, accurate, and in a usable format.

Refer to JP 2-01, Joint and National Intelligence Support to Military Operations, for additional information on RFIs.

Tasking, Collection, Processing, Exploitation, and Dissemination (TCPED) Architecture

The DOD's global connectivity enables commanders to task assigned or attached ISR sensors or assets and submit collection and production requirements directly to other ISR or IC activities.

For more information on TCPED, refer to JP 2-01, Joint and National Intelligence Support to Military Operations.

B. Threat Detection and Characterization

Some threats in cyberspace are detected by intelligence sources and others during the course of military maneuver.

Detection
The activities in cyberspace of a sophisticated threat may be difficult to detect. Unlike actions in the physical domains, which are often detected by the presence of military equipment or other types of observables, threat actions in cyberspace may not be easily distinguishable from legitimate network activity. Detecting of activities in cyberspace is critical for enabling effective CO.

Characterization
Because the DOD cyberspace missions are categorized based on the commander's intent and because friendly forces are often uncertain of a threat's actual intent, threat activities in cyberspace are referred to more generically. Threat actions in cyberspace are generally referred to as malicious cyberspace activity. If known details of adversary activity support more precise categorization, specific threat actions may qualify as cyberspace attack if they have created noticeable denial effects or cyberspace exploitation if the adversary has only maneuvered for collection or enabling purposes.

Analysis and Attribution
Due to the characteristics of the physical network, logical network, and cyber-persona layers of cyberspace, attribution of malicious cyberspace activity to a specific person, criminal organization, non-state threat, or even a responsible nation-state can be exceptionally difficult. Although attribution is not necessarily required for self-defense, the difficulty of attribution, along with the possibility that an apparent threat may actually be an attempt at misdirection, is one of the principal reasons DCO-RA mission planning may be more difficult than planning for response to conventional attack. The risks of a defensive response against the wrong threat, particularly a nation-state or a target within an unwitting nation-state where the attack originated, are weighed against strategic objectives and the consequences of making an attribution mistake. Working effectively within these constraints requires unique skills on the part of all-source intelligence analysts to understand the context of the threat activity. They use skills like analyzing deception techniques, anonymity techniques, virtual representations and avatars, and other artifacts of the logical network and cyber-persona layers to characterize activities with the requisite degree of confidence required to enable an effective response.

C. Intelligence Gain/Loss (IGL)

Another planning concern is that maneuver and fires in red and gray cyberspace could potentially compromise intelligence collection activities sources and methods. To the maximum extent practicable, an IGL assessment is required prior to executing such actions. The IGL assessment can be complicated by the array of non-DOD USG and multinational partners operating in cyberspace. JFCs use IGL analysis to weigh the risks of conducting the CO versus achieving the desired objective via other methods.

D. Warning Intelligence

Cyberspace threat intelligence includes all-source analysis to factor in political, military, and technical warning intelligence. Adversary cyberspace actions may occur separate from, and well in advance of, related activities in the physical domains. Additionally, cyberspace threat sensors may recognize malicious activity with only a very short time available to respond. These factors make the inclusion of all-source intelligence analysis very important for effectively assessing adversaries' intentions in cyberspace.

E. Open-Source Intelligence (OSINT)

All-source intelligence analysis of cyberspace sources should take advantage of the information available from OSINT, including Internet social media and other nontraditional sources of information. The constantly evolving sphere of open-source activity offers the opportunity to add useful data to all-source analysis. But this constantly changing landscape of media and the low "signal to noise" ratio of data available in cyberspace also complicate the intelligence collection problem, requiring active collection management to stay abreast of these sources.

F. ISR in Cyberspace

ISR in cyberspace is an activity that synchronizes and integrates the planning and operation of sensors; assets; and processing, exploitation, and dissemination systems in direct support of current and future operations. This is an integrated intelligence and operations function. ISR in cyberspace focuses on gathering tactical and operational information and on mapping enemy and adversary networks to support military planning. To facilitate the optimum utilization of all available ISR assets, an ISR concept of the operations (CONOPS) should be developed in conjunction with the command's planning effort. The ISR CONOPS should be based on the collection strategy and ISR execution planning and should be developed jointly by the joint force intelligence directorate of a joint staff and the operations directorate of a joint staff. The ISR CONOPS documents the synchronization, integration, and operation of ISR resources in direct support of current and future operations. It outlines the capability to task, collect, process, exploit, and disseminate accurate and timely information that provides the awareness necessary to successfully plan and conduct operations. It addresses how all available ISR collection assets and associated processing, exploitation, and dissemination infrastructure, including multinational and commercial assets, will be used to satisfy the joint force's anticipated collection tasks. It also requires appropriate deconfliction and personnel that are trained and certified to a common standard with the IC.

IV. Targeting

The purpose of targeting is to integrate and synchronize fires (the use of weapon systems or other actions to create a specific lethal or nonlethal effect on a target) into joint operations. Targeting is the process of selecting and prioritizing targets and matching the appropriate response to them, considering operational requirements and capabilities. Integrating and synchronizing planning, execution, and assessment are pivotal to the success of joint targeting. The overall joint targeting cycle and target development process described in JP 3-60, Joint Targeting, apply generally to targeting in support of CO. In addition, the coordination required by Chairman of the Joint Chiefs of Staff Manual (CJCSM) 3139.01, (U) Review and Approval Process for Cyberspace Operations, for certain OCO and DCO-RA missions is unique to CO and applies to many aspects of the joint targeting cycle. Therefore, CO planners and decision makers often use a targeting process specifically adapted to the circumstance. Three fundamental aspects of CO require consideration in the targeting processes: recognizing cyberspace capabilities are a viable option for engaging some designated targets; understanding a CO option may be preferable in some cases, because it may offer low probability of detection and/or no associated physical damage; and higher-order effects on targets in cyberspace may impact elements of the DODIN, including retaliation for attacks attributed to the joint force. Additionally, some characteristics unique to the cyberspace components of targets and to cyberspace capabilities are described below.

See pp. 4-29 to 4-34, Targeting (D3A), for related discussion from FM 3-12.

A. Target Access

Cyberspace forces develop access to targets or target elements in cyberspace by using cyberspace exploitation actions. This access can then be used for various purposes, ranging from information collection to maneuver and to targeting nomination. Not all accesses are equally useful for military operations. For instance, the level of access required to collect information from an entity may not be sufficient to create a desired effect. Developing access to targets in or through cyberspace follows a process which can often take significant time. In some cases, remote access is not possible, and close proximity may be required. All target access efforts in cyberspace require coordination with the IC for deconfliction IAW national policy and to illuminate potential IGL concerns. If direct access to the target is unavailable or undesired, sometimes a similar or partial effect can be created by indirect access using a related target that has higher-order effects on the desired target. Some denial of service cyberspace attacks leverage this type of indirect access.

B. Target Nomination and Synchronization

CO use standard target nomination processes, but target folders should include unique cyberspace aspects (e.g., hardware and software configurations, IP address, cyber-persona applications) of the target. Development of this data is imperative to understand and characterize how elements targetable through cyberspace are relevant to the commander's objective. This data also allows the planner to match an appropriate cyberspace capability against a particular target. Component commanders, national agencies, supporting commands, and/or the JFC planning staff nominate targets to the targeting staff for development and inclusion on the joint target list (JTL). Once placed on the JTL, JFCs in receipt of an EXORD with relevant objectives and ROE can engage the target with organic assets (if within a component commander's assigned area of operations) or nominate the target to CDRUS-CYBERCOM for action by other joint force components and other organizations.

Refer to JP 3-60, Joint Targeting, and CJCSI 3370.01, Target Development Standards, for additional details on vetting, validation, and joint targeting working groups.

Targeting In and Through Cyberspace

Ref: JP 3-12, Cyberspace Operations (Jun '18), p. IV-9.

Planning and targeting staffs develop and select targets in and through cyberspace based on the commander's objectives rather than on the capabilities available to achieve them. The focus is on creating effects that accomplish targeting-related tasks and objectives, not on using a particular cyberspace capability simply because it is available. Targets that can be accessed in cyberspace are developed, vetted, and validated within the established targeting process. Although targets paired with cyberspace capabilities can often be engaged with no permanent damage, due to the interconnectedness of cyberspace, the effects of CO may cross geographical boundaries and, if not carefully planned, may have unanticipated effects. As a result, engaging targets in and through cyberspace requires close coordination within DOD and with interagency and multinational partners. Every target has distinct intrinsic or acquired characteristics (i.e., physical, functional, cognitive, environmental, and temporal) that form the basis for detection, location, and identification; for determining target value within the target system; and for classification for future surveillance, analysis, strike, and assessment. The challenge in targeting for CO is to identify, correlate, coordinate, and deconflict multiple activities occurring across the physical network, logical network, and cyber-persona layers. This requires a C2 capability that can operate at the tempo of CO and can rapidly integrate impacted stakeholders.

Physical Network Layer Target Features

The physical network layer is the medium where the data travels. It includes wired (e.g., land and undersea cable) and wireless (e.g., radio, radio-relay, cellular, satellite) transmission means. It is a point of reference for determining geographic location and the applicable legal framework.

Logical Network Layer Target Features

The logical network layer provides an alternate view of the target, abstracted from its physical location, and referenced from its logical position in cyberspace. This position is often represented through a network address (e.g., IP address). It depicts how nodes in the physical domains address and refer to one another to form entities in cyberspace. The logical network layer is the first point where the connection to the physical domains may be lost. Targeting in the logical layer requires the logical identity and logical access to the target to have a direct effect.

Cyber-Persona Layer Target Features

The cyber-persona layer, the aggregate of an individual's or group's online identity(ies), and an abstraction of logical network layer data, holds important implications for joint forces in terms of positive target identification and affiliation and activity attribution. Cyber-personas are created to group information together about targeted actors in order to organize analysis, engagement, and intelligence reporting. Because cyber-personas can be complex, with elements in many virtual locations but often not linked to a single physical location or form, sufficient intelligence collection and analysis capabilities are required for the joint forces to gain insight and situational awareness required to enable effective targeting of a cyber-persona. Ultimately, cyber-personas will be linked to features that will be engaged in either the logical or physical network layers.

See pp. 4-29 to 4-34, Targeting (D3A), for related discussion from FM 3-12.

C. Time-Sensitive Targets (TSTs)

A TST is a validated target of such high priority to friendly forces that the commander designates it for immediate engagement because it poses (or will soon pose) a threat to friendly forces or is a highly lucrative, fleeting target. TSTs are normally engaged dynamically. However, to be successfully engaged, they require considerable planning and preparation within the joint targeting cycle. Engaging TSTs in cyberspace is difficult in most situations, because they are likely to cross-AORs and require detailed joint, interagency, and/or multinational planning efforts.

Being prepared to engage a TST in cyberspace requires coordination between cyberspace planners, operators, and the supported commander early in the planning phase, to increase the likelihood that adequate flexibility and access is available should a fleeting opportunity arise. In addition, JFCs should establish procedures to quickly promulgate strike orders for TSTs in cyberspace. Successful prosecution of TSTs in cyberspace requires a well-organized and well-rehearsed process for sharing sensor data and target information, identifying suitable strike assets, obtaining mission approval, and rapidly deconflicting cyberspace capability employment. Performing as much advanced coordination and decision making as possible, based on the types of TSTs expected and the nature of the mission, is the key to success.

Refer to JP 3-60, Joint Targeting, for additional information on joint targeting, and JP 2-01, Joint and National Intelligence Support to Military Operations, for additional information on intelligence operations.

V. Command and Control (C2) of Cyberspace Forces

Clearly established command relationships are crucial for ensuring timely and effective employment of forces, and CO require unity of command and unity of effort. However, the complex nature of CO, where cyberspace forces can be simultaneously providing actions at the global level and at the theater or JOA level, requires adaptations to traditional C2 structures. Joint forces principally employ centralized planning with decentralized execution of operations. CO require constant and detailed coordination between theater and global operations, creating a dynamic C2 framework that can adapt to the constant changes, emerging threats, and unknowns. Certain CO functions, including protection of the DODIN's global networks and pursuit of global cyberspace threats, lend themselves to centralized planning and execution to meet multiple, near-instantaneous requirements for response. Centrally controlled CO should be integrated and synchronized with the CCDR's regional or local CO, conducted by forces assigned or attached to the CCDR, or in support of the CCDR. For these reasons, there may be times when C2 of forces executing simultaneous global CO and theater CO is conducted using supported/supporting command relationships under separate, but synchronized, chains of command. CO are integrated and synchronized by the supported commander into their CONOPS, detailed plans and orders, and specific joint operations.

A. C2 for Global CO

CDRUSCYBERCOM is the supported commander for transregional and global CO and manages day-to-day global CO even while he or she is the supporting commander for one or more geographic or functional CCDR's operations. For a specific CO mission, the supported/supporting command relationships are established in an EXORD, OPORD, or establishing directive. A supported relationship for CO does not exempt either command from coordinating response options with affected commanders prior to conducting an operation. Regardless of the approach employed for any particular operation, unless otherwise specified by the President or SecDef, C2 for CO are implemented IAW existing CJCS C2 EXORD and other relevant orders to help ensure effective coordination and synchronization of joint forces and to provide a common construct for JFCs to execute their mission within a global context. JFHQ-DODIN centrally coordinates and directs global DODIN operations and DCO-IDM when these operations have the potential to impact the integrity and operational readiness of multiple DOD components. Although execution of many actions may be decentralized, CDRUSCYBERCOM is the supported commander for CO to secure, operate, and defend the DODIN and, when ordered, to defend other US critical cyberspace assets, systems, and functions. As the DODIN continues to migrate towards a common architecture standard, routine cyberspace security actions for global networks will continue shifting to centralized locations, such as a global enterprise operations center.

B. C2 for CO Supporting CCMDs

CCDRs are supported for CO in their AOR or for their transregional responsibilities, with CDRUSCYBERCOM supporting as necessary. These CO comprise actions intended to have effects localized within a GCC's AOR or a functional CCMD's transregional responsibilities. These could be cyberspace security and defense actions internal to a theater DODIN segment or external actions, such as cyberspace exploitation or cyberspace attack against a specific enemy capability. In addition to the theater segments of global networks, CCMD-level DODIN operations and DCO-IDM include the protection of stand-alone and tactical networks and computers used exclusively by the CCMD. For example, CCMD-level maneuvers in cyberspace include activities to reposition capabilities to enhance threat detection in specified areas, focus cyberspace forces activity in areas linked to specific operational branches and sequels to keep the adversary at risk, or activate stand-by tactical cyberspace capabilities to transition friendly C2 to more secure locations. Such CO maneuvers are vital when a CCDR's systems are under attack to the degree that subsets of the DODIN are degraded, compromised, or lost. In such operations, the supported CCDR coordinates, through their USCYBERCOM CO-IPE, with their associated enterprise operation center, supported by JFHQ-DODIN and DISA, to restore the affected cyberspace. The supported CCDR also integrates, synchronizes, and normally directs CO actions in red and gray cyberspace, including fires, with other lethal and nonlethal effects, for which they may use assigned, attached, or supporting cyberspace forces. CCDRs develop and coordinate their requirements for such effects with the USCYBERCOM CO-IPE, for deconfliction and prioritized execution. When a CCDR establishes a subordinate force (e.g., a joint task force), the cyberspace unit(s) assigned to support that force are determined by the CCDR's mission requirements in coordination with CDRUSCYBERCOM.

C. C2 Distinctives for Routine and Crisis/Contingency

The CJCS has established two models for C2 of CO depending upon the prevailing circumstances, one for routine CO and one for crisis/contingency CO.

These C2 relationships are described and depicted graphically in Figure IV-1 and Figure IV-2, on the following pages (pp. 1-50 to 1-51).

Mission-Tailored Force Package (MTFP)

A MTFP is a USCYBERCOM-tailored support capability comprised of assigned CO forces, additional CO support personnel, and cyberspace capabilities, as required. When directed, USCYBERCOM establishes a tailored force to support specific CCMD crisis or contingency mission requirements beyond the capacity of forces available for routine support. Each MTFP is task-organized and provided to the supported CCDR for the duration of the crisis/contingency operation or until redeployed by CDRUSCYBERCOM in coordination with the supported CCDR.

C2 Distinctives for Routine and Crisis/Contingency CO

Ref: JP 3-12, Cyberspace Operations (Jun '18), pp. IV-12 to IV-15.

The CJCS has established two models for C2 of CO, depending upon the prevailing circumstances. The relationships are described below and depicted graphically in Figure IV-1 and Figure IV-2.

Routine CO

The following relationships guide the C2 of cyberspace forces during normal operating conditions, when no crisis or contingency is in effect:

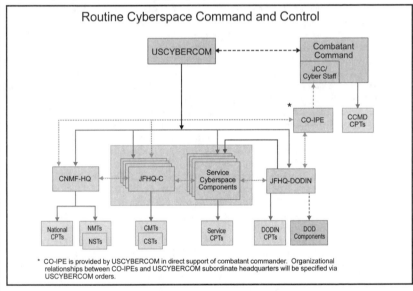

Ref: JP 3-12, Cyberspace Operations (Jun '18), fig. IV-1. Routine Cyberspace Command and Control.

USCYBERCOM C2 Relationships

1. CDRUSCYBERCOM has COCOM of all GFMIG-assigned cyberspace forces.

2. CDRUSCYBERCOM has support relationships with all other CCDRs.

3. CNMF commander has OPCON of NMTs/NSTs and national CPTs.

4. JFHQ-C commanders have OPCON of CMTs/CSTs.

5. SCC commanders have OPCON of Service CPTs and other forces attached by CDRUSCYBERCOM (e.g., CSSPs).

6. JFHQ-DODIN commander has OPCON of DODIN CPTs.

7. JFHQ-DODIN commander has tactical control (TACON) of SCC commands for DODIN operations and DCO-IDM only.

8. JFHQ-DODIN commander has DACO, delegated from CDRUSCYBERCOM, over all DOD components for global DODIN operations and DCOIDM.

9. SCC commanders have DACO, delegated from CDRUSCYBERCOM, over all related Service components for DODIN operations and DCO-IDM.

CCMD C2 Relationships

1. CCDRs have COCOM of assigned cyberspace forces.
2. CCDRs have OPCON of CCMD CPTs.
3. SecDef establishes support relationships between CCDRs for CO.
4. JFHQ-C commanders support more than one CCDR using the general support model.
5. USCYBERCOM CO-IPEs provide direct support to CCDRs.

Crisis/Contingency CO

When a cyberspace-related crisis or contingency is in effect, the routine relationships carry over, with these additional caveats:

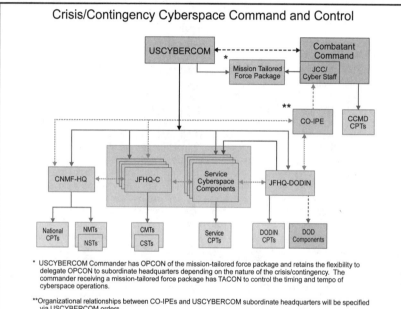

* USCYBERCOM Commander has OPCON of the mission-tailored force package and retains the flexibility to delegate OPCON to subordinate headquarters depending on the nature of the crisis/contingency. The commander receiving a mission-tailored force package has TACON to control the timing and tempo of cyberspace operations.

**Organizational relationships between CO-IPEs and USCYBERCOM subordinate headquarters will be specified via USCYBERCOM orders.

1. USCYBERCOM commander retains OPCON of any cyberspace forces USCYBERCOM provides to support a CCDR for crisis/contingency operations.

2. When directed, CCDRs receiving forces from USCYBERCOM for crisis/contingency operations (e.g., a mission-tailored force package [MTFP]) have TACON of those forces.

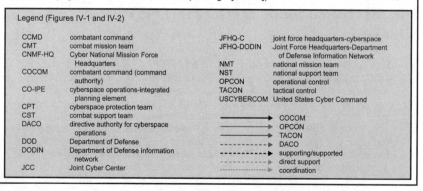

VI. Synchronization of Cyberspace Operations

The pace of CO requires significant pre-operational collaboration and constant vigilance after initiation, for effective coordination and deconfliction throughout the OE. Keys to this synchronization are maintaining cyberspace situational awareness and assessing the potential impacts to the joint force of any planned CO, including the protection posture of the DODIN, changes from normal network configuration, or observed indications of malicious activity. The timing of planned CO should be determined based on a realistic assessment of their ability to create effects and support operations throughout the OE. This may require use of cyberspace capabilities in earlier phases of an operation than the use of other types of capabilities. Effective planners and operators understand how other operations within the OE may impact the CO. For example, the joint force uses fire support coordination measures in air, land, and maritime operations to facilitate the rapid engagement of targets and simultaneously provide safeguards for friendly forces. CO deconfliction and coordination efforts with other operations should include similar measures.

A. Deconfliction

For CO, deconfliction is the act of coordinating the employment of cyberspace capabilities to create effects with applicable DOD, interagency, and multinational partners to ensure operations do not interfere, inhibit, or otherwise conflict with each other. The commander's intended effects in cyberspace, and the capabilities planned to create these effects, require deconfliction with other commands and agencies that may have equities in the same area of cyberspace. This critical step is managed from multiple aspects. From a purely technical perspective, it can be shown that two cyberspace capabilities can either interoperate without interference in the same environment or they cannot. However, from an operational risk perspective, even if multiple capabilities can operate without interference, it may not be wise to use them together. For example, the effect of one capability may draw the adversary's attention on the target system in a way that jeopardizes another previously unnoticed US or mission partner capability. Technical deconfliction uses the results of technical assurance evaluations and includes detailed interoperability analysis of each capability and the cyberspace aspects of the OE. CDRUSCYBERCOM is the DOD focal point for interagency deconfliction of all actions proposed for OCO and DCO-RA missions. Commander, JFHQ-DODIN, is the focal point for interagency deconfliction of global DODIN operations and DCO-IDM activities which may affect more than one DOD component. The timelines required for analysis and coordination should be considered and included in the plan. Interagency coordination often takes longer than concomitant DOD coordination. CO may also require deconfliction and synchronization with integrated joint special technical operations (IJSTO). Information and processes related to IJSTO and its contribution to CO can be obtained from the IJSTO planners at CCMD or Service component HQ.

B. Electromagnetic Spectrum (EMS) Factors

EMS Dependencies

Advancements in technology, including an ever-increasing shift to mobile technologies, have created a progressively complex EMS portion of the OE. This has significant implications for CO. The JFC uses joint EMS operations to coordinate elements of CO, space operations, electronic warfare (EW), navigation warfare, various forms of EMS-dependent information collection, and C2. Although these activities can be integrated with other information-related capabilities (IRCs) as part of information operations synchronization, the offensive aspects of CO, space operations, and EW operations are often conducted under different specific authorities. Likewise, some IRCs enabled by CO, such as MISO and MILDEC, have their own execution approval process. Therefore, synchronizing IRCs that use the EMS is a complex process that requires significant foresight and awareness of the various applicable policies. Planners should also maintain awareness of their operational dependencies on mobile devices and wireless networks, including cellular, wireless local area networks, Global Positioning System, and other commercial and military uses of the EMS. Plans that assume access to the EMS for effects in cyberspace should consider contingencies for when bandwidth or interference issues preclude access to the required portion of the EMS.

Fires in and through the EMS

Cyberspace attack, EA, and offensive space control (OSC) are deconflicted to maximize the impact of each type of fires. Uncoordinated EA may significantly impact EMS-enabled cyberspace attack actions, and vice-versa. Depending upon power levels, the geographic terrain in which they are used, and the nature of the system being targeted, unintended effects of EA and OSC could also occur outside of a local commander's OA, just as higher-order effects of CO may be possible outside the OA. The JFC and staff may need to comply with different coordination requirements for the various types of fires that depend upon the EMS, forwarding requests for execution as early in the planning process as possible to comply with US law and to facilitate effective and timely effects. To minimize overlap, the primary responsibility for cyberspace attack coordination between USCYBERCOM and the joint force resides with the applicable JFHQ-C and USCYBERCOM CO-IPEs in coordination with the CCMD CO staff. Refer to respective doctrine and policy documents of supported IRCs for specifics on their authorities.

C. Integration of Cyberspace Fires

Cyberspace attack capabilities, although they can be used in a stand-alone context, are generally most effective when integrated with other fires. Some examples of integrating cyberspace fires are: disruption of enemy air defense systems using EMS-enabled cyberspace attack, insertion of messages into enemy leadership's communications, degradation/disruption of enemy space-based and ground-based precision navigation and timing systems, and disruption of enemy C2. Effects in cyberspace can be created at the strategic, operational, or tactical level, in any phase of the military operation, and coordinated with lethal fires to create maximum effect on target. Integrated fires are not necessarily simultaneous fires, since the timing of cyberspace attack effects may be most advantageous when placed before or after the effects of lethal fires. Each engagement presents unique considerations, depending upon the level and nature of the enemy's dependencies upon cyberspace. Supporting cyberspace fires may be used in a minor role, or they can be a critical component of a mission when used to enable air, land, maritime, space, and special operations. Forces operating lethal weapons and other capabilities in the physical domains cannot use cyberspace fires to best advantage unless they clearly understand the type and timing of planned effects in cyberspace. Properly prepared and timed cyberspace fires can create effects that cannot be created any other way. Poorly timed fires in cyberspace can be useless, or even worse, interfere with an otherwise effective mission.

D. Risk Concerns

JFCs should continuously seek to minimize risks to the joint force, as well as to friendly and neutral nations, societies, and economies, caused by use of cyberspace. Coordinated joint force operations benefit from the use of various cyberspace capabilities, including unclassified Web sites and Web applications used for communication efforts with audiences internal and external to DOD. Forward-deployed forces use the Internet, mobile phones, and instant messaging for logistics and morale purposes, including communication with friends and family. These uses of cyberspace are targeted by myriad actors, from foreign nations to malicious insiders. The JFC works with JFHQ-DODIN and the Services, as well as with assigned cyberspace forces, to limit the threat to the DODIN and mission partners' cyberspace. Several areas of significant risk exist for the JFC:

Insider threats are a significant concern to the joint force. Because insiders have a trusted relationship with access to the DODIN, the effects of their malicious or careless activity can be far more serious than those of external threat actors. Any user who does not closely follow cybersecurity policy can become an insider threat. Malicious insiders may exploit their access at the behest of foreign governments, terrorist groups, criminal elements, unscrupulous associates, or on their own initiative. Whether malicious insiders are committing espionage, making a political statement, or expressing personal disgruntlement, the consequences for DOD and national security can be devastating. JFCs use risk mitigation measures for this threat, such as reinforcing training of the joint force to be alert for suspicious insider activity and use of two-person controls on particularly sensitive hardware, software, or data.

Internet-based capabilities, including e-mail, social networking, Web sites, and cloud-based repositories, are used for both official and unofficial purposes and pose continuously evolving security risks that are not fully understood. The security risks of Internet-based capabilities are often obscured, and our ability to mitigate these risks is limited, due to the commercial ownership of the majority of the supporting information systems or sites. These cyberspace and information security concerns, combined with bandwidth requirements of Internet applications, create an imperative for the commander to be aware of and actively manage the impact of official and unofficial use of Internet-based capabilities.

Cross-domain (network) solutions that connect systems operating at different classification levels can provide significant operational value to the JFC but complicate cryptographic and other security support considerations and should be included as a planning consideration. Cross-domain solutions are often required in multinational operations and at the tactical level. The pace of operations and increasing demand for information from commanders and their staffs can sometimes pressure end-users into using poor security practices. Likewise, emergent tasking for information sharing has sometimes caused network managers to build ad hoc links over existing commercial infrastructure or connect non-DOD US and partner cyberspace without adequate security controls. The security risk of these behaviors is significant. US-CYBERCOM, through JFHQ-DODIN, works with JFCs to develop appropriate technical solutions and detailed security policies to address the operational requirements without adding unnecessary risk. Planners should include requirements for early coordination so the security features included are appropriate for the commander's needs.

VII. Assessment of Cyberspace Operations

Assessment measures progress of the joint force toward mission accomplishment. Commanders continuously assess the OE and the progress of CO and compare them to their vision and intent. Measuring this progress toward the end state, and delivering timely, relevant, and reliable feedback into the planning process to adjust operations during execution, involves deliberately comparing the forecasted effects of CO with actual outcomes to determine the overall effectiveness of cyberspace force employment. More specifically, assessment helps the commander determine progress toward attaining the desired end state, achieving objectives, or performing tasks.

The assessment process for external CO missions begins during planning and includes measures of performance (MOPs) and measures of effectiveness (MOEs) of fires and other effects in cyberspace, as well as their contribution to the larger operation or objective. Historically, combat assessment has emphasized the battle damage assessment (BDA) component of measuring physical and functional damage, but this approach does not always represent the most complete effect, particularly with respect to CO. CO effects are often created outside the scope of battle and often do not create physical damage. Assessing the impact of CO effects requires typical BDA analysis and assessment of physical, functional, and target system components. However, the higher-order effects of cyberspace actions are often subtle, and assessment of second- and third-order effects can be difficult. Therefore, assessment of fires in and through cyberspace frequently requires significant intelligence collection and analysis efforts. Incorporating pre-strike prediction and post-strike assessment for CO into the existing joint force staff processes increases the likelihood that all objectives are met.

See following page (p. 1-56) for further discussion of MOPs/MOEs.

Assessment of CO at the Operational Level

The operational-level planner is concerned with the accumulation of tactical effects into an overall operational effect. At the operational level, planning and operations staffs develop objectives and desired effects for the JFC to assign to subordinates. Subordinate staffs use the assigned operational objectives to develop tactical-level objectives, tasks, and subordinate targeting objectives and effects and to plan tactical actions and MOPs/MOEs for those actions. Individual tactical actions typically combine with other tactical actions to create operational-level effects; however, they can have operational or strategic implications. Usually, the summation of tactical actions in an operational theater is used to conduct an operational-level assessment principally operation assessments (see JP 3-0, Joint Operations, and JP 5-0, Joint Planning), which in turn supports the strategic-level assessment (as required). Operational MOPs/MOEs avoid tactical information overload by providing commanders a shorthand method of tracking tactical actions and maintaining situational awareness. MOPs and MOEs are clearly definable and measurable, are selected to support and enhance the commander's decision process, and guide future actions that achieve objectives and attain end states.

CO often involve multiple commanders. Additionally, with CO typically conducted as part of a larger operation, assessment of CO is usually done in the context of supporting the overarching objectives. Therefore, CO assessments require close coordination within each staff and across multiple commands. Coordination and federation of the assessment efforts may require prior arrangements before execution. CO planners submit assessment requests as early as possible and provide sufficient justification to support priority allocation of relevant collection capabilities, including those outside of cyberspace.

Refer to JP 5-0, Joint Planning, for a detailed description of assessment. Refer to JP 3-60, Joint Targeting, and Defense Intelligence Agency Publication 2820-4-03, Battle Damage Assessment (BDA) Quick Guide, for more information on the assessment process related to targeting, BDA, and munitions effectiveness assessment.

Measures of Effectiveness (MOEs) / Measures of Performance (MOPs)

Ref: JP 3-12, Cyberspace Operations (Jun '18), pp. IV-12 to IV-15.

The assessment process for external CO missions begins during planning and includes measures of performance (MOPs) and measures of effectiveness (MOEs) of fires and other effects in cyberspace, as well as their contribution to the larger operation or objective.

Measures of Effectiveness (MOEs)
MOEs are used to assess changes in targeted system behavior or in the OE. They measure progress toward the attainment of an end state, achievement of an objective, or creation of an effect. Data gathered on the target from its pre-mission state through access, execution, and possibly long-term post-operations analysis may enable later, more comprehensive assessment, including that of higher-order effects. MOEs generally reflect a trend or show progress toward or away from a measurable threshold. While MOEs may be harder to derive than MOP for a discrete task, they are nonetheless essential to effective assessment. For example, a MOE for a cyberspace attack action might be a meaningful reduction in the throughput of enemy data traffic or their shift to a more interceptable means of communication. Assessment of CO takes place both inside and outside of cyberspace. For instance, an OCO mission to disrupt electric power might be assessed through visual observation to determine that the power is actually out.

Measures of Performance (MOPs)
MOPs are criteria for measuring task performance or accomplishment. MOPs are generally quantitative and are used in most aspects of combat assessment, which typically seeks specific quantitative data or a direct observation of an event to determine accomplishment of tactical tasks. An example of a MOP for a cyberspace exploitation action might be gaining a required access or emplacing a cyberspace capability on a targeted system.

Development of operational-level MOPs/MOEs for CO is still an emerging aspect of operational art. In some cases, activities in cyberspace alone have operational-level effects; for example, the use of a cyberspace attack to bring down or corrupt the enemy HQ network could very well reverberate through the entire JOA. A CO option may be preferable in some scenarios if its effects are temporary or reversible. In such cases, accurate assessment requires the ability to effectively track the current status of the potentially changing effect using MOE indicators.

Refer to JFODS5-1: The Joint Forces Operations & Doctrine SMARTbook (Guide to Joint, Multinational & Interorganizational Operations). Updated for 2019, topics include joint doctrine fundamentals (JP 1), joint operations (JP 3-0 w/Chg 1), an expanded discussion of joint functions, joint planning (JP 5-0), joint logistics (JP 4-0), joint task forces (JP 3-33), joint force operations (JPs 3-30, 3-31, 3-32 & 3-05), multinational operations (JP 3-16), interorganizational cooperation (JP 3-08), & more!

V. Interorganizational & Multinational

Ref: JP 3-12, Cyberspace Operations (Jun '18), pp. IV-23 to IV-26.

I. Interorganizational Considerations

When appropriate, JFCs coordinate and integrate their CO with interagency partners during planning and execution. Effective integration of interagency considerations is vital to successful military operations, especially when the joint force conducts shaping, stability, and transition to civil authority activities. Just as JFCs and their staffs consider how the capabilities of other USG components and NGOs can be leveraged to assist in accomplishing military missions and broader national strategic objectives, JFCs should also consider the capabilities and priorities of interagency partners in planning and executing CO. In collaboration with interagency representatives, JS, and USCYBERCOM, JFCs should coordinate with interagency partners during CO planning to help ensure appropriate agreements exist to support their plans.

At the national level, the National Security Council, with its policy coordination committees and interagency working groups, advises and assists the President on all aspects of national security policy. OSD and JS, in consultation with the Services and CCMDs, coordinate interagency support required to support the JFC's plans and orders. While supported CCDRs are the focal points for interagency coordination in support of operations in their AORs, interagency coordination with supporting commanders is also important. For integration into their operational-level estimates, plans, and operations, commanders should only consider interagency capabilities and capacities that interagency partners can realistically commit to the effort.

Military leaders work with the other members of the national security team to promote unified action. A number of factors can complicate the coordination process, including various agencies' different and sometimes conflicting policies, overlapping legal authorities, roles and responsibilities, procedures, and decision-making processes for CO. A supported commander develops interagency coordination requirements and mechanisms for each OPLAN. The JFC's staff requires a clear understanding of military CO capabilities, requirements, operational limitations, liaison, and legal considerations. Additionally, planners should understand the nature of this relationship and the types of CO support interagency partners can provide. In the absence of a formal interagency command structure, JFCs are required to build consensus to achieve unity of effort. Robust liaison facilitates understanding, coordination, and mission accomplishment.

Interagency command relationships, lines of authority, and planning processes vary greatly from those of DOD. Interagency management techniques often involve committees, steering groups, and/or interagency working groups organized along functional lines. During joint operations, use of a JIACG provides the CCDR and subordinate JFCs with an increased capability to coordinate with other USG departments and agencies.

Refer to JFODS5-1: The Joint Forces Operations & Doctrine SMARTbook (Guide to Joint, Multinational & Interorganizational Operations). Updated for 2019, topics include joint doctrine fundamentals (JP 1), joint operations (JP 3-0 w/Chg 1), an expanded discussion of joint functions, joint planning (JP 5-0), joint logistics (JP 4-0), joint task forces (JP 3-33), joint force operations (JPs 3-30, 3-31, 3-32 & 3-05), multinational operations (JP 3-16), interorganizational cooperation (JP 3-08), & more!

(Joint Cyberspace Operations) V. Interorganizational & Multinational 1-57

II. Multinational Considerations

Collective security is a strategic objective of the US, and joint planning is frequently accomplished within the context of planning for multinational operations. There is no single doctrine for multinational action, and each alliance or coalition develops its own protocols and plans. US planning for joint operations accommodates and complements such protocols and plans for potential use of US cyberspace forces to protect MNF networks. JFCs also anticipate and incorporate mission partner planning factors, such as their domestic laws, regulations, and operational limitations on the use of various cyberspace capabilities and tactics.

When working within an MNF, each nation and Service can expect to be tasked by the commander with the mission(s) most suited to their particular capability and capacity. For example, a CPT supporting a CCMD could be tasked, with the agreement of all nations involved, to investigate and mitigate the effects of malicious cyberspace activity on a multinational network. CO planning, coordination, and execution items that require consideration when an MNF operation or campaign plan is developed include:

- National agendas of the PNs on an MNF may differ significantly from those of the US, creating potential difficulties in determining the CO objectives.

- Differing national standards and foreign laws, as well as interpretation of international laws pertaining to operations in cyberspace, may affect their ability to participate in certain CO.

- Nations without established CO doctrine may need to be advised of the potential benefits of CO and assisted in integrating CO into the planning process.

- Nations in an MNF often require approval for the CO portion of plans and orders from higher authority, which may impede CO implementation. This national-level approval requirement increases potential constraints and restraints upon the participating national forces and further lengthens the time required to gain approval for their participation. Commanders and planners should be proactive in seeking to understand PNs' laws, policies, and other matters that might affect their use of CO and anticipate the additional time required for approval through parallel national command structures.

- Security restrictions may prevent full disclosure of individual CO plans and orders between multinational partners; this may complicate cyberspace synchronization efforts. Therefore, the JFC's staff should seek approval for sharing required information among partners and then issue specific guidance on the release of classified US material to the MNF as early as possible during planning. Likewise, once these information-sharing restrictions are identified by each nation, policy should be established and mechanisms put in place to encourage appropriate CO-related information sharing across the force.

- To effectively conduct multinational operations, mission partners require appropriate access to systems, services, and information. Emerging standards for the technologies and applications applied to DODIN segments used in a joint environment are designed to allow seamless and secure interaction with multinational partners. Until such technology is widespread, the US joint force strives to provide necessary and appropriate access and support at the lowest appropriate security classification level on the infrastructure they have available.

- Responsibility for cyberspace security and cyberspace defense actions to protect multinational networks should be made clear before the network is activated. If responsibility for these actions is to be shared amongst PNs, explicit agreements, including expectations and limitation of action of each partner, should be in place. Unless otherwise agreed, US cyberspace forces or other DOD personnel protect DODIN segments of multinational networks.

I. Cyberspace and the Electromagnetic Spectrum

Ref: FM 3-12, Cyberspace Operations and Electromagnetic Warfare (Aug '21), chap. 1.

Cyberspace operations and electromagnetic warfare (EW) play an essential role in the Army's conduct of unified land operations as part of a joint force and in coordination with unified action partners. **Cyberspace operations** are the employment of cyberspace capabilities where the primary purpose is to achieve objectives in or through cyberspace (JP 3-0). **Electromagnetic warfare (EW) is** a military action involving the use of electromagnetic and directed energy to control the **electromagnetic spectrum** or to attack the enemy (JP 3-85).

See pp. 2-17 to 2-26 for discussion of cyberspace operations and chap. 3 for discussion of electromagnetic warfare (EW).

Electromagnetic Spectrum (EMS)

The electromagnetic spectrum (EMS) is a maneuver space essential for facilitating control within the operational environment (OE) and impacts all portions of the OE and military operations. Based on specific physical characteristics, the EMS is organized by frequency bands, including radio waves, microwaves, infrared radiation, visible light, ultraviolet radiation, x-rays, and gamma rays.

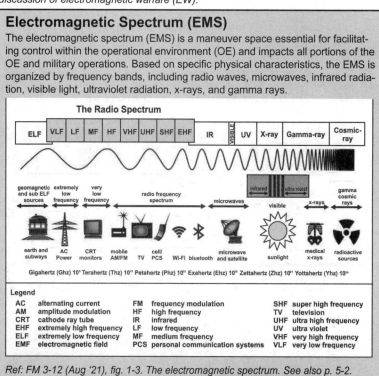

Ref: FM 3-12 (Aug '21), fig. 1-3. The electromagnetic spectrum. See also p. 5-2.

Cyberspace is one of the five domains of warfare and uses a portion of the electromagnetic spectrum (EMS) for operations, for example, Bluetooth, Wi-Fi, and satellite transport. Therefore, cyberspace operations and EW require frequency assignment, management, and coordination performed by spectrum management operations.

Spectrum management operations consist of four key functions—spectrum management, frequency assignment, host-nation coordination, and policy adherence. Spectrum management operations include preventing and mitigating frequency conflicts and electromagnetic interference (EMI) between friendly forces and host nations during Army operations. *See chap. 5, Spectrum Management Operations.*

I. Cyberspace and the Electromagnetic Spectrum (EMS)

Cyberspace and the EMS are critical for success in today's operational environment (OE). U.S. and adversary forces alike rely heavily on cyberspace and EMS-dependent technologies for command and control, information collection, situational understanding, and targeting. Achieving relative superiority in cyberspace and the EMS gives commanders an advantage over adversaries and enemies. By conducting cyberspace operations and EW, commanders can limit adversaries' available courses of action, diminish their ability to gain momentum, degrade their command and control, and degrade their ability to operate effectively in the other domains.

Commanders must leverage cyberspace and EW capabilities using a combined arms approach to seize, retain, and exploit the operational initiative. Effective use of cyberspace operations and EW require commanders and staffs to conduct cyberspace electromagnetic activities (CEMA). Cyberspace electromagnetic activities is the process of planning, integrating, and synchronizing cyberspace operations and electromagnetic warfare in support of unified land operations (ADP 3-0). By integrating and synchronizing cyberspace operations and EW, friendly forces gain an information advantage across multiple domains and lines of operations.

See facing page (Logic Chart, fig. 1-1) for an illustration of how cyberspace operations and EW contribute to Army operations.

Army's reliance on networked systems and weapons necessitates highly trained forces to protect warfighting systems and networks dependent upon access to cyberspace and the EMS. Cyberspace and the EMS are heavily congested due to the high volume of friendly, neutral, and adversary use, and contested due to adversary actions.

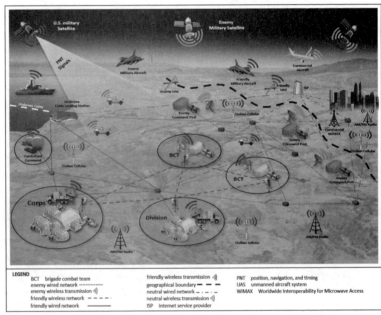

Ref: FM 3-12 (Aug '21), fig. 1-4. Congestion in cyberspace and the electromagnetic spectrum.

Cyberspace Operations & Electromagnetic Warfare (EW) Logic Chart

Ref: FM 3-12, Cyberspace Operations and Electromagnetic Warfare (Aug '21), fig. 1-1.

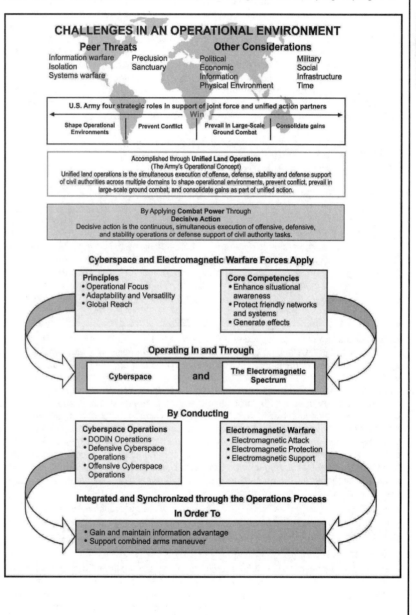

(Cyberspace Operations) I. Cyberspace & the EMS 2-3 *

A. Operational Environment (OE) Overview

Ref: FM 3-12, Cyberspace Operations and Electromagnetic Warfare (Aug '21), pp. 1-4 to 1-5.

An **operational environment** is a composite of the conditions, circumstances, and influences that affect the employment of capabilities and bear on the decisions of the commander (JP 3-0). Conditions in cyberspace and the EMS often change rapidly and can positively or negatively impact a commander's ability to achieve mission objectives. Friendly, neutral, adversary, and enemy actions in cyberspace and the EMS can create near-instantaneous effects on the battlefield or in garrison. Given the global nature of cyberspace and the EMS, these actions can impact a commander's OE even though the actions may originate or terminate beyond that OE. Cyberspace and EW effects also cross through and impact multiple domains simultaneously. For these reasons, commanders must gain and maintain an in-depth understanding of the OE that extends beyond the land domain to the multi-domain extended battlefield to seize, exploit, and retain operational initiative.

Operational Initiative

Operational initiative is the setting of tempo and terms of action throughout an operation (ADP 3-0). By gaining and maintaining positions of relative advantage, including information advantage in and through cyberspace and the EMS, commanders can seize and retain the operational initiative. To gain and maintain information advantage, commanders must account for the temporal nature of information and the temporary nature of many cyberspace and EW effects. On average, the relative operational advantage that a commander can gain from a piece of information or from a cyberspace or EW effect degrades over time. This means that a commander who takes action first, on average, will obtain a greater information advantage from a similar piece of information or effect than a commander who acts later. In this way, the commander who can sense, understand, decide, act, and assess faster than an opponent will generally obtain the greatest information advantage.

Commanders can use cyberspace and EW capabilities to gain enhanced situational awareness and understanding of the enemy through reconnaissance and sensing activities. These reconnaissance and sensing activities can augment and enhance the understanding a commander gains from information collection and intelligence processes. Commanders can also use cyberspace and EW capabilities to decide and act faster than an adversary or enemy. By protecting friendly information systems and signals from disruption or exploitation by an adversary or enemy, a commander can ensure command and control and maintain tactical and operational surprise. Conversely, a commander might use cyberspace and EW capabilities to slow or degrade an enemy's decision-making processes by disrupting enemy sensors, communications, or data processing. To make effective use of cyberspace and EW capabilities to achieve an information advantage, a commander must plan early to integrate cyberspace operations and EW actions fully into the overall scheme of maneuver.

See following pages (pp. 2-6 to 2-7) for discussion of the cyberspace domain.

The Multi-Domain Extended Battlefield

Ref: FM 3-0, Operations (Oct '17), pp. 1-6 to 1-8.

The interrelationship of the air, land, maritime, space, and the information environment (including cyberspace) requires a cross-domain understanding of an OE. Commanders and staffs must understand friendly and enemy capabilities that reside in each domain. From this understanding, commanders can better identify windows of opportunity during operations to converge capabilities for best effect. Since many friendly capabilities are not organic to Army forces, commanders and staffs plan, coordinate for, and integrate joint and other unified action partner capabilities in a multi-domain approach to operations.

A **multi-domain approach** to operations is not new. Army forces have effectively integrated capabilities and synchronized actions in the air, land, and maritime domains for decades. Rapid and continued advances in technology and the military application of new technologies to the space domain, the EMS, and the information environment (particularly cyberspace) require special consideration in planning and converging effects from across all domains.

See p. 2-16 for further discussion. Refer to TRADOC PAM 525-3-1, The U.S. Army in Multi-Domain Operations (Dec '18) for further discussion.

Information Environment

The information environment is the aggregate of individuals, organizations, and systems that collect, process, disseminate, or act on information (JP 3-13). The information environment is not separate or distinct from the OE but is inextricably part of it. Any activity that occurs in the information environment simultaneously occurs in and affects one or more of the physical domains. Most threat forces recognize the importance of the information environment and emphasize information warfare as part of their strategic and operational methods.

The information environment is comprised of three dimensions: physical, informational, and cognitive. The physical dimension includes the connective infrastructure that supports the transmission, reception, and storage of information.

Across the globe, information is increasingly available in near-real time. The ability to access this information, from anywhere, at any time, broadens and accelerates human interaction across multiple levels, including person to person, person to organization, person to government, and government to government. Social media, in particular, enables the swift mobilization of people and resources around ideas and causes, even before they are fully understood. Disinformation and propaganda create malign narratives that can propagate quickly and instill an array of emotions and behaviors from anarchy to focused violence. From a military standpoint, information enables decision making, leadership, and combat power; it is also key to seizing, gaining, and retaining the initiative, and to consolidating gains in an OE. Army commanders conduct information operations to affect the information environment.

Space Domain

The space domain is the space environment, space assets, and terrestrial resources required to access and operate in, to, or through the space environment (FM 3-14). Space is a physical domain like land, sea, and air within which military activities are conducted. Proliferation of advanced space technology provides more widespread access to space-enabled technologies than in the past. Adversaries have developed their own systems, while commercially available systems allow almost universal access to some level of space enabled capability with military applications. Army forces must be prepared to operate in a denied, degraded and disrupted space operational environment (D3SOE).

Cyberspace and the Electromagnetic Spectrum (EMS)

Cyberspace is a global domain within the information environment consisting of interdependent networks of information technology infrastructures and resident data, including the Internet, telecommunications networks, computer systems, and embedded processors and controllers. Cyberspace is an extensive and complex global network of wired and wireless links connecting nodes that permeate every domain. Networks cross geographic and political boundaries connecting individuals, organizations, and systems around the world. Cyberspace is socially enabling, allowing interactivity among individuals, groups, organizations, and nation-states.

See following pages (pp. 2-6 to 2-7) for discussion of the cyberspace domain.

B. Cyberspace Domain

Ref: FM 3-12, Cyberspace Operations and Electromagnetic Warfare (Aug '21), pp. 1-5 to 1-7. See pp. 1-2 to 1-3 for related discussion from JP 3-12.

Cyberspace is a global domain within the information environment consisting of the interdependent networks of information technology infrastructures and resident data, including the Internet, telecommunications networks, computer systems, and embedded processors and controllers (JP 3-12). Cyberspace operations require the use of links and nodes located in other physical domains to perform logical functions that create effects in cyberspace that then permeate throughout the physical domains using both wired networks and the EMS.

The use of cyberspace is essential to operations. The Army conducts cyberspace operations and supporting activities as part of both Army and joint operations. Because cyberspace is a global communications and data-sharing medium, it is inherently joint, inter-organizational, multinational, and often a shared resource, with signal and intelligence maintaining significant equities. Friendly, enemy, adversary, and host-nation networks, communications systems, computers, cellular phone systems, social media websites, and technical infrastructures are all part of cyberspace.

To aid the planning and execution of cyberspace operations, cyberspace is sometimes visualized in three layers. These layers are interdependent, but each layer has unique attributes that affect operations. Cyberspace operations generally traverse all three layers of cyberspace but may target effects at one or more specific layers. Planners must consider the challenges and opportunities presented by each layer of cyberspace as well as the interactions amongst the layers. Figure 1-2 on page 1-6 depicts the relationship between the three cyberspace layers. The three cyberspace layers are—

- The physical network layer.
- The logical network layer.
- The cyber-persona layer.

See pp. 1-2 to 1-3 for related discussion from JP 3-12.

Physical Network Layer

The physical network layer consists of the information technology devices and infrastructure in the physical domains that provide storage, transport, and processing of information within cyberspace, to include data repositories and the connections that transfer data between network components (JP 3-12). Physical network components include the hardware and infrastructure such as computing devices, storage devices, network devices, and wired and wireless links. Components of the physical network layer require physical security measures to protect them from damage or unauthorized access, which, if left vulnerable, could allow a threat to gain access to both systems and critical data.

Every physical component of cyberspace is owned by a public or private entity. The physical layer often crosses geo-political boundaries and is one of the reasons that cyberspace operations require multiple levels of joint and unified action partner coordination. Cyberspace planners use knowledge of the physical location of friendly, neutral, and adversary information technology systems and infrastructures to understand appropriate legal frameworks for cyberspace operations and to estimate impacts of those operations. Joint doctrine refers to portions of cyberspace, based on who owns or controls that space, as either blue, gray, or red cyberspace (refer to JP 3-12). This publication refers to these areas as friendly, neutral, or enemy cyberspace respectively.

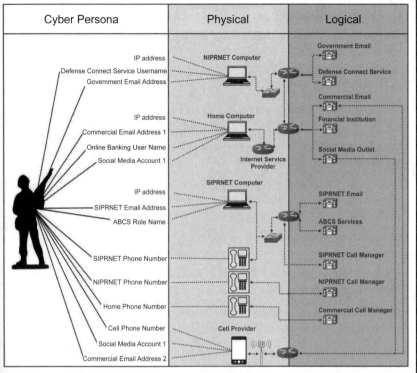

Ref: FM 3-12 (Aug '21), fig. 1-2. Relationship between the cyberspace network layers.

Logical Network Layer

The logical network layer consists of those elements of the network related to one another in a way that is abstracted from the physical network, based on the logic programming (code) that drives network components (i.e., the relationships are not necessarily tied to a specific physical link or node, but to their ability to be addressed logically and exchange or process data) (JP 3-12). Nodes in the physical layer may logically relate to one another to form entities in cyberspace not tied to a specific node, path, or individual. Web sites hosted on servers in multiple physical locations where content can be accessed through a single uniform resource locator or web address provide an example. This may also include the logical programming to look for the best communications route, instead of the shortest physical route, to provide the information requested.

Cyber-Persona Layer

The cyber-persona layer is a view of cyberspace created by abstracting data from the logical network layer using the rules that apply in the logical network layer to develop descriptions of digital representations of an actor or entity identity in cyberspace, known as a cyber-persona (JP 3-12). Cyber-personas are not confined to a single physical or logical location and may link to multiple physical and logical network layers. When planning and executing cyberspace operations, staffs should understand that one actor or entity (user) may have multiple cyber-personas, using multiple identifiers in cyberspace. These various identifiers can include different work and personal emails and different identities on different Web forums, chatrooms, and social network sites. For example, an individual's account on a social media website, consisting of the username and digital information associated with that username, may be just one of that individual's cyber-personas.

(Cyberspace Operations) I. Cyberspace & the EMS 2-7 *

C. Operational & Mission Variables

Ref: Adapted from FM 3-12, Cyberspace & Electronic Warfare Operations (Apr '17), pp. 1-18 to 1-19.

Commanders and staffs use the operational and mission variables to help build their situational understanding. They analyze and describe an operational environment in terms of eight interrelated operational variables: political, military, economic, social, information, infrastructure, physical environment, and time (PMESII-PT). Upon receipt of a mission, commanders filter information categorized by the operational variables into relevant information with respect to the mission. They use the mission variables, in combination with the operational variables, to refine their understanding of the situation and to visualize, describe, and direct operations. The mission variables are mission, enemy, terrain and weather, troops and support available, time available, and civil considerations (METT-TC).

See pp. 4-2 to 4-8 for related discussion of the military decisionmaking process (MDMP) as related to cyberspace and electronic warfare operations.

Cyberspace and the Operational Variables (PMESII-PT)

Commanders and staffs continually analyze and describe the operational environment in terms of eight interrelated operational variables: political, military, economic, social, information, infrastructure, physical environment, and time. Each variable applied to an analysis of designated cyberspace can enable a more comprehensive understanding of the operational environment. The analysis describes the planning, preparation, execution, and assessment activities for both the wired and EMS portions cyberspace operations. The following are operational variable example questions specific to networks and nodes—

P - Political
What networks and nodes require the most emphasis on security and defense to enable the functioning of the government?

M - Military
Where are networks and nodes utilized by enemy and adversary actors to enable their activities?

E - Economic
What networks and nodes require the most emphasis on security and defense to enable commerce and other economic-related activities?

S - Social
What network nodes enable communication with the host nation population for the purpose of providing information or protecting them from potential negative effects caused by military operations in cyberspace?

I - Information
What is the nature of the data transiting cyberspace that influences or otherwise affects military operations?

I - Infrastructure
What networks and nodes enable critical infrastructure and key resource capabilities and supporting supervisory control and data acquisition systems?

P - Physical Environment
How are wireless networks affected by the electromagnetic environment which includes terrain and weather?

T - Time
What are the optimal times to create effects to support the overarching mission?

Cyberspace and the Mission Variables (METT-TC)

The analysis of mission variables specific to cyberspace operations enables Army forces to integrate and synchronize cyberspace capabilities to support Army operations. Mission variables describe characteristics of the area of operations. The mission variables are mission, enemy, terrain and weather, troops and support available, time available, and civil considerations. For cyberspace operations, mission variables provide an integrating framework upon which critical questions can be asked and answered throughout the operations process. The questions may be specific to either the wired portion of cyberspace, the EMS, or both. The following is a list of the mission variables example questions—

M - Mission
Where can we integrate elements of cyberspace operations to support the unit mission? What essential tasks could be addressed by the creation of one or more effects by cyberspace operations?

E - Enemy
How can we leverage information collection efforts regarding threat intentions, capabilities, composition, and disposition in cyberspace? What enemy vulnerabilities can be exploited by cyberspace capabilities?

T - Terrain and Weather
What are the opportunities and risks associated with the employment of cyberspace operations capabilities when terrain and weather may cause adverse impacts on supporting information technology infrastructures?

T - Troops and Support Available
What resources are available (internal and external) to integrate, synchronize, and execute cyberspace operations? What is the process to request, receive, and integrate these resources?

T - Time Available
How can we synchronize OCO and related desired effects with the scheme of maneuver within the time available for planning and execution?

C - Civil Considerations
How can we employ cyberspace operations without negative impacts on noncombatants?

Refer to BSS6: The Battle Staff SMARTbook, 6th Ed. for further discussion. BSS6 covers the operations process (ADP 5-0); commander's activities; Army planning methodologies; the military decisionmaking process and troop leading procedures (FM 7-0 w/Chg 2); integrating processes (IPB, information collection, targeting, risk management, and knowledge management); plans and orders; mission command, C2 warfighting function tasks, command posts, liaison (ADP 6-0); rehearsals & after action reviews; and operational terms and military symbols (ADP 1-02).

II. Trends and Characteristics

Ref: FM 3-12, Cyberspace Operations and Electromagnetic Warfare (Aug '21), pp. 1-8 to 1-12.

The rapid proliferation of cyberspace and EMS capabilities has further congested an already challenging OE. In addition to competing with threat actors in cyberspace and the EMS, U.S. forces also encounter challenges resulting from neutral actors. Such neutral systems as commercial aircraft and airports, Worldwide Interoperability for Microwave Access, and commercial cellular infrastructures contribute to continuing congestion in cyberspace and the EMS.

Several key trends and characteristics impact a commander's ability to use cyberspace and the EMS. Such trends and characteristics include—

- Congested environments.
- Contested environments.
- Threats.
- Hazards.
- Terrain.

A. Congested Environments

Both cyberspace and the EMS are increasingly congested environments that friendly, neutral, and threat actors use to transmit and process large amounts of information. Since 2000, the Army's use of networked information systems in almost every aspect of operations has increased tenfold. Neutral and threat actors have similarly expanded their use of cyberspace and the EMS for a wide range of military and non-military purposes.

B. Contested Environments

As cyberspace and the EMS continue to become more congested, the capabilities of state and non-state actors to contest U.S. advantages in both areas have also expanded. State and non-state threats use a wide range of advanced technologies that may represent relatively inexpensive ways for a small or materially disadvantaged adversary to pose a significant threat to the United States. The application of low-cost cyberspace capabilities can provide an advantage against a technology-dependent nation or organization and an asymmetric advantage to those who could not otherwise effectively oppose U.S. military forces.

C. Threats

For every operation, threats are a fundamental part of an OE. A threat is any combination of actors, entities, or forces that have the capability and intent to harm United States forces, United States national interests, or the homeland (ADP 3-0). Threat is an umbrella term that includes any actor with the potential to harm the United States or its interests. Threats include—

- **Enemy**. An enemy is a party identified as hostile against which the use of force is authorized (ADP 3-0). An enemy is also called a combatant and treated as such under the laws of war. Enemies will employ various advanced technologies to attack Army forces in cyberspace and EMS to disrupt or destroy the ability to conduct operations or collect information that will give friendly forces a strategic, operational, or tactical advantage.

- **Adversary**. An adversary is a party acknowledged as potentially hostile to a friendly party and against which the use of force may be envisaged (JP 3-0). Though an adversary is not treated as a combatant, the goal is still to prevent and deter conflict by keeping their activities within a desired state of cooperation and competition.

- **Peer Threat**. A peer threat is an adversary or enemy able to effectively oppose U.S. forces world-wide while enjoying a position of relative advantage in a specific region

(ADP 3-0), including cyberspace and the EMS. Peer threats often have cyberspace and EW capabilities that are comparable to U.S. capabilities. Peer threats may employ these capabilities across the competition continuum to collect intelligence, delay the deployment of U.S. forces, degrade U.S. capabilities, and disrupt U.S. operations. Peer threats have electromagnetic attack (EA) capabilities such as telecommunications and EMS jamming equivalent to or better than U.S. forces. Peer threats can conduct advanced cyberspace attacks, including denial-of-service, various forms of phishing, eavesdropping, and malware.

- **Hybrid Threat.** A hybrid threat is the diverse and dynamic combination of regular forces, irregular forces, or criminal elements unified to achieve mutually benefitting effects (ADP 3-0). Commanders and staffs must understand that the diversity of a hybrid threat complicates operations since hostility is coming from multiple actors operating from various geographical territories. A hybrid threat complicates the United States' efforts to identify, characterize, attribute, and respond to threats in cyberspace and the EMS.

- **Organized Crime or other Non-State, Illegitimate Organizations.** These organizations often make sophisticated malware available for purchase or free, allowing even unsophisticated threat actors to acquire advanced capabilities at little to no cost. Because of the low barriers to entry and the potentially high payoff, the United States can expect an increasing number of adversaries to use cyberspace capabilities to attempt to negate U.S. advantages in military capability.

- **Insider Threat.** An insider threat is a person with placement and access who intentionally causes loss or degradation of resources or capabilities or compromises the ability of an organization to accomplish its mission through espionage, providing support to international terrorism, or the unauthorized release or disclosure of information about the plans and intentions of United States military forces (AR 381-12). Insider threats may include spies within or working with U.S. forces, as well as personnel who may be unaware of their actions either through deception or third party manipulation. Insider threats present unique challenges because they are trusted individuals with authorized access to Army capabilities and sensitive operational information. Insider threats may include spies within or working with U.S. forces.

Note. Law enforcement and counterintelligence capabilities also operate in cyberspace during their efforts to neutralize criminal activities. Countering insider threats falls primarily within the purview of these organizations and outside the authorized activities of the cyberspace forces. However, information discovered in the course of authorized cyberspace operations may aid these other organizations.

D. Hazards

A hazard is a condition with the potential to cause injury, illness, or death of personnel, damage to or loss of equipment or property, or mission degradation (JP 3-33). Disruption to cyberspace's physical infrastructure often occurs due to operator errors, industrial accidents, and natural disasters. These unpredictable events may have just as significant impact on operations as the actions of enemies. Recovery from accidents and hazardous incidents may require significant coordination external to the DOD or the temporary reliance on backup systems with which operators may be less familiar.

Electromagnetic energy can also impact the operational capability of military forces, equipment, systems, and platforms. Various hazards from electromagnetic energy include electromagnetic environmental effects, electromagnetic compatibility issues, EMI, electromagnetic pulse, and electromagnetic radiation hazards.

Electromagnetic radiation hazards include hazards of electromagnetic radiation to personnel; hazards of electromagnetic radiation to ordnance; hazards of electromagnetic radiation to fuels; and natural phenomena effects such as space weather, lightning, and precipitation static.

III. Core Competencies & Fundamentals

Ref: FM 3-12, Cyberspace Operations and Electromagnetic Warfare (Aug '21), pp. 1-3 to 1-4.

A. Core Competencies

Cyberspace forces and EW professionals are organized, trained, and equipped to provide the following core competencies that deliver essential and enduring capabilities to the Army—

- Enable situational understanding.
- Protect friendly personnel and capabilities.
- Deliver effects.

Create Understanding

Cyberspace forces execute cyberspace intelligence, surveillance, and reconnaissance in and through the information environment to identify and understand adversary networks, systems, and processes. This information enables commanders to understand adversary capabilities and vulnerabilities, thereby enhancing the commanders' ability to prioritize and deliver effects.

EW professionals surveil the EMS to collect combat information used to characterize adversary use of the EMS and understand the integration of adversary emitter systems arrays at echelon. This information enables understanding friendly vulnerabilities and threat capabilities while allowing commanders to prioritize and deliver effects.

Protect Friendly Personnel and Capabilities

Cyberspace forces defend networks, warfighting platforms, capabilities, and data from ongoing or imminent malicious cyberspace activity. By protecting critical networks and systems, cyberspace forces help maintain the Army's ability to conduct operations and project power across all domains.

EW forces, in coordination with the G-6 or S-6 and in support of the commander's directive, implement and enhance measures to protect friendly personnel, facilities, warfighting platforms, capabilities, and equipment from adverse effects in the EMS. EW forces recommend measures to mask or control friendly emissions from enemy detection and deny adversaries the ability to locate and target friendly formations. EW forces detect and mitigate enemy attacks in or through the EMS to maintain the Army's ability to conduct operations and project power across all domains.

Deliver Effects

Cyberspace forces deliver cyberspace effects against adversary networks, systems, and weapons. These effects enhance the Army's ability to conduct operations, reduce adversary combat power, and project power across all domains.

EW professionals deliver effects in the EMS against adversary networks, systems, and weapons. These actions reduce adversary combat power, protect friendly forces, and enhance friendly forces and weapons' lethality.

B. Fundamental Principles

Fundamental principles are basic rules or assumptions of central importance that guide how cyberspace and EW professionals' approach and conduct cyberspace operations and EW. These fundamental principles are—

- Operational focus.
- Adaptability and versatility.
- Global reach.

Operational Focus
Cyberspace and EW forces execute missions in support of a commander's overarching operational design. When properly integrated and synchronized as part of a combined arms approach, cyberspace and EW capabilities can produce layered dilemmas for the adversary in multiple domains and enhance relative combat power. To accomplish this, cyberspace and EW staff must collaborate across all warfighting functions.

Adaptability and Versatility
Cyberspace and EW forces conduct operations using capabilities that are adaptable to a variety of mission requirements. Cyberspace and EW capabilities vary in both the size of the force employed and the magnitude or scope of effects created. Depending on mission requirements, cyberspace and EW capabilities may be used as primary or supporting efforts for decisive, shaping or sustaining operations.

Global Reach
The nature of the cyberspace domain increases the operational reach of cyberspace and EW forces. Combat mission force(s) and EW professionals deliver strategic, operational, or tactical effects worldwide from remote, co-located, or forward operating positions.

An operational environment is a composite of the conditions, circumstances, and influences that affect the employment of capabilities and bear on the decisions of the commander (JP 3-0). Conditions in cyberspace and the EMS often change rapidly and can positively or negatively impact a commander's ability to achieve mission objectives. Friendly, neutral, adversary, and enemy actions in cyberspace and the EMS can create near-instantaneous effects on the battlefield or in garrison. Given the global nature of cyberspace and the EMS, these actions can impact a commander's OE even though the actions may originate or terminate beyond that OE. Cyberspace and EW effects also cross through and impact multiple domains simultaneously. For these reasons, commanders must gain and maintain an in-depth understanding of the OE that extends beyond the land domain to the multi-domain extended battlefield to seize, exploit, and retain operational initiative.

Operational initiative is the setting of tempo and terms of action throughout an operation (ADP 3-0). By gaining and maintaining positions of relative advantage, including information advantage in and through cyberspace and the EMS, commanders can seize and retain the operational initiative. To gain and maintain information advantage, commanders must account for the temporal nature of information and the temporary nature of many cyberspace and EW effects. On average, the relative operational advantage that a commander can gain from a piece of information or from a cyberspace or EW effect degrades over time. This means that a commander who takes action first, on average, will obtain a greater information advantage from a similar piece of information or effect than a commander who acts later. In this way, the commander who can sense, understand, decide, act, and assess faster than an opponent will generally obtain the greatest information advantage.

Commanders can use cyberspace and EW capabilities to gain enhanced situational awareness and understanding of the enemy through reconnaissance and sensing activities. These reconnaissance and sensing activities can augment and enhance the understanding a commander gains from information collection and intelligence processes. Commanders can also use cyberspace and EW capabilities to decide and act faster than an adversary or enemy. By protecting friendly information systems and signals from disruption or exploitation by an adversary or enemy, a commander can ensure command and control and maintain tactical and operational surprise. Conversely, a commander might use cyberspace and EW capabilities to slow or degrade an enemy's decision-making processes by disrupting enemy sensors, communications, or data processing. To make effective use of cyberspace and EW capabilities to achieve an information advantage, a commander must plan early to integrate cyberspace operations and EW actions fully into the overall scheme of maneuver.

IV. Contributions to the Warfighting Functions

Ref: FM 3-12, Cyberspace Operations and Electromagnetic Warfare (Aug '21), pp. 1-12 to 1-15.

This section describes how cyberspace operations and EW support the warfighting functions. It specifies the types of cyberspace operations and EW missions and actions that contribute to the various tasks related to each warfighting function.

Command and Control

Commanders rely heavily on cyberspace and the EMS for command and control. At corps and below, the network in the command-and-control system is the Department of Defense information network-Army (DODIN-A). The Department of Defense information network-Army is an Army-operated enclave of the DODIN that encompasses all Army information capabilities that collect, process, store, display, disseminate, and protect information worldwide (ATP 6-02.71). Signal forces establish, manage, secure, and defend the DODIN-A by conducting Department of Defense information network operations and maintaining cybersecurity compliance to prevent intrusions into the DODIN-A. EW supports command and control through electromagnetic protection (EP) to eliminate or mitigate the negative impact of friendly, neutral, enemy, or naturally occurring EMI on command-and-control systems. The frequency assignment and deconfliction tasks of spectrum management operations support EP. Such EP tasks include—emission control, mitigating electromagnetic environmental effects, electromagnetic compatibility, electromagnetic masking, preemptive countermeasures, and electromagnetic warfare reprogramming. These tasks require integration with spectrum management operation for frequency management and deconfliction.

Movement and Maneuver

Cyberspace operations and EW enhance friendly forces commanders' movement and maneuver by disrupting adversary command and control, reducing adversary and increasing friendly situational awareness, and negatively affect the adversary's ability to make sound decisions. Due to the range and reach of cyberspace capabilities, cyberspace forces are often able to support friendly maneuver in close areas while simultaneously supporting deep area operations.

DODIN operations support movement and maneuver by establishing secure tactical networks that allow communications with friendly forces conducting operations laterally in close and deep areas, in addition to communications with higher headquarters in the rear area. Units use the DODIN-A as the primary means of communication during movement and maneuver. Satellite communications, combat net radios, and wired networks are elements of the DODIN-A used to synchronize operations, collaborate, understand the environment, and coordinate fires. The network enables near real-time updates to the common operational picture. The upper and lower tiers of the DODIN-A connect headquarters to subordinate, adjacent, and higher headquarters and unified action partners.

Offensive cyberspace operations (OCO) in coordination with other forms of fires also support movement and maneuver by opening avenues necessary to disperse and displace enemy forces. Synchronizing OCO with other fires sets conditions that enable maneuver to gain or exploit positions of relative advantage.

EW assets support movement and maneuver by conducting operations to degrade, neutralize, or destroy enemy combat capabilities in the EMS. Defensive EA protects friendly forces from enemy attacks during movement and maneuver by denying the enemy the use of the EMS. Using friendly EA to counter radio-controlled devices, such as improvised explosive devices, drones, robots, or radio-guided munitions is an example of defensive EA. During defensive EA, EW assets conduct operations to degrade, neutralize, or destroy enemy combat capabilities in the EMS. EW assets conduct defensive EA by employing EA capabilities such as counter radio-controlled improvised explosive device electronic warfare and devices used for aircraft survivability. Offensive EA supports movement and maneuver by projecting power within the time and tempo of the scheme of maneuver. Electromagnetic jamming, electromagnetic intrusion, and electromagnetic probing are examples of offensive EA. Electromagnetic support (ES) supports movement and maneuver by providing combat information for a situational understanding of the OE.

Intelligence

Cyberspace operations, EW, and intelligence mutually identify the cyberspace and EMS aspects of the OE to provide recommendations for friendly courses of action during the military decision-making process. Cyberspace and EW forces support information collection that may be used by intelligence professionals. Conversely, intelligence operations provide products that enhance understanding of the OE, enable targeting, and support defense in cyberspace and the EMS. It is critical that information acquired through cyberspace operations and EW is standardized and reported to the intelligence community. Intelligence supports cyberspace operations through the intelligence process, intelligence preparation of the battlefield (IPB), and information collection. Intelligence at all echelons supports cyberspace operations and EW planning, and helps measure performance and effectiveness through battle damage assessment. Cyberspace planners leverage intelligence analysis, reporting, and production capabilities to understand the OE, develop plans and targets, and support operations throughout the operations process. In the context of cyberspace and the EMS, the OE includes network topology overlays that graphically depict how information flows and resides within the operational area and how the network transports data in and out of the area of interest.

Fires

OCO and EA tasks are part of the fires warfighting function. Cyberspace forces employ cyberspace attacks to deny, degrade, disrupt, and destroy or otherwise affect enemies' cyberspace or information-dependent capabilities. EW personnel employ EA to degrade, and neutralize the enemies' ability to use the EMS. Cyberspace and EW effects transcends beyond cyberspace and the EMS and may result in second-and-third-order effects that could impact the other physical domains. Army cyberspace and EW effects applied against enemy capabilities and weapon systems deny their ability to communicate, track, or target. EW also supports fires by enabling lethal fires through the employment of ES to search for, identify, and locate or localize sources of radiated electromagnetic energy used by the enemy for targeting. Defensive EA can support fires through the deployment of decoys or noise to mask friendly fires networks.

Sustainment

Cyberspace operations support sustainment through DODIN operations and defensive cyberspace operations (DCO). Sustainment organizations, functions, systems, and sustainment locations are highly dependent on DODIN operations. DODIN operations establish the necessary communications to conduct sustainment functions. Cyberspace forces defend sustainment systems when adversaries breach cybersecurity measures of networks and systems from threat cyberspace attacks. EW supports sustainment through EP and ES, ensuring freedom of action for DODIN operations in and through the EMS for continued sustainment support. Management, coordination, and deconfliction of frequencies in the EMS are functions of spectrum management operations.

Protection

DCO-IDM and EP tasks, in addition to the cyberspace security tasks of DODIN operations, are part of the protection warfighting function. DODIN operations, DCO-IDM, EP, and defensive EA support protection by securing and defending the DODIN-A. Cyberspace forces conduct DCO-IDM to detect, characterize, counter, and mitigate ongoing or imminent threats to the DODIN-A. DODIN operations and DCO-IDM also enable other protection tasks by providing secured communications for area security, police operations, personnel recovery, air and missile defense, and detention operations. EP involves actions to protect personnel, facilities, and equipment from friendly, neutral, or enemy use of the EMS. EP includes measures to protect friendly personnel and equipment in a contested and congested electromagnetic operational environment (EMOE).

Adversaries continue to develop sophisticated weapons and networked systems that project power through or depend on cyberspace and the EMS. The Army employs cyberspace and EW capabilities as part of a joint and combined arms approach to defeat threat activities in cyberspace and the EMS, protect friendly forces, and enable friendly freedom of action across the conflict continuum. Army cyberspace and EW forces apply the following core competencies and underlying fundamental principles to ensure friendly forces gain and maintain positions of relative advantage.

V. Conflict and Competition

Army forces face continuous competition and conflict in cyberspace and the EMS from threats intending to diminish friendly capabilities. Commanders must seek and exploit opportunities for success in cyberspace and the EMS wherever and whenever authorized.

A. Competition Continuum

Cyberspace operations, EW, and spectrum management operations take place across the competition continuum. The competition continuum describes a world of enduring competition conducted through a mixture of cooperation, competition below armed conflict, and armed conflict Superiority in cyberspace and the EMS enables U.S. forces to conduct operations to achieve the goals and accomplish the objectives assigned to them by the President and Secretary of Defense. Though U.S. forces may conduct cyberspace operations and EW during competition below the level of armed conflict, they are critical enablers to combat power when conducting large-scale combat operations during armed conflict. Competition below armed conflict consists of situations in which joint forces take actions outside of armed conflict against a strategic actor in pursuit of policy objectives.

Spectrum management operations fulfill a crucial within the CEMA construct. Spectrum management operations take place across the entire competition continuum and ensure proper coordination of EMS activities spanning the entirety of military operations.

B. Multi-Domain Extended Battlefield

The enemy seeks to employ capabilities to create effects in multiple domains to counter U.S. interests and impede friendly operations. Threat actors will conduct activities in the information environment, space, and cyberspace to influence U.S. decision makers and disrupt the deployment of friendly forces. Land-based threats will attempt to impede joint force freedom of action across the air, land, maritime, space, and cyberspace domains. They will disrupt the EMS, sow confusion, and challenge the legitimacy of U.S. actions. Understanding how threats can present multiple dilemmas to Army forces in all domains helps Army commanders identify (or create), seize, and exploit their opportunities. Implementing operations security (OPSEC) is critical to protecting essential friendly information technology infrastructures, command and control, and targeting systems. Operations security is a capability that identifies and controls critical information, indicators of friendly force actions attendant to military operations and incorporates countermeasures to reduce the risk of an adversary exploiting vulnerabilities (JP 3-13.3).

See also pp. 2-4 to 2-5.

C. Positions of Relative Advantage (in Cyberspace and the Electromagnetic Spectrum)

The Army conducts cyberspace operations and EW to attain positions of relative advantage in cyberspace and the EMS, to establish information superiority. A position of relative advantage is a location or the establishment of a favorable condition within the area of operations that provides the commander with temporary freedom of action to enhance combat power over an enemy or influence the enemy to accept risk and move to a position of disadvantage (ADP 3-0).

Chap 2
II. Cyberspace Operations

Ref: FM 3-12, Cyberspace Operations and Electromagnetic Warfare (Aug '21), chap. 2.

Cyberspace operations and electromagnetic warfare (EW) can benefit from synchronization with other Army capabilities using a combined arms approach to achieve objectives against enemy forces. Cyberspace operations and EW can provide commanders with positions of relative advantage in the multi-domain fight. Effects that bleed over from the cyberspace domain into the physical domain can be generated and leveraged against the adversary. A cyberspace capability is a device or computer program, including any combination of software, firmware, or hardware, designed to create an effect in or through cyberspace (JP 3-12).

Electromagnetic Spectrum Superiority
Electromagnetic spectrum superiority is the degree of control in the electromagnetic spectrum that permits the conduct of operations at a given time and place without prohibitive interference, while affecting the threat's ability to do the same (JP 3-85). Electromagnetic warfare (EW) creates effects in the EMS and enables commanders to gain EMS superiority while conducting Army operations. EW capabilities consist of the systems and weapons used to conduct EW missions to create lethal and non-lethal effects in and through the EMS.
See chap. 3, Electromagnetic Warfare (EW), for further discussion.

I. Cyberspace Operations

The joint force and the Army divide cyberspace operations into three categories based on the portion of cyberspace in which the operations take place and the type of cyberspace forces that conduct those operations. Each of type of cyberspace operation has varying associated authorities, approval levels, and coordination considerations. An Army taxonomy of cyberspace operations is depicted in figure 2-1, below. The three types of cyberspace operations are—

Cyberspace Operations

 DODIN Operations

 Defensive Cyberspace Operations (DCO)

 Offensive Cyberspace Operations (OCO)

The Army conducts DODIN operations on internal Army and DOD networks and systems using primarily signal forces. The Army employs cyberspace forces to conduct DCO which includes two further sub-divisions—DCO-IDM and defensive cyberspace operations-response actions (DCO-RA). Cyberspace forces conduct DCO-IDM within the DODIN boundary, or on other friendly networks when authorized, in order to defend those networks from imminent or ongoing attacks. At times cyberspace forces may also take action against threat cyberspace actors in neutral or adversary networks in defense of the DODIN or friendly networks. These types of actions, called DCO-RA, require additional authorities and coordination measures. Lastly, cyberspace forces deliberately target threat capabilities in neutral, adversary, and enemy-held portions of cyberspace by conducting OCO. Cyberspace forces may include joint forces from the DOD cyber mission forces or Army-retained cyberspace forces.

See pp. 2-27 to 2-36 for discussion of cyberspace forces.

A. Department of Defense Information Network Operations (DODIN)

The Department of Defense information network is the set of information capabilities and associated processes for collecting, processing, storing, disseminating, and managing information on demand to warfighters, policy makers, and support personnel, whether interconnected or stand-alone. Also called DODIN (JP 6-0). This includes owned and leased communications and computing systems and services, software (including applications), data, security services, other associated services, and national security systems. Department of Defense information network operations are operations to secure, configure, operate, extend, maintain, and sustain Department of Defense cyberspace to create and preserve the confidentiality, availability, and integrity of the Department of Defense information network. Also called DODIN operations (JP 3-12). DODIN operations provide authorized users at all echelons with secure, reliable end-to-end network and information system availability. DODIN operations allow commanders to effectively communicate, collaborate, share, manage, and disseminate information using information technology systems.

Signal forces install tactical networks, conduct maintenance and sustainment activities, and security evaluation and testing. Signal forces performing DODIN operations may also conduct limited DCO-IDM. Since both cyberspace security and defense tasks are ongoing, standing orders for DODIN operations and DCO-IDM cover most cyberspace security and initial cyberspace defense tasks.

The Army secures the DODIN-A using a layered defense approach. Layered defense uses multiple physical, policy, and technical controls in to guard against threats on the network. Layering integrates people, technology, and operational capabilities to establish security barriers across multiple layers of the DODIN-

A. Various types of security barriers include—
- Antivirus software.
- Firewalls.
- Anti-spam software.
- Communications security.
- Data encryption.
- Password protection.
- Physical and technical barriers.
- Continuous security training.
- Continuous network monitoring.

Security barriers are protective measures against acts that may impair the effectiveness of the network, and therefore the mission command system. Additionally, layering includes perimeter security, enclave security, host security, physical security, personnel security, and cybersecurity policies and standards. Layering protects the cyberspace domain at the physical, logical, and administrative control levels.

B. Defensive Cyberspace Operations (DCO)

Defensive cyberspace operations are missions to preserve the ability to utilize blue cyberspace capabilities and protect data, networks, cyberspace-enabled devices,

Cyberspace Operations (Missions & Actions)

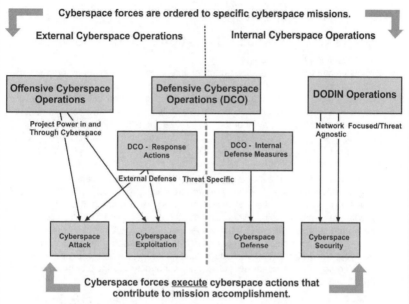

Ref: FM 3-12, Cyberspace Operations and Electromagnetic Warfare (Aug '21), fig. 2-2. Cyberspace operations missions and actions.

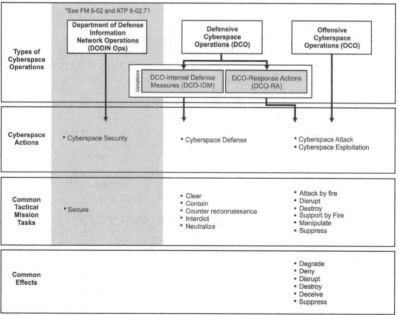

Ref: FM 3-12, Cyberspace Operations and Electromagnetic Warfare (Aug '21), fig. 2-1. Cyberspace operations taxonomy.

(Cyberspace Operations) II. Cyberspace Operations 2-19

and other designated systems by defeating on-going or imminent malicious cyberspace activity (JP 3-12). The term blue cyberspace denotes areas in cyberspace protected by the United States, its mission partners, and other areas the Department of Defense may be ordered to protect. DCO are further categorized based on the location of the actions in cyberspace as—

Defensive Cyberspace Operations-Internal Defensive Measures (DCO-IDM)

Defensive cyberspace operations-internal defensive measures are operations in which authorized defense actions occur within the defended portion of cyberspace (JP 3-12). DCO-IDM is conducted within friendly cyberspace. DCO-IDM involves actions to locate and eliminate cyber threats within friendly networks. Cyberspace forces employ defensive measures to neutralize and eliminate threats, allowing reestablishment of degraded, compromised, or threatened portions of the DODIN. Cyberspace forces conducting DCO-IDM primarily conduct cyberspace defense tasks, but may also perform some tasks similar to cyberspace security.

Cyberspace defense includes actions taken within protected cyberspace to defeat specific threats that have breached or are threatening to breach cyberspace security measures and include actions to detect, characterize, counter, and mitigate threats, including malware or the unauthorized activities of users, and to restore the system to a secure configuration. (JP 3-12). Cyberspace forces act on cues from cybersecurity or intelligence alerts of adversary activity within friendly networks. Cyberspace defense tasks during DCO-IDM include hunting for threats on friendly networks, deploying advanced countermeasures, and responding to eliminate these threats and mitigate their effects.

Defensive Cyberspace Operations-Response Actions (DCO-RA)

Defensive cyberspace operation-response actions are operations that are part of a defensive cyberspace operations mission that are taken external to the defended network or portion of cyberspace without permission of the owner of the affected system (JP 3-12). DCO-RA take place outside the boundary of the DODIN. Some DCO-RA may include actions that rise to the level of use of force and may include physical damage or destruction of enemy systems. DCO-RA consist of conducting cyberspace attacks and cyberspace exploitation similar to OCO. However, DCO-RA use these actions for defensive purposes only, unlike OCO that is used to project power in and through cyberspace.

Decisions to conduct DCO-RA depend heavily on the broader strategic and operational contexts such as the existence or imminence of open hostilities, the degree of certainty in attribution of the threat; the damage the threat has or is expected to cause, and national policy considerations. DCO-RA are conducted by national mission team(s) and require a properly coordinated military order, coordination with interagency and unified action partners, and careful consideration of scope, rules of engagement, and operational objectives.

C. Offensive Cyberspace Operations (OCO)

Offensive cyberspace operations are missions intended to project power in and through cyberspace (JP 3-12). Cyberspace forces conduct OCO outside of DOD networks to achieve positions of relative advantage through cyberspace exploitation and cyberspace attack actions in support of commanders' objectives. Commanders must integrate OCO within the combined arms scheme of maneuver throughout the operations process to achieve optimal effects.

The Army provides cyberspace forces trained to perform OCO across the range of military operations to the joint force. Army forces conducting OCO do so under the authority of a joint force commander. Refer to Appendix C for information on integrating with unified action partners. Joint forces may provide OCO support to corps and below Army commanders in response to requests through the joint targeting process. Refer to Appendix D for more information on joint cyberspace forces. Targets for cyberspace effects may require extended planning time, extended approval time, as well as synchronization and deconfliction with partners external to the DOD. Chapter 4 covers targeting considerations in detail.

II. Cyberspace Actions

Execution of these cyberspace operations entails one or more specific tasks, which joint cyberspace doctrine refers to as cyberspace actions (refer to JP 3-12), and the employment of one or more cyberspace capabilities. Figure 2-2 on page 2-6 depicts the relationships between the types of cyberspace operations and their associated actions, the location of those operations in cyberspace, and the forces that conduct those operations. The four cyberspace actions are—

A. Cyberspace Security

Cyberspace security is actions taken within protected cyberspace to prevent unauthorized access to, exploitation of, or damage to computers, electronic communications systems, and other information technology, including platform information technology, as well as the information contained therein, to ensure its availability, integrity, authentication, confidentiality, and nonrepudiation (JP 3-12). These preventive measures include protecting the information on the DODIN, ensuring the information's availability, integrity, authenticity, confidentiality, and nonrepudiation. Cyberspace security is generally preventative in nature, but also continues throughout DCO-IDM and incident responses in instances where a cyberspace threat compromises the DODIN. Some common types of cyberspace security actions include—

- Password management.
- Software patching.
- Encryption of storage devices.
- Mandatory cybersecurity training for all users.
- Restricting access to suspicious websites.
- Implementing procedures to define the roles, responsibilities, policies, and administrative functions for managing DODIN operations.

B. Cyberspace Defense

Cyberspace defense are actions taken within protected cyberspace to defeat specific threats that have breached or are threatening to breach cyberspace security measures and include actions to detect, characterize, counter, and mitigate threats, including malware or the unauthorized activities of users, and to restore the system to a secure configuration. (JP 3-12)

C. Cyberspace Exploitation

Cyberspace exploitation consists of actions taken in cyberspace to gain intelligence, maneuver, collect information, or perform other enabling actions required to prepare for future military operations (JP 3-12). These operations must be authorized through mission orders and are part of OCO or DCO-RA actions in gray or red cyberspace that do not create cyberspace attack effects, and are often intended to remain clandestine. Cyberspace exploitation includes activities to support operational preparation of the environment for current and future operations by gaining and maintaining access to networks, systems, and nodes of military value; maneuvering to positions of advantage within cyberspace; and positioning cyberspace capabilities to facilitate follow-on actions. Cyberspace exploitation actions are deconflicted with other United States Government departments and agencies in accordance with national policy.

D. Cyberspace Attack

Ref: FM 3-12, Cyberspace Operations and Electromagnetic Warfare (Aug '21), p. 2-7.

Cyberspace attack actions taken in cyberspace that create noticeable denial effects (i.e., degradation, disruption, or destruction) in cyberspace or manipulation that leads to denial effects in the physical domains (JP 3-12). A cyberspace attack creates effects in and through cyberspace and may result in physical destruction. Modification or destruction of cyberspace capabilities that control physical processes can lead to effects in the physical domains. Some illustrative examples of common effects created by a cyberspace attack include—

Deny
To prevent access to, operation of, or availability of a target function by a specified level for a specified time (JP 3-12). Cyberspace attacks deny the enemy's ability to access cyberspace by hindering hardware and software functionalities for a specific duration of time.

Degrade
To deny access to, or operation of, a target to a level represented as a percentage of capacity. Level of degradation is specified. If a specific time is required, it can be specified (JP 3-12).

Disrupt
To completely but temporarily deny access to, or operation of, a target for a period of time. A desired start and stop time are normally specified. Disruption can be considered a special case of degradation where the degradation level is 100 percent (JP 3-12). Commanders can use cyberspace attacks that temporarily but completely deny an enemy's ability to access cyberspace or communication links to disrupt decision making, ability to organize formations, and conduct command and control. Disruption effects in cyberspace are usually limited in duration.

Destroy
To completely and irreparably deny access to, or operation of, a target. Destruction maximizes the time and amount of denial. However, destruction is scoped according to the span of a conflict, since many targets, given enough time and resources, can be reconstituted (JP 3-12). Commanders can use cyberspace attacks to destroy hardware and software beyond repair where replacement is required to restore system function. Destruction of enemy cyberspace capabilities could include irreversible corruption to system software causing loss of data and information, or irreparable damage to hardware such as the computer processor, hard drive, or power supply on a system or systems on the enemy's network.

Manipulate
Manipulation, as a form of cyberspace attack, controls or changes information, information systems, and/or networks in gray or red cyberspace to create physical denial effects, using deception, decoying, conditioning, spoofing, falsification, and other similar techniques. It uses an adversary's information resources for friendly purposes, to create denial effects not immediately apparent in cyberspace (JP 3-12). Commanders can use cyberspace attacks to manipulate enemy information or information systems in support of tactical deception objectives or as part of joint military deception. Refer to FM 3-13.4 for information on Army support to military deception.

Note. Cyberspace attacks are types of fires conducted during DCO-RA and OCO actions and are limited to cyber mission force(s) engagement. They require coordination with other United States Government departments and agencies and careful synchronization with other lethal and non-lethal effects through established targeting processes.

III. Interrelationship with Other Operations

This section describes the relationship that cyberspace operations and EW have with other operations. It discusses how cyberspace operations and EW mutually support intelligence operations, space operations, and information operations.

A. Intelligence Operations

As an operation, intelligence is (1) the product resulting from the collection, processing, integration, evaluation, analysis, and interpretation of available information concerning foreign nations, hostile or potentially hostile forces or elements, or areas of actual or potential operations; (2) the activities that result in the production; and (3) the organizations engaged in such activities (JP 2-0). Intelligence at all echelons supports the planning of cyberspace operations and EW and assists with defining measures of performance and effectiveness. Intelligence also assists the fires support element in developing the high payoff target (HPT) list, and collaborating with the CEMA section to ensure the high payoff target list includes enemy cyberspace and EW-related targets. Intelligence also plays a crucial part in assisting the fires support element in continued target development, including forwarding targets to the joint task force (JTF) headquarters for assessment as potential targets for the joint targeting list.

Information collection supports cyberspace operations and EW by collecting information to satisfy commander's critical information requirement(s) (CCIRs) and staff members' information requirements (IRs) regarding friendly, neutral, and enemy cyberspace and EMS capabilities, activities, disposition, and characteristics within the OE. Information collection also drives capability development. A robust intelligence package is imperative to understanding the target space, developing tools and having meaningful effects in cyberspace. There are four tasks and missions nested in information collection: intelligence operations, reconnaissance, surveillance, and security operations (See Chapter 4).

Information obtained by information collection drives the IPB process. Through the IPB process, the G-2 or S-2 analyzes operational and mission variables in an area of interest to determine their effect on operations. These variables affect how friendly forces will conduct cyberspace operations and EW within the assigned AO. Conversely, cyberspace operations and EW also contribute to intelligence by supporting information collection. Cyberspace operations and EW capabilities collect combat information to answer CCIRs and IRs for situational awareness and targeting.

SIGINT, cyberspace operations, and EW may overlap during operations in the EMS. For this reason, effective integration of SIGINT, cyberspace, EW, and spectrum management operations extends well beyond simple coordination. Effective integration requires both deconfliction and identification of windows of opportunity among these operations. This integration requires close staff collaboration, detailed procedural controls, and various technical channels. See Chapter 4 for additional details.

The intelligence staff also identifies adversary and enemy key terrain as part of the IPB process. Cyberspace operations use the concept of key terrain as a model to identify critical aspects of the cyberspace domain. Identified key terrain in cyberspace is subject to actions the controlling combatant (friendly, enemy, or adversary) deems advantageous such as defending, exploiting, and attacking. Key terrain in cyberspace corresponds to nodes, links, processes, or assets in cyberspace, whether part of the physical, logical, or cyber-persona layer. Key terrain in cyberspace may include—

- Locations in cyberspace in which friendly forces can gather intelligence.
- Locations in cyberspace that support network connectivity.
- Entry points to friendly networks that require priorities for defense.
- Locations in cyberspace that friendly forces require access for essential functions or capabilities.

B. Space Operations

Cyberspace and space operations are interdependent. Access to the space domain is critical to cyberspace operations, especially DODIN operations, enabling global end-to-end network connectivity. In the Army, the space domain is only accessible through space operations. Conversely, space capabilities such as navigation warfare, offensive space control, and defensive space control are dependent on operations conducted in space, cyberspace, and the EMS. This interrelationship is critical, and addressing the interdependencies between the three must be managed throughout the operations process.

Both cyberspace operations and EW can affect space operations. Ground control systems that control satellites rely on networked computers to maintain orbital parameters and direct onboard sensors, particularly to maintain stable orbits; radios transmit computer commands to the satellites. Computer code sent directly to satellites in orbit can potentially allow remote control of the system, preventing others' access to onboard sensors or communications systems. Adversaries could similarly enter ground control systems and issue alternative orders to satellites to move them out of position or shut off critical systems. Because satellites routinely receive commands using radio frequencies, an adversary might attempt to shut off sensors or directly gain control of the spacecraft, rather than trying to issue orders through a ground control system.

All space operations rely on the EMS for command and control, sensing, and information distribution. The vital nature of space operations in multi-domain operations requires close coordination with other EMS activities associated with spectrum management operations to ensure proper prioritization, integration, synchronization, and deconfliction. The G-2 or S-2 uses information gathered through space-based intelligence, surveillance, and reconnaissance to assist the commander and staff with attaining situational awareness and understanding of the OE.

Navigation warfare is the deliberate defensive and offensive action to assure and prevent positioning, navigation, and timing information through coordinated employment of space, cyberspace, and electromagnetic warfare operations (JP 3-14). A navigation warfare attack denies threat actors a global navigation satellite system through various methods, including OCO, space operations, and EA. Global navigation satellite system is the general term used to describe any space-based system providing positioning, navigation, and timing (PNT) information worldwide (for example, Global Positioning System). Navigation warfare effectiveness requires synchronization of space operations, cyberspace operations, and EW capabilities with lethal and nonlethal attack actions to create desired effects. EW must be synchronized with space operations to understand the impacts of navigation warfare operations, deny adversary access to global navigation satellite system information, and protect friendly spectrum-dependent devices using specific frequencies within the EMS.

Refer to FM 3-14 for more information on navigation warfare.

The space domain consists of three segments: space, link, and ground. The space segment is the operational area corresponding with the space domain and comprises satellites in both geosynchronous and non-geosynchronous Earth orbit. The link segment consists of signals connecting ground and space segments through the EMS. The ground segment consists of ground-based facilities and equipment supporting command and control of space assets, ground-based processing equipment, earth terminals, user equipment, space situational awareness sensors, and the interconnectivity between the facilities and equipment. Earth terminals include all multi-Service ground, shipborne, submarine, and airborne satellite terminals that establish connectivity to the satellites in the space segment. The three space domain segments rely heavily on cyberspace operations to protect networking and information technologies and infrastructures while depending on the EMS to conduct operations between the space, link, and ground segments.

Cyberspace operations contribute to space operations by protecting friendly networks that leverage the global navigation satellite system while targeting similar enemy and adversary capabilities. Additionally, cyberspace operations establish network connectivity between ground-based facilities and equipment throughout the space domain's ground segment. EW supports navigation warfare by denying the enemy access to global navigation satellite system information while protecting friendly space capabilities operating in the EMS.

Integrating cyberspace operations, EW, and space operations enable commanders and staffs at each level to synchronize capabilities and effects. Space-based capabilities (space segment) enable distributed and global cyberspace operations. Cyberspace and space-based capabilities provide responsive and timely support that allows commanders to project combat power from the highest echelons down to the tactical level. Synchronization with spectrum management operations is necessary to ensure the availability of resources in the EMS and to prevent spectrum conflicts.

Refer to FM 3-14 for more information about space operations.

C. Information Operations (IO)

Information operations are the integrated employment, during military operations, of information-related capabilities in concert with other lines of operations to influence, disrupt, corrupt, or usurp the decision-making of adversaries and potential adversaries while protecting our own (JP 3-13). Information operations (IO) integrate and synchronize information-related capabilities to create effects in and through the information environment and deliver an operational advantage to the commander. IO optimize the information element of combat power and support and enhance all other elements to gain operational advantage over a threat. IO consist of three interrelated efforts that work in tandem and overlap each other. These three efforts are—

- A commander-led staff planning and synchronization effort.
- A preparation and execution effort carried out by information-related capabilities units, IO units, or staff entities in concert with the IO working group.
- An assessment effort that is carried out by all involved.

When commanders employ cyberspace and EW capabilities to create desirable conditions within the OE, they synchronize these actions through IO. Commanders use cyberspace operations and EW to gain a strategic advantage in cyberspace and the EMS. Cyberspace and EW capabilities support operations by enabling the ability to share information among friendly forces or affecting the enemy's ability to use cyberspace and the EMS.

Cyberspace operations and EW effects influence, disrupt, corrupt, or manipulate the decision-making cycle of threat actors. Cyberspace operations support operations through OCO or DCO-RA by creating denial or manipulation effects to degrade, disrupt, or destroy the enemy's cyberspace capability or change enemy information, information systems, or networks. EW supports operations through EA by degrading, neutralizing, or destroying enemy capability to use the EMS. EW also supports operations through EP actions by concealing or manipulating friendly EMS signatures, to degrade or deceive enemy sensors or targeting systems. When integrated and synchronized with other capabilities, cyberspace operations and EW can help commanders set favorable conditions for information advantage, whether in cyberspace, the EMS, or other domains.

Cyberspace operations and EW can also create cognitive effects by impacting physical components of enemy capabilities. For example, affecting the ability of an enemy's fires network through a cyberspace attack or EA may deny or create doubt about their ability to use artillery effectively. Similarly, restricting the enemy's ability to use cyberspace or EMS at critical points can affect enemy judgments when exercising command and control. Synchronizing defensive EW and cyberspace operations with other capabilities can also disrupt a threat's ability to make decisions while ensuring friendly forces freedom of action.

Cyberspace operations and EW synchronized through the operations process and targeting can provide commanders additional ways and means to—

- Affect threat capabilities that inform or influence decision making.
- Affect threat capabilities for command and control, movement and maneuver, fires, intelligence, communications, and information warfare.
- Affect threat capabilities to target and attack friendly command and control and related decision support systems.
- Affect threat capabilities that distribute, publish, or broadcast information designed to persuade relevant actors to oppose friendly operations.
- Enable military deception directed against threat decision making, intelligence and information gathering, communications, dissemination, and command and control capabilities.
- Enable friendly OPSEC to protect critical information.
- Enable friendly influence activities, such as military information support operations, to improve or sustain positive relations with foreign actors in and around the operational area and to degrade threat influence over the same.
- Protect friendly information, technical networks, and decision-making capabilities from exploitation by enemy and adversary information warfare assets.

See pp. 0-10 to 0-16 for further discussion of information operations.

Refer to INFO1: The Information Operations & Capabilities SMARTbook (Guide to Information Operations & the IRCs). INFO1 chapters and topics include information operations (IO defined and described), information in joint operations (joint IO), information-related capabilities (PA, CA, MILDEC, MISO, OPSEC, CO, EW, Space, STO), information planning (information environment analysis, IPB, MDMP, JPP), information preparation, information execution (IO working group, IO weighted efforts and enabling activities, intel support), fires & targeting, and information assessment.

III. Army Organizations & Command and Control

Ref: FM 3-12, Cyberspace Operations and Electromagnetic Warfare (Aug '21), chap. 3.

Army maneuver commanders use cyberspace operations and EW to understand the OE, support decision-making, and affect adversaries. Maneuver commanders at the brigade combat team level and above rely on assigned CEMA sections to leverage Army and joint cyberspace and EW capabilities. During joint operations, a corps or division designated as a JTF headquarters or a joint force headquarters combines its spectrum management chief with its CEMA section to establish an electromagnetic spectrum operations (EMSO) cell to support the joint electromagnetic spectrum operations cell (JEMSOC). Numerous Army and joint organizations contribute forces and capabilities for use in cyberspace operations and EW. Commanders at corps and below should possess a general understanding of the roles and responsibilities of these organizations and how they interact with the units' CEMA sections.

I. United States Army Cyber Command

ARCYBER operates and defends Army networks and delivers cyberspace effects against adversaries to defend the nation. ARCYBER rapidly develops and deploys cyberspace capabilities to equip our force for the future fight against a resilient, adaptive adversary. ARCYBER also integrates intelligence, fires, space, psychological operations, strategic communications, public affairs, special technical operations, cyberspace operations, electromagnetic warfare, and information operations to allow Army commanders a decisional advantage during competition and conflict.

ARCYBER protects DODIN-A through DCO-IDM and DODIN operations. Commander, ARCYBER, is also the commander of joint force headquarters-cyber (JFHQ-C [Army]). In this role, Commander, ARCYBER, possesses the capability to conduct OCO to attack and exploit the enemy upon authorization from Unites States Cyber Command (USCYBERCOM). ARCYBER is the Army's point of contact for reporting and assessing cyber incidents and events involving suspected adversary activity. The United States Army Network Enterprise Technology Command (NETCOM) and the regional cyber center act as the chief action arms, having been delegated operational control and directive authority for cyberspace operations by ARCYBER for DODIN operations over all Army networks. ARCYBER serves as the Army's principal cybersecurity service provider and provides program oversight while NETCOM and the regional cyber centers act as the principal executors of the program. Units assigned to ARCYBER are—

- NETCOM.
- 1st Information Operations Command (Land).
- 780th Military Intelligence Brigade.
- Cyber protection brigade.
- 915th Cyber Warfare Battalion.

II. Army Information Warfare Operations Center

The Army Information Warfare Operations Center serves as ARCYBER's hub for coordinating, integrating, synchronizing, and tracking cyberspace operations, electromagnetic warfare (EW), IO, and answering intelligence requirements in support of national, regional, and Army directives. The Army Information Warfare Operations Center maintains global and regional situational awareness and understanding while executing mission command of all assigned or allocated Army cyber and IO forces.

The Army Information Warfare Operations Center is composed of personnel with information-related capabilities expertise (IO, cyber, EW, psychological operations [forces], public affairs, civil affairs, military deception, United States Space Command and special technical operations), to include representatives from all staff functions and embeds from partner organizations. The Army Information Warfare Operations Center is responsible for integrating information-related capabilities across the staff into the command's current operations and plans processes. Additionally, the Army Information Warfare Operations Center —

- Receives reports from subordinate commands.
- Prepares reports required by higher headquarters.
- Processes requests for support (RFS).
- Publishes operation orders (OPORDs) and cyber tasking orders (CTOs).
- Consolidates Commander's critical information requirements.
- Answers requests for information from higher HQs, CCMDs, other Services and agencies.
- Assesses the overall progress of ongoing operations.

III. Cyberspace Electromagnetic Activities at Corps and Below

CEMA sections are assigned to the G-3 or S-3 within corps, divisions, BCTs, and combat aviation brigades. Commanders are responsible for ensuring that CEMA sections integrate cyberspace operations and EW into their concept of operations. The CEMA section involves key staff members in the CEMA working group to assist in planning, development, integration, and synchronization of cyberspace operations and EW.

Note. The structure of the CEMA section is similar at all corps and below echelons. However, 1st IO Command may augment a corps' CEMA section to provide increased capabilities for synchronizing and integrating cyberspace operations and EW with IO.

A. Commander's Role

Commanders direct the continuous integration of cyberspace operations and EW within the operations process, whether in a tactical environment or at home station. By leveraging cyberspace operations and EW as part of combined arms approach, commanders can sense, understand, decide, act, and assess faster than the adversary assesses and achieve a decisional advantage in multiple domains during operations.

Commanders should—

- Include cyberspace operations and EW within the operations process.
- Continually enforce cybersecurity standards and configuration management.
- Understand, anticipate, and account for cyberspace and EW effects, capabilities, constraints, and limitations, including second and third order effects.
- Understand the legal and operational authorities to affect threat portions of cyberspace or EMS.
- Understand the implications of cyberspace operations and EW operations on the mission and scheme of maneuver.
- Understand how the selected course of action (COA) affects the prioritization of

resources to their portion of the DODIN-A.
- Leverage effects in and through cyberspace and the EMS to support the concept of operations.
- Develop and provide intent and guidance for actions and effects inside and outside of the DODIN-A.
- Identify critical missions or tasks in phases to enable identification of key terrain in cyberspace.
- Ensure active collaboration across the staff, subordinate units, higher headquarters, and unified action partners that enable a shared understanding of cyberspace and the EMS.
- Approve high-priority target lists, target nominations, collection priorities, and risk mitigation measures.
- Ensure the synchronization of cyberspace operations and EW with other lethal and nonlethal fires to support the concept of operations.
- Oversee the development of cyberspace operations and EW-related home-station training.

B. Cyberspace Electromagnetic Activities (CEMA) Section

The CEMA section plans, coordinates, and integrates OCO, DCO and EW in support of the commander's intent. The CEMA section collaborates with numerous staff sections to ensure unity of effort in meeting the commander's total operational objectives such as collaborating with the G-2 or S-2 to attain situation awareness and understanding of friendly, enemy, and neutral actors operating within the AO. The CEMA section is responsible for providing regular updates to the commander and staff on OCO and other supported operations conducted in the AO. The CEMA section is responsible for synchronizing and integrating cyberspace operations and EW with the operations process and through other integrating processes. Personnel assigned to the CEMA section are the—

- CEWO.
- Cyber warfare officer.
- EW technician.
- EW sergeant major (corps) or EW NCOIC (division).
- EW noncommissioned officer (NCO).
- CEMA spectrum manager.

See following pages (pp. 2-30 to 2-31) for an overview and further discussion.

C. Cyberspace Electromagnetic Activities (CEMA) Working Group

The CEMA section leads the CEMA working group. The CEMA working group is not a formal working group that requires dedicated staff members from other sections. When needed, the CEWO uses a CEMA working group to assist in synchronizing and integrating cyberspace operations and EW into the concept of operations. The CEMA section normally collaborates with key stakeholders during staff meetings established as part of the unit's battle rhythm and throughout the operations process. Membership in the CEMA working group will vary based on mission requirements.

If scheduled, the CEMA working group must be integrated into the staff's battle rhythm. The CEMA working group is responsible for coordinating horizontally and vertically to support operations and assist the fires support element throughout the

Cyberspace Electromagnetic Activities (CEMA) Section

Ref: FM 3-12, Cyberspace Operations and Electromagnetic Warfare (Aug '21), pp. 3-5 to 3-8.

The CEMA section plans, coordinates, and integrates OCO, DCO and EW in support of the commander's intent. The CEMA section collaborates with numerous staff sections to ensure unity of effort in meeting the commander's total operational objectives such as collaborating with the G-2 or S-2 to attain situation awareness and understanding of friendly, enemy, and neutral actors operating within the AO. The CEMA section is responsible for providing regular updates to the commander and staff on OCO and other supported operations conducted in the AO. The CEMA section is responsible for synchronizing and integrating cyberspace operations and EW with the operations process and through other integrating processes. Personnel assigned to the CEMA section are the—

Cyber Electromagnetic Warfare Officer (CEWO)

The CEWO is the commander's designated staff officer responsible for integrating, coordinating, and synchronizing actions in cyberspace and the EMS. The CEWO is responsible for understanding all applicable classified and unclassified cyberspace and spectrum-related policies to assist the commander with planning, coordinating, and synchronizing cyberspace operations, EW, and CEMA. A commander that has been delegated electromagnetic attack control authority from higher headquarters may further delegate it to the CEWO. Refer to ATP 3-12.3 for specific roles and responsibilities of the CEWO. Tasks for which the CEWO is responsible include—

- Advising the commander on effects in cyberspace (including associated rules of engagement, impacts, and constraints) in coordination with the staff judge advocate.
- Advising the commander of mission risks presented by possible cyberspace and EW vulnerabilities and adversary capabilities.
- Analyzing the OE to understand how it will impact operations within cyberspace and the EMS.
- Developing and maintaining the consolidated cyberspace and EW target synchronization matrix and recommending targets for placement on the units' target synchronization matrix.
- Assisting the G-2 or S-2 with the development and management of the electromagnetic order of battle.
- Serving as the electromagnetic attack control authority for EW missions when directed by the commander.
- Advising the commander on how cyberspace and EW effects can impact the OE.
- Receiving and integrating cyberspace and EW forces and associated capabilities into operations.
- Coordinating with higher headquarters for OCO and EW support on approved targets.
- Recommending cyberspace operations and EW-related CCIRs.
- Preparing and processing all requests for cyberspace and EW support.
- Overseeing the development and implementation of cyberspace operations and EW-related home-station training.
- Providing employment guidance and direction for organic and attached cyberspace operations and EW assets.
- Tasking authority for all assigned EW assets.

Cyber Warfare Officer (Corps And Brigade) or Cyber Operations Officer (Division)

The cyber warfare officer (corps and brigade) or cyber operations officer (division) assists the CEWO with integrating and synchronizing cyberspace operations into the operations process and provides insight into cyberspace capabilities. The cyber warfare officer or cyber operations officer collaborates with the CEWO in vetting and processing potential targets received from subordinate units for OCO effects. The cyber warfare officer or cyber operations officer—

- Assists the CEWO in the integration, coordination, and synchronization of cyberspace operations and EW with operations.
- Provides the CEWO with information on the effects of cyberspace operations, including associated rules of engagement, impacts, and constraints used to advise the commander.
- Assists the CEWO with developing and maintaining a consolidated cyberspace target synchronization matrix and assists in nominating OCO-related targets for approval by the commander.
- Assists the CEWO in monitoring and assessing measures of performance and effectiveness while maintaining updates on cyberspace operation's effects on the OE.
- Assists the CEWO in requesting and coordinating for OCO support while integrating received cyber mission forces into operations.
- Coordinates with unified action partners for cyberspace capabilities that complement or increase the unit's cyberspace operations posture.
- Coordinates cyberspace operations with the G-2 or S-2 and the G-6 or S-6.
- Develops and implements cyberspace operations-related home station training.

Electromagnetic Warfare Technician (EWT)

The Electromagnetic Warfare Technician (EWT) is a critical asset to the CEMA section and the EW platoon as they serve as the resident technical and tactical expert across all echelons. The EWT assist in the accomplishment of mission objectives by coordinating, integrating, and synchronizing CEMA effects to exploit and gain an advantage over adversaries and enemies in both cyberspace and the electromagnetic spectrum (EMS), while simultaneously denying and degrading adversary and enemy use of the same.

See pp. 3-11 to 3-16 for discussion of EW key personnel and duties from ATP 3-12.3.

Electromagnetic Warfare Sergeant Major (Corps) or NCOIC (Division)

The EW sergeant major or NCOIC is the CEWO's senior enlisted advisor for EW. The EW sergeant major or NCOIC assists the CEWO and cyber warfare officer with integrating, coordinating, and cyberspace operations and EW with operations.

See pp. 3-11 to 3-16 for discussion of EW key personnel and duties from ATP 3-12.3.

Electromagnetic Warfare Noncommissioned Officer (EW NCO)

The EW NCO manages the availability and employment of EW assets assigned to the unit.

See pp. 3-11 to 3-16 for discussion of EW key personnel and duties from ATP 3-12.3.

Cyberspace Electromagnetic Activities Spectrum Manager

The CEMA spectrum manager assists the CEMA section in the planning, coordination, assessment, and implementation of EW through frequency management. The CEMA spectrum manager defines the EMOE for the CEMA section.

See chap. 5, Spectrum Management Operations.

execution of an operation. Generally, the CEMA working group is comprised of staff representatives with equities in CEMA, and typically include—
- The G-2 or S-2.
- The G-6 or S-6.
- The IO officer or representative.
- The G-6 or S-6 spectrum manager.
- The fire support officer or a fires support element representative.
- The staff judge advocate.
- The Protection Officer.

IV. Staff and Support at Corps and Below

During the operations process and associated integrating processes, cyberspace operations and EW require collaborative and synchronized efforts with other key staff. The G-6 or S-6 oversees DODIN operations, and the G-6 or S-6 spectrum manager collaborates with the CEMA spectrum manager to synchronize spectrum management operations with EW. The G-2 or S-2 manages the integration and synchronization of the IPB process and information collection. The IO officer oversees the integration and synchronization of information-related capabilities for IO. The staff judge advocate advises the commander on the legality of operations.

A. Assistant Chief of Staff, Intelligence

The G-2 or S-2 provides intelligence to support CEMA. The G-2 or S-2 facilitates understanding the enemy situation and other operational and mission variables. The G-2 or S-2 staff provides direct or indirect support to cyberspace operations and EW through information collection, enabling situational understanding, and supporting targeting and IO. The G-2 or S-2 further supports CEMA by—
- Assessing CEMA intelligence and plans while overseeing information collection and analysis to support the IPB, target development, enemy COA estimates, and situational awareness.
- Continually monitoring intelligence operations and coordinating intelligence with supporting higher, lateral, and subordinate echelons.
- Coordinating SIGINT.
- Coordinating for intelligence and local law enforcement support to enhance cyberspace security.
- Leading the IPB and developing IPB products.
- Overseeing the development and management of the electromagnetic order of battle.
- Providing all-source intelligence to CEMA.
- Coordinating with the G-3 or S-3 and fires support element to identify high-value target(s) from the high-payoff target list for each friendly COA.
- Coordinating with the intelligence community to validate threat-initiated cyberspace attack or EA activities in the OE.
- Requesting intelligence support and collaborating with the intelligence community and local law enforcement to gather intelligence related to threat cyberspace operations and EW in the OE.
- Providing information and intelligence on threat cyberspace and EW characteristics that facilitate situational understanding and supports decision making.
- Coordinating with Air Force Combat Weather Forecasters for information on the terrain and weather variables for situational awareness.

- Ensuring information collection plans and operations support CEMA target development, target update requirements, and combat assessment.
- Developing requests for information and collection for information requirements that exceed the unit's organic intelligence capabilities.
- Collecting, processing, storing, displaying, and disseminating cyberspace operations and EW relevant information throughout the operations process and through command and control systems.
- Consolidating all high-value target(s) on a high-payoff target list.
- Providing input for guarded frequencies from the intelligence community.
- Providing the CEMA section and G-6 or S-6 prioritized EMS usage requirements for intelligence operations.
- Participating as a member of the CEMA working group.
- Assisting the CEMA spectrum manager in mitigating EMI and resolving EMS deconfliction and assisting with determining the source of unacceptable EMI.

B. Assistant Chief of Staff, Signal

In collaboration with the joint force and unified action partners (as appropriate), the G-6 or S-6 staff directly or indirectly supports cyberspace operations by conducting DODIN operations. G-6 or S-6 is the primary staff representative responsible for spectrum management operations. The G-6 or S-6 staff supports CEMA by—

- Establishing the tactical portion of the DODIN-A, known as the tactical network, at theater army and below.
- Conducting DODIN operations activities, including cyberspace security, to meet the organization's communications requirements.
- Assisting in developing the cyberspace threat characteristics specific to enemy and adversary activities and related capabilities within friendly networks, and advising on cyberspace operations COAs.
- Conducting cyberspace security risk assessments based on enemy or adversary tactics, techniques, and procedures, identifying vulnerabilities to crucial infrastructure that may require protection measures that exceed the unit's capabilities and require DCO-IDM support.
- Participating in the CEMA working group.
- Providing a common operational picture of the DODIN for planning purposes and situational awareness.
- Providing subject matter expertise regarding wired and wireless networks.
- Ensuring security measures are configured, implemented, and monitored on the DODIN-A based on threat reports.
- Overseeing spectrum management operations.
- Implementing layered security by employing tools to provide layered cyberspace security and overseeing security training throughout the organization.
- Coordinating with the regional cyber center to ensure the unit understands and meets compliance of all cyberspace operations policies and procedures within the region.
- Requesting satellite and gateway access through the regional satellite communications support center.
- Coordinating with regional hub node to establish network connectivity and access services.

C. G-6 or S-6 Spectrum Manager

The G-6 or S-6 spectrum manager coordinates EMS usage for various communications and electronic systems and resources. The G-6 or S-6 spectrum manager supports CEMA by—

- Coordinating spectrum resources for the organization.
- Coordinating for spectrum usage with higher headquarters, host nations, and international agencies as necessary.
- Coordinating frequency allocation, assignment, and usage.
- Coordinating spectrum resources for communications assets used for deception operations.
- Coordinating with the higher headquarters' spectrum manager to mitigate EMI identified in the unit's portion of EMOE.
- Seeking assistance from the higher the headquarters' spectrum managers for a resolution to unresolvable internal EMI.
- Participating in the CEMA working group.
- Assisting the CEMA spectrum manager with deconflicting friendly EMS requirements with planned EW, cyberspace operations, and information collection.
- Collaborating with the CEMA spectrum manager to ensure the integration and synchronization of spectrum management operations with EW.

D. Information Operations Officer (Corps & Division) or Representative (Bde & Below)

The IO officer or representative leads the unit's IO element. The IO officer or representative contributes to the IPB by identifying and evaluating threats targeted actors in the AO.

See facing page for further discussion.

E. Fires Support Element

The fires support element plans, coordinates, integrates, synchronizes, and deconflicts current and future fire support to meet the commander's objectives. Fire support coordination may include collaboration with joint forces and unified action partners. The fires support element coordinates with the CEMA section to synchronize, plan, and execute cyberspace attacks and EA as part of the targeting process. The fires support element support CEMA by—

- Leading the targeting working group and participating in the targeting board chaired by the commander.
- Assisting the G-2 or S-2 with synchronizing the information collection plan with cyberspace operations, EW, and other fires.
- Collaborating with the CEMA section and the G-2 or S-2 in developing and managing the high-payoff target list, target selection standards, attack guidance matrix, and targeting synchronization matrix, all of which include cyberspace attack and EA-related targets.
- FM 3-12 provides additional bullets for FSE support to CEMA.

F. Staff Judge Advocate

The staff judge advocate is the field representative of the Judge Advocate General and the primary legal adviser to the commander. The staff judge advocate also advises the CEMA working group concerning operational law, and the legality of cyberspace operations and EW, particularly those cyberspace and EW tasks that may affect noncombatants. The staff judge advocate is the unit's subject matter expert on the law of war, rules of engagement, the protection of noncombatants, detainee operations, and fiscal and contract law, providing commanders and staff with essential input on plans, directives, and decisions related to lethal and nonlethal targeting.

Information Operations Officer (Corps & Division) or Representative (Bde & Below)

Ref: FM 3-12, Cyberspace Operations and Electromagnetic Warfare (Aug '21), pp. 3-10 to 3-11.

The IO officer or representative leads the unit's IO element. The IO officer or representative contributes to the IPB by identifying and evaluating threats targeted actors in the AO. The IO officer or representative leads the planning, synchronization, and employment of information-related capabilities not managed by a capability owner or proponent. The IO officer or representative coordinates with the CEMA section with integrating cyberspace operations and EW into IO.

- Leading the IO working group.
- Identifying the most effective information-related capabilities to achieve the commander's objectives.
- Synchronizing cyberspace operations and EW with other information-related capabilities to achieve the commander's objectives in the information environment.
- Assessing the risk-to-mission and risk-to-force associated with employing cyberspace operations, EW, and other information-related capabilities in collaboration with the CEMA section.
- Identifying information-related capabilities gaps not resolvable at the unit level.
- Coordinating with Army, other Services, or joint forces for information-related capabilities to augment the unit's shortfalls.
- Providing information, as required, in support of OPSEC at the unit level.
- Collaborating with the CEMA section to employ cyberspace manipulation and EA deception tasks in support of military deception.
- Assessing the effectiveness and making plan modifications to employed information-related capabilities.
- Developing products that describe all military and civilian communications infrastructures and connectivity links in the AO in coordination with the G-2 or S-2.
- Locating and describing all EMS systems and emitters in the EMOE in coordination with the G-2 or S-2, CEMA section, and other information-related capabilities owners.
- Identifying network vulnerabilities of friendly, neutral, and threat forces in coordination with the G-2 or S-2, CEMA section, and other information-related capabilities owners.
- Providing understanding of information-related conditions in the OE in coordination with the G-2 or S-2.
- Participating in the military decision-making process and developing IO-related IRs.
- Participating member of the CEMA working group.
- Integrating IO into the unit's targeting process.
- Integrating non-organic information-related capabilities into operations.
- Ensuring IO-related information is updated in the common operational picture.
- Collaborating with the fire support coordinator for lethal and non-lethal effects.

Refer to INFO1: The Information Operations & Capabilities SMARTbook (Guide to Information Operations & the IRCs). INFO1 chapters and topics include information operations (IO defined and described), information in joint operations (joint IO), information-related capabilities (PA, CA, MILDEC, MISO, OPSEC, CO, EW, Space, STO), information planning (information environment analysis, IPB, MDMP, JPP), information preparation, information execution (IO working group, IO weighted efforts and enabling activities, intel support), fires & targeting, and information assessment.

V. Electromagnetic Warfare (EW) Organizations

Ref: FM 3-12, Cyberspace Operations and Electromagnetic Warfare (Aug '21), pp. 3-3 to 3-4. See also 3-11 to 3-16, EW key personnel.

Electromagnetic Warfare (EW) Platoon

EW platoons are located in the military intelligence company of a brigade combat team's brigade engineer battalion. An EW platoon consists of three EW teams with the capability to provide EW support during close operations. Though the CEMA section aligns EW and cyberspace operations with the operations process, they must collaborate with the BCT's S-2 to task the military intelligence company for deploying EW platoon assets in support of assigned EW missions.

The EW platoon performs electromagnetic reconnaissance to identify and locate enemy emitters and spectrum-dependent devices within assigned AO using sensors. Data and information attained through electromagnetic reconnaissance provide the commander with critical combat information. This data and information also supports electromagnetic battle management by providing continuous situational awareness to the CEMA spectrum manager to develop and update the common operational picture of the EMOE. An EW platoon can also conduct EA to degrade and neutralize enemy spectrum-dependent devices.

When given electromagnetic attack control authority from the JTF headquarters, the JFLCC may further delegate electromagnetic attack control authority to subordinate Army commanders. Electromagnetic attack control authority is a broader evolution of jamming control authority that enables subordinate commanders with the authority to transmit or cease transmission of electromagnetic energy. Electromagnetic attack control authority allows commanders to control EA missions conducted throughout their AO within the constraints of their higher headquarters. Before receiving electromagnetic spectrum coordinating authority, commanders should ensure they have situational awareness of the EMOE, operational control of EW capabilities, and the ability to monitor and estimate EW transmission activities within their AO to determine corrective actions when necessary. Commanders should also ensure that EW missions are thoroughly vetted to ensure deconfliction with friendly spectrum dependent devices. The G-6 spectrum management chief or the G-6 or S-6 spectrum manager is responsible for performing electromagnetic battle management for the unit.

EW platoons reprogram all assigned EW equipment according to system impact messages received from Service equipment support channels that include recommendations to respond to identified threat changes. Commanders may require an EW platoon to make immediate changes to their tactics to regain or improve EW equipment performance.

Intelligence, Information, Cyber, Electromagnetic Warfare, and Space Detachment (I2CEWS)

The I2CEWS detachment is a battalion-sized unit assigned to a multi-domain task force and includes an enhanced CEMA section. The I2CEWS provides cyberspace operations and EW support to an Army Service Component Command, theater army, or the JTF conducting long-range precision joint strikes during multi-domain operations. The I2CEWS is composed of four companies consisting of cyberspace forces with the capability to perform Service-level DCO-IDM and EW operators capable of delivering EA effects throughout the MDTFs assigned AO.

The I2CEWS has organic sensing and intelligence, information, and space operations assets that, when integrated and synchronized with DCO-IDM and EW, allows Army forces to simultaneously defend their assigned portion of the DODIN-A while disrupting, denying, and degrading enemy EMS capabilities. The I2CEWS is structured to meet the continually changing OE in which joint operations are being conducted collaboratively and simultaneously in multiple domains.

Chap 2: IV. Integration through the Operations Process

Ref: FM 3-12, Cyberspace Operations and Electromagnetic Warfare (Aug '21), chap. 4.

At corps and below, the planning, synchronization, and integration of cyberspace operations and EW are conducted by the CEMA section, in collaboration with key staff members that make up the CEMA working group. The CEMA section is an element of the G-3 or S-3 and works closely with members of the CEMA working group to ensure unity of effort to meet the commander's objectives.

I. The Operations Process

The operations process includes the major command and control activities performed during operations: planning, preparing, executing, and continuously assessing the operation (ADP 5-0).

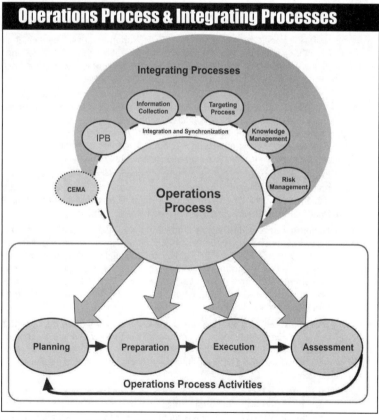

Ref: FM 3-12 (Aug '21), fig. 4-1. The operations process and integrating processes.

The operations process is the Army's framework for the organization and implementation of command and control. The CEMA working group enables the commander with the ability to understand cyberspace and the EMOE.

(Cyberspace Operations) IV. Integration through the Operations Process 2-37

Commanders, staff, and subordinate headquarters use the operations process to organize efforts, integrate the warfighting functions across multiple domains, and synchronize forces to accomplish missions. Army forces plan, prepare, execute, and assess cyberspace operations and EW in collaboration with joint forces and unified action partners as required. Army commanders and staffs will likely coordinate or interact with joint forces to facilitate cyberspace operations and EW. For this reason, commanders and staff should have an awareness of joint planning systems and processes that enable cyberspace operations and EW.

A. Planning

Planning is the art and science of understanding a situation, envisioning a desired future, and laying out effective ways of bringing that future about (ADP 5-0). Commanders apply the art of command and the science of control to ensure cyberspace operations and EW support the concept of operations. Whether cyberspace operations and EW are planned and directed from higher headquarters or requested from tactical units, timely staff actions and commanders' involvement coupled with continued situational awareness of cyberspace and the EMS are critical for mission success.

See chap. 4, Cyberspace and EW Planning.

B. Preparation

Preparation consists of those activities performed by units and Soldiers to improve their ability to execute an operation (ADP 5.0). Preparation activities include initiating information collection, DODIN operations preparation, rehearsals, training, and inspections. Preparation requires the commander, staff, unit, and Soldiers' active engagement to ensure the force is ready to execute operations.

Preparation activities typically begin during planning and continue into execution. At corps and below, subordinate units' that are task-organized to employ cyberspace operations and EW capabilities (identified in the OPLAN or OPORD) conduct preparation activities to improve the force's opportunity for success during operations. Commanders drive preparation activities through leading and assessing. Using the following preparation functions, commanders and staff can—

- **Improve situational understanding.** Commanders, staff, and subordinate units continue to refine knowledge of cyberspace and the EMOE within the assigned AO, including the improved insight on how the use of cyberspace and the EMS could affect operations across multiple domains.

- **Develop a shared understanding of the plan.** Commanders, staff, and tasked subordinate units develop a shared understanding of the plan (described in the OPLAN or OPORD) by conducting home-station training and combat training center(s). These training events provide the perfect opportunity for subordinate commanders, leaders, and Soldiers to execute the developed plan in a controlled environment and to identify issues in the developing plan that require modification.

- **Train and become proficient in critical tasks.** Through rehearsals and training, subordinate units gain and refine skills in those individual and collective tasks essential to the success of cyberspace operations and EW. Commanders also allocate training time for anticipated and unanticipated events and circumstances.

- **Integrate the force.** Commanders allocate preparation time to put the new task-organized force into effect. Integrating the force includes detaching units, moving cyberspace and EW assets, and receiving and integrating new units and Soldiers into the force. Task-organized forces require preparation time to learn the gaining unit's policies and standards and to understand their role in the overall plan. The gaining unit requires time to assess the task-organized forces' cyberspace and EW capabilities and limitations and integrate new capabilities.

- **Ensure the positioning of forces and resources.** Positioning and task organization occur concurrently. Commanders ensure cyberspace and EA assets consist of the right personnel and equipment using pre-operations checks while ensuring those assets are in the right place at the right time.

C. Execution

Execution is the act of putting a plan into action by applying combat power to accomplish the mission (ADP 5-0). The commander, staff, and subordinate commander's focus on translating decisions made during planning and preparing into actions. Commanders conduct OCO and EA to project combat power throughout cyberspace and the EMS, conduct DCO and EP to protect friendly forces and systems, and conduct reconnaissance through cyberspace and the EMS to gather combat information for continuing situational awareness.

Commanders should understand that detailed planning provides a reasonable forecast of execution but must also be aware that situations may change rapidly in cyberspace and the EMOE. During execution, commanders take concerted action to seize, retain, and exploit operational initiative while accepting risk.

Operational initiative is the setting or tempo and terms of action throughout an operation (ADP 3-0). By presenting the enemy with multiple cross-domain dilemmas, including cyberspace and the EMS, commanders force the enemy to react continuously, driving the enemy into positions of disadvantage.

Commanders can use cyberspace attacks and EA to force enemy commanders to abandon their preferred courses of action and make costly mistakes. Commanders retain the initiative by synchronizing cyberspace attacks and EA as fires combined with other elements of combat power to apply unrelenting pressure on the enemy using continuously changing combinations of combat power at a tempo an enemy cannot effectively counter.

Commanders and staff continue to use information collection and electromagnetic reconnaissance assets to identify enemy attempts to regain the initiative. Information collected can be used to readjust targeting priorities and fire support plans, including cyberspace attacks and EA, to keep adversaries on the defensive.

Once friendly forces seize the initiative, they immediately exploit it through continued operations to accelerate the enemy's defeat. Defeat is to render a force incapable of achieving its objective (ADP 3-0). Commanders can use cyberspace attacks and EA to disrupt enemy attempts to reconstitute forces and exacerbate enemy disorganization by targeting adversary command and control and sensing nodes.

D. Assessment

Assessment is the determination of the progress toward accomplishing a task, creating an effect, or achieving an objective (JP 3-0). The commander and staff continuously assess cyberspace operations and EW to determine if they have resulted in the desired effect. Assessment activities support decision making by ascertaining the progress of the operation to develop and refine plans.

Assessment both precedes and guides the other activities of the operations process, and there is no single way to conduct it. Commanders develop an effective assessment plan built around the unique challenges of the operations.

Refer to BSS6: The Battle Staff SMARTbook, 6th Ed. for further discussion. BSS6 covers the operations process (ADP 5-0); commander's activities; Army planning methodologies; the military decisionmaking process and troop leading procedures (FM 7-0 w/Chg 2); integrating processes (IPB, information collection, targeting, risk management, and knowledge management); plans and orders; mission command, C2 warfighting function tasks, command posts, liaison (ADP 6-0); rehearsals & after action reviews; and operational terms and military symbols (ADP 1-02).

II. Integrating Processes

Ref: FM 3-12, Cyberspace Operations and Electromagnetic Warfare (Aug '21), pp. 4-4 to 4-21.

Commanders and staff integrate warfighting functions and synchronize the force to adapt to changing circumstances throughout the operations process. The CEMA section aligns cyberspace operations and EW with the operations process and its associated integrating processes to identify threats in cyberspace and the EMS, to target and attack enemy cyberspace and EMS enabled systems, and to support the warfighting functions.

The operations process is the principal essential activity conducted by a commander and staff. The commander and staff integrate and synchronize CEMA with five key integrating processes throughout the operations process (see figure 4-1). These integrating processes are—

A. Intelligence Preparation of the Battlefield (IPB)

To integrate and synchronize the tasks and missions of information collection, the G-2 or S-2 leads the staff through the IPB process. Intelligence preparation of the battlefield is the systematic process of analyzing the mission variables of enemy, terrain, weather, and civil considerations in an area of interest to determine their effect on operations (ATP 2-01.3). IPB assists in developing an in-depth understanding of relevant aspects of the OE, including threats.

Integrating the IPB process into the operations process is essential in supporting the commander's ability to understand the OE and visualize operations throughout the operations process. Integrating the IPB process and the operations process is an enabler that allows commanders to design and conduct operations continuously. Integrating the IPB process and the operations process provides the information and intelligence required to plan, prepare, execute, and assess operations.

See pp. 4-a to 4-4 for full discussion of IPB.

B. Information Collection

Information collection is an activity that synchronizes and integrates the planning and employment of sensors and assets as well as the processing, exploitation, and dissemination systems in direct support of current and future operations (FM 3-55). These sensors and assets may include cyberspace operations and EW assets conducting cyberspace exploitation operations, electromagnetic probing, and electromagnetic reconnaissance for information collection.

Information collection is the acquisition of information and the provision of this information to processing elements. Information collection integrates the intelligence and operations staff functions with a focus on answering the CCIRs, and IRs that assists the commander and staff in shaping the OE and conducting operations. The commander drives information collection coordinated by the staff and led by the G-2 or S-2. The following are the steps of information collection:

- Plan requirements and assess collection.
- Task and direct collection.
- Execute collection.

Information collection enables the commander to understand and visualize the operation. Information collection identifies gaps in information that require aligning intelligence assets with cyberspace exploitation, electromagnetic reconnaissance, and electromagnetic probing to collect data on those gaps. The decide and detect steps of targeting also rely heavily on information collection. Enemy cyberspace capabilities identified through information collection assist the CEMA working group in identifying potential targets and key terrain in cyberspace.

See p. 2-44 for further discussion of information collection.

C. Targeting

Targeting is the process of selecting and prioritizing targets and matching the appropriate response to them, considering operational requirements and capabilities (JP 3-0). A target is an entity or object that performs a function for the adversary considered for possible engagement or other actions. (JP 3-60).

When targeting for cyberspace effects, the physical network layer is the medium through which all digital data travels. The physical network layer includes wired (land and undersea cable), and wireless (radio, radio-relay, cellular, satellite) transmission means. The physical network layer is a point of reference used during targeting to determine the geographic location of an enemy's cyberspace and EMS capabilities.

When targeting, planners may know the logical location of some targets without knowing their physical location. The same is true when defending against threats in cyberspace. Defenders may know the logical point of origin for a threat without necessarily knowing the physical location of that threat. Engagement of logical network layer targets can only occur with a cyberspace capability.

See pp. 4-29 to 4-34, Targeting (D3A).

D. Risk Management

Risk management is the process to identify, assess, and control risks and make decisions that balance risk cost with mission benefits (JP 3-0) and an element of command and control. Risk is the exposure of someone or something valued to danger, harm, or loss, and is inherent in all operations. The commander and staff conduct risk management throughout the operations process to identify and mitigate risks associated with hazards that can cause friendly and civilian casualties, damage or destruction of equipment, or otherwise impact mission effectiveness. Aspects of cyberspace defense and security operations and EP missions include risk mitigation measures as part of risk management.

Risk management is integrated into planning activities and continues throughout the operations process. Risk management consists of the following steps:

- Identify the hazards.
- Assess the hazards.
- Develop controls and make risk decisions.
- Implement controls.
- Supervise and evaluate.

The CEMA section, as with all staff elements, incorporate risk management into cyberspace operations and EW-related running estimates and recommendations to mitigate risk. The G-3/S-3 coordinates risk management amongst all staff elements during the operations process.

See following pages (pp. 2-42 to 2-43) for discussion of risks in cyberspace and the EMS.

E. Knowledge Management

Knowledge management is the process of enabling knowledge flow to enhance shared understanding, learning, and decision making (ADP 6-0). The four components of knowledge management are people, processes, tools, and organizations. Knowledge management facilitates the transfer of knowledge among the commander, staff, and forces to build and maintain situational awareness and enhance organizational performance. Through knowledge management, information gets to the right personnel at the right time to facilitate decision making.

During knowledge management, the necessary cyberspace operations and EW-related information and tools from higher headquarters are provided to the CEMA working group in a timely enough manner to make decisions during mission analysis and COA development. Through the knowledge management process, cyberspace operations and EW-related intelligence received through information collection and IO is disseminated for decision making by the CEMA working group.

III. Risks In Cyberspace and the EMS

Ref: FM 3-12, Cyberspace Operations and Electromagnetic Warfare (Aug '21), pp. 4-18 to 4-20.

Risk is inherent in all military operations. When commanders accept risks, they create opportunities to seize, retain, and exploit the initiative and achieve decisive results. The willingness to incur risks is often the key to exposing an enemy's weaknesses that the enemy considers beyond friendly reach. Commanders assess and mitigate risks continuously throughout the operations process. Many risks to the DODIN-A come from enemies, adversaries, and insiders. Some threats are well equipped and well trained, while some are novices using readily available and relatively inexpensive equipment and software. Army users of the DODIN are trained on basic cyberspace security, focusing on the safe use of information technology and understanding common threats in cyberspace.

Risk management is the Army's primary decision-making process for identifying hazards and controlling risks. The process applies to all types of operations, tasks, and activities, including cyberspace operations. The factors of mission, enemy, terrain and weather, troops and support available, time available, and civil considerations provide a standardized methodology for addressing both threat and hazard-based risks. Risks associated with cyberspace operations fall into four major categories—

A. Operational Risks

Operational risks pertain to the consequences that cyberspace and EMS threats pose to mission effectiveness. Operational consequences are the measure of cyberspace attack and EA effectiveness. Cyberspace intrusions or attacks, and likewise in the EMS, can compromise networks, systems, and data, which can result in operational consequences such as injury or death of personnel, damage to or loss of equipment or property, degradation of capabilities, mission degradation, or even mission failure. Exfiltration of data from Army networks by the enemy can undermine the element of surprise and result in loss of initiative. Enemy or adversary forces may conduct cyberspace and EMS attacks to exposed friendly networks and capabilities, compromising future cyberspace attacks and cyberspace exploitation missions.

Friendly forces conducting cyberspace operations and EW encounter many operational risks. Commander and staff consider cascading effects because of employing cyberspace attacks and EA. The CEMA section ensures that the commander and staff understand the characteristics of the various cyberspace and EW capabilities and their associated effects. The CEMA section informs the commander and staff of the reversibility of effects resulting from cyberspace attacks and EA to understand that some effects are irreversible at the operator level. Attaining an understanding of the characteristics, cascading effects, and reversibility effects provide a commander with situational awareness and in determining the acceptable risks when conducting cyberspace operations and EW.

It is essential to consider risk management when conducting OCO and EA that could reveal friendly locations and intentions to an adversary prematurely. Some OCO or EA effects have a one-time use and once utilized cannot be effectively used again. OCO and EA may also create cascading effects that could hinder other operations.

Personal electronic device(s) such as smartwatches, smartphones, tablets, laptops, and gaming systems can be a significant OPSEC vulnerability to friendly cyberspace and EW capabilities. The CEWO gathers understanding surrounding risks associated with PEDs from the G2 or S2 and OPSEC and makes recommendations to the commander regarding their usage in the organization.

B. Technical Risks

Technical risks exist when there are exploitable vulnerabilities in systems on the DODIN-A, and there are threats that can exploit those vulnerabilities. Nearly every technical system within the Army is networked, resulting in a vulnerability in one system compromising other connected systems, creating a shared vulnerability. These potentially vulnerable networked systems and components directly impact the Army's ability to conduct operations. DCO mitigates risks by defending against specified cyberspace attacks, thereby denying the enemy's ability to take advantage of technical vulnerabilities that could disrupt operations.

Robust information systems engineering disciplines result in chain risk management, security, counterintelligence, intelligence, and hardware and software assurance that assist the leaders with managing technical risk. Friendly forces examine the technical risks when conducting cyberspace attacks to avoid making friendly networks vulnerable to enemy cyberspace counterattacks.

C. Policy Risks

Policy risk pertains to authorities, legal guidance, and international law. Policies address cyberspace boundaries, authorities, and responsibilities. Commanders and decision makers must perform risk assessments and consider known probable cascading and collateral effects due to overlapping interests between military, civil, government, private, and corporate activities on shared networks in cyberspace. Policies, the United States Code (USC), the Uniform Code of Military Justice, regulations, publications, operation orders, and standard operating procedures all constitute a body of governance for making decisions about activities in cyberspace.

Policy risk includes considering international norms and practices, the effect of deviating from those norms, and potential shifts in international reputation because of the effects resulting from a cyberspace operation. Cyberspace attacks can be delivered through networks owned, operated, and geographically located within the sovereignty of multiple governments. EA can also deliver effects that impact frequencies in the spectrum owned and operated by commercial, government, and other neutral users. Therefore, it is vital to consider the legal, cultural, and political costs associated with using cyberspace and the EMS as avenues of approach.

Policy risks occur where policy fails to address operational necessity. For example, a policy emplaced that limits cyberspace operations, which results in low levels of collateral effects, can result in a unit constrained to cyberspace attacks that will not result in the desired outcomes necessary for mission success. A collateral effects analysis to meet policy limits is distinct from the proportionality and necessity analysis required by the law of war. Even if a proposed cyberspace operation is permissible after a collateral effect's analysis, the proposed cyberspace operation or EW mission must include a legitimate military objective that is also permissible under the law of war.

Policy risk applies to risk management under civil or legal considerations. An OCO or EA mission may pose a risk to host nation civilians and non-combatants in an OE where a standing objective is to minimize collateral damage. During a mission, it may be in the Army's best interest for host nation populations to be able to perform day-to-day activities. Interruptions of public networks may present hazards to the DODIN-A and pose dangers to Army forces because of social impacts that lead to riots, criminal activity, and the emergence of insurgent opportunists seeking to exploit civil unrest.

C. Operations Security Risks

Both cyberspace and the EMS provides a venue for OPSEC risks. The Army depends on cyberspace security programs and training to prevent or mitigate OPSEC risks. Commanders emphasize and establish OPSEC programs to minimize the risks. OPSEC measures include actions and information on the DODIN and non-DODIN information systems and networks. All personnel are responsible for protecting sensitive and critical information. EP denies unauthorized access to information that an enemy intercept in the EMS through electromagnetic security operations.

For more information on OPSEC, refer to AR 530-1 and ATP 3-13.3.

Information Collection

Ref: FM 3-12, Cyberspace Operations and Electromagnetic Warfare (Aug '21), pp. 4-9 to 4-10. See also p. 2-40.

The focus when executing information collection is to collect data that answers CCIRs and IRs for analysis during the IPB process. The G-2 or S-2 executes collection by conducting—

Intelligence Operations. Intelligence operations are the tasks undertaken by military intelligence units through the intelligence disciplines to obtain information to satisfy validated requirements (ADP 2-0). Through intelligence operations, the G-2 or S-2 attains information regarding threat capabilities, activities, disposition, and characteristics. Intelligence operations use multiple intelligence disciplines to collect information regarding cyberspace and the EMS to satisfy CCIRs and IRs. However, knowledge attained from the other intelligence disciplines may also provide cyberspace and EMS related insight. In addition to gathering information on peer and near-peer threats through SIGINT, criminal intelligence collects information on cyberspace and EMS-related illegal activities conducted throughout the assigned AO.

Reconnaissance. Reconnaissance is a mission undertaken to obtain, by visual observation or other detection methods, information about the activities and resources of an enemy or adversary, or to secure data concerning the meteorological, hydrographic, or geographic characteristics of a particular area (JP 2-0). Reconnaissance produces information about the assigned AO. Through reconnaissance, the G-2 or S-2 can collect information regarding such mission and operational variables as terrain characteristics, enemy and friendly obstacles to movement, and the disposition of enemy forces and civilians. Combined employment of three methods of reconnaissance (dismounted, mounted, and aerial) can result in the location and type(s) of friendly, civilian, and threat cyberspace and EW capabilities operating in the assigned AO. Upon request, the CEMA section supports the G-2 or S-2's reconnaissance efforts by employing EW assets to conduct electromagnetic reconnaissance to collect information in the EMS and request OCO support to conduct cyberspace exploitation in cyberspace.

Surveillance. Surveillance is the systematic observation of aerospace, cyberspace, surface, or subsurface areas, places, persons, or things by visual, aural, electronic, photographic, or other means (JP 3-0). Surveillance involves observing an area to collect information and monitoring civilians and threats in a named area of interest or target area of interest. Surveillance may be autonomous or part of a reconnaissance mission. Collecting information in cyberspace and the EMS as part of a surveillance mission is also called network surveillance. Network Surveillance is the observation of organizational, social, communications, cyberspace, or infrastructure connections and relationships (FM 2-0). Network surveillance can also include detailed information on connections and relationships among individuals, groups, and organizations, and the role and importance of aspects of physical or virtual infrastructure.

Security Operations. Security operations are those operations performed by commanders to provide early and accurate warning of enemy operations, to provide the forces being protected with time and maneuver space within which to react to the enemy and to develop the situation to allow commanders to effectively use their protected forces (ADP 3-90). Early and accurate warnings provide friendly forces with time and maneuverability to react and create an opportunity for the commander to employ force protection measures. Cyberspace defense, cyberspace security, and EP include actions that allow early detection and mitigation of threats in cyberspace and the EMS. Additionally, ES missions conduct electromagnetic reconnaissance to attain information about the disposition of enemy threats in the EMS and modify security efforts.

Chap 3

I. Electromagnetic Warfare (EW)

Ref: FM 3-12, Cyberspace Operations and Electromagnetic Warfare (Aug '21), pp. 2-8 to 2-15.

I. Electromagnetic Warfare (EW)*

Electromagnetic Warfare (EW) is military action involving the use of electromagnetic and directed energy to control the electromagnetic spectrum or to attack the enemy. EW consists of three functions: electromagnetic attack, electromagnetic protection, and electromagnetic support.

Modern militaries rely on communications equipment using broad portions of the electromagnetic spectrum (EMS) to conduct military operations allowing forces to talk, transmit data, and provide navigation and timing information, and command and control troops worldwide. They also rely on the EMS for sensing and awareness of the OE. The Army conducts electromagnetic warfare (EW) to gain and maintain positions of relative advantage within the EMS. The Army's contribution to electromagnetic spectrum operations is accomplished by integrating and synchronizing EW and spectrum management operations.

Electromagnetic Warfare (EW)

 Electromagnetic Attack (EA)

 Electromagnetic Protection (EP)

 Electromagnetic Support (ES)

 Electromagnetic Warfare Reprogramming

The three divisions often mutually support each other in operations. For example, radar-jamming EA can serve a protection function for friendly forces to penetrate defended airspace; it can also prevent an adversary from having a complete operating picture.

---

*Editor's Note: In keeping with doctrinal terminology changes in JP 3-85, Joint Electromagnetic Spectrum Operations (May '20) and FM 3-12, Cyberspace Operations and Electromagnetic Warfare (Aug '21), the term "electronic warfare (EW)" has been updated to "electromagnetic warfare (EW)". Lilkewise, the EW divisions have been updated as "electromagnetic attack (EA), electromagnetic protection (EP), and electromagnetic support (ES)." For purposes of the CYBER1 SMARTbook, EW/EA/EP/ES acronyms and terms will remain the same as presented in the original cited and dated source -- for example, ATP 3-12.3, Electronic Warfare Techniques (Jul '19). Readers should anticipate that as those specific references are updated/revised, so will the terms.

(Electromagnetic Warfare) I. EW Overview 3-1 *

A. Electromagnetic Attack (EA)

Army forces conduct both offensive and defensive EA to fulfill the commander's objectives in support of the mission. EA projects power in and through the EMS by implementing active and passive actions to deny enemy capabilities and equipment, or by employing passive systems to protect friendly capabilities. Electromagnetic attack is a division of electromagnetic warfare involving the use of electromagnetic energy, directed energy, or antiradiation weapons to attack personnel, facilities, or equipment with the intent of degrading, neutralizing, or destroying enemy combat capability and considered a form of fires (JP 3-85). EA requires systems or weapons that radiate electromagnetic energy as active measures and systems that do not radiate or re-radiate electromagnetic energy as passive measures.

Offensive EA

Offensive EA prevents or reduces an enemy's effective use of the EMS by employing jamming and directed energy weapon systems against enemy spectrum-dependent systems and devices. Offensive EA systems and capabilities include—

- Jammers.
- Directed energy weaponry.
- Self-propelled decoys.
- Electromagnetic deception.
- Antiradiation missiles.

Defensive EA

Defensive EA protects against lethal attacks by denying enemy use of the EMS to target, guide, and trigger weapons that negatively impact friendly systems. Defensive EA supports force protection, self-protection and OPSEC efforts by degrading, neutralizing, or destroying an enemy's surveillance capabilities against protected units. Defensive EA systems and capabilities include—

- Expendables (flares and active decoys).
- Jammers.
- Towed decoys.
- Directed energy infrared countermeasure systems.
- Radio controlled improvised explosive device (RCIED) systems.
- Counter Unmanned Aerial Systems (C-UAS).

Electromagnetic Attack (EA) Effects

EA effects available to the commander include—

- **Destroy.** Destruction makes the condition of a target so damaged that it can neither function nor be restored to a usable condition in a timeframe relevant to the current operation. When used in the EW context, destruction is the use of EA to eliminate targeted enemy personnel, facilities, or equipment (JP 3-85).
- **Degrade.** Degradation reduces the effectiveness or efficiency of an enemy EMS-dependent system. The impact of degradation may last a few seconds or remain throughout the entire operation (JP 3-85).
- **Disrupt.** Disruption temporarily interrupts the operation of an enemy EMS dependent system (JP 3-85).
- **Deceive.** Deception measures are designed to mislead the enemy by manipulation, distortion, or falsification of evidence to induce them to react in a manner prejudicial to their interests. Deception in an EW context presents enemy operators and higher-level processing functions with erroneous inputs, either directly through the sensors themselves or through EMS-based networks such as voice communications or data links (JP 3-85).

II. Electromagnetic Warfare Taxonomy

Ref: FM 3-12, Cyberspace Operations and Electromagnetic Warfare (Aug '21), p. 2-8.

Electromagnetic warfare (EW) consists of three distinct divisions: electromagnetic attack (EA), electromagnetic protection (EP), and electromagnetic support (ES).

Divisions of Cyberspace Operations	Electromagnetic Attack (EA)	Electromagnetic Protection (EP)	Electromagnetic Support (ES)
Types of Operations	Attack personnel, facilities, or equipment	Protect friendly Electromagnetic Spectrum (EMS)-dependent capabilities	• Intercept • Identify • Locate • Evaluate
Types of Enabling Operations	• Reconnaissance • Enemy Attack	Preemptive Protection	• Situational Understanding • Combat Information • Targeting • Intelligence Preparation of Battlespace (IPB) Development
Common Tactical Mission Tasks	• Employing Directed Energy Weaponry • Electromagnetic Pulse • Reactive Countermeasures • Deception Measures • Electromagnetic Intrusion • Electromagnetic Jamming • Electromagnetic Probing • Meaconing	• Deconflict Electromagnetic Environmental Effects • Ensure Electromagnetic Compatibility • Electromagnetic Hardening • Emission Control • Electromagnetic Masking • Preemptive Countermeasures • Electromagnetic Security • Conduct Wartime Reserve Modes	• Conduct Electromagnetic Reconnaissance • Threat Warning • Direction Finding
Common Effects	• Disrupt • Degrade • Neutralize • Destroy • Deceive	• Deception • Denial • Disrupt • Neutralize	• Exploit • Detect

Ref: FM 3-12 (Aug '21), fig. 2-3. Electromagnetic warfare taxonomy.

Electromagnetic Attack (EA) Tasks

Ref: FM 3-12, Cyberspace Operations and Electromagnetic Warfare (Aug '21), pp. 2-9 to 2-11. See also pp. 3-21 to 3-28 for electronic attack techniques from ATP 3-12.3.

EA has the unique potential to affect enemy use of the EMS and attack the enemy through the EMS. Other offensive options can affect enemy use of the EMS but are likely to cause collateral damage outside the EMS, whereas EA uses the EMS for its effects. Concurrently, EA's potential to cause EMS fratricide necessitates caution and coordination in its employment.

EA tasks include—
- Employing directed energy weaponry.
- Electromagnetic pulse.
- Reactive countermeasures.
- Deception measures.
- Electromagnetic intrusion.
- Electromagnetic jamming.
- Electromagnetic probing.
- Meaconing.

Directed Energy. Directed energy is an umbrella term covering technologies that relate to the production of a beam of concentrated electromagnetic energy or atomic or subatomic particles. (JP 3-85). Directed energy becomes a directed energy weapon when used to conduct EA. A directed-energy weapon is a weapon or system that uses directed energy to incapacitate, damage, or destroy enemy equipment, facilities, and/ or personnel (JP 3-85). EA involving the use of directed-energy weapons is called directed-energy warfare. Directed-energy warfare is military action involving the use of directed-energy weapons, devices, and countermeasures (JP 3-85). The purpose of directed-energy warfare is to disable, cause direct damage, or destroy enemy equipment, facilities, or personnel. Another use for directed-energy warfare is to determine, exploit, reduce, or prevent hostile use of the EMS by neutralization or destruction.

Electromagnetic Pulse. Electromagnetic pulse is a strong burst of electromagnetic radiation caused by a nuclear explosion, energy weapon, or by natural phenomenon, that may couple with electrical or electronic systems to produce damaging current and voltage surges (JP 3-85). The effects of an electromagnetic pulse can extend hundreds of kilometers depending on the height and power output of the electromagnetic pulse burst. A high-altitude electromagnetic pulse can generate destructive effects over a continent-sized area. The most affected portion of the EMS by electromagnetic pulse or high-altitude electromagnetic pulse is the radio spectrum. Electromagnetic energy produced by an electromagnetic pulse excludes the highest frequencies of the optical (infrared, visible, ultraviolet) and ionizing (X and gamma rays) ranges. An indirect impact of an electromagnetic pulse or high-altitude electromagnetic pulse includes electrical fires caused by the overheating of electrical systems and components.

Reactive Countermeasures. EA includes reactive countermeasures as a response to an enemy attack in the EMS. Response to enemy attack may include employing radio frequency countermeasures, such as flares and chaff, in disrupting enemy systems and weapons, such as precision-guided or radio-controlled weapons, communications equipment, and sensor systems. Radio frequency countermeasures are any device or technique employing radio frequency materials or technology that is intended to impair the effectiveness of enemy activity, particularly with respect to precision guided and sensor systems (JP 3-85). Chaff is radar confusion reflectors, consisting of thin, narrow metallic strips of various lengths and frequency responses, which are used to reflect echoes for confusion purposes (JP 3-85). Reactive countermeasures may provoke the employment of directed energy weaponry or electromagnetic pulse and can include the use of lethal fires. Army forces can disrupt enemy guided weapons and sensor systems by deploying passive and active electro-optical-infrared countermeasures that include smokes, aerosols, signature suppressants, decoys, pyrotechnics, pyrophoric, laser jammers, high-energy lasers, and directed infrared energy.

Deception Measures. Deception measures are designed to mislead the enemy by manipulation, distortion, or falsification of evidence to induce them to react in a manner prejudicial to their interests. Electromagnetic deception uses misleading information by injecting false data into the adversary's EMS-dependent voice and data networks to inhibit the effectiveness of intelligence, surveillance, and reconnaissance sensor systems. EW uses the EMS to deceive a threat's decision loop, making it difficult to establish an accurate perception of Army forces' objective reality. EW supports all deceptions plans, both Joint military deception and tactical deception, using electromagnetic deception measures and scaling appropriately for the desired effect. Electromagnetic deception measures provide misleading signals in electromagnetic energy, for example by injecting false signals into a threat's sensor systems such as radar. Commander's authority to plan and execute deception integrated with electromagnetic deception measures may be limited by separate EW authorities and rules of engagement.

Electromagnetic Intrusion. Electromagnetic intrusion is the intentional insertion of electromagnetic energy into transmission paths in any manner, with the objective of deceiving operators or of causing confusion (JP 3-85). An example of electromagnetic intrusion is injecting false or misleading information into an enemy's radio communications, acting as the enemy's higher headquarters. Electromagnetic intrusion can also create deception or confusion in a threat aircraft's intelligent flight control system, compromising the intelligent flight control system' neural network and the pilot's ability to maintain control.

Electromagnetic Jamming. Electromagnetic jamming is the deliberate radiation, reradiation, or reflection of electromagnetic energy for the purpose of preventing or reducing an enemy's effective use of the electromagnetic spectrum, with the intent of degrading or neutralizing the enemy's combat capability (JP 3-85). Targets subjected to jamming may include radios, navigational systems, radars, and satellites. Electromagnetic jamming can disrupt a threat aircraft's intelligent flight control system by jamming its sensors, denying its ability to obtain navigational or altitude data crucial to flight performance. Electromagnetic jamming can also prevent or reduce the effectiveness of an enemy's integrated air defense system by jamming its anti-aircraft sensors used for targeting.

Electromagnetic Probing. Electromagnetic probing is the intentional radiation designed to be introduced into the devices or systems of an adversary for the purpose of learning the functions and operational capabilities of the devices or systems (JP 3-85). Electromagnetic probing involves accessing an enemy's spectrum-dependent devices to obtain information about the targeted devices' functions, capabilities, and purpose. Electromagnetic probing may provide information about threat capabilities and their ability to affect or detect friendly operations. Army forces may conduct overt electromagnetic probing to elicit a response from an enemy, exposing their location.

Meaconing. Meaconing consists of receiving radio beacon signals and rebroadcasting them on the same frequency to confuse navigation. Meaconing stations cause inaccurate bearings to be obtained by aircraft or ground stations (JP 3-85).

B. Electromagnetic Protection (EP)

Ref: FM 3-12, Cyberspace Operations and Electromagnetic Warfare (Aug '21), pp. 2-11 to 2-13. See also pp. 3-29 to 3-34 for electronic protection techniques from ATP 3-12.3.

Electromagnetic protection is the division of electromagnetic warfare involving actions taken to protect personnel, facilities, and equipment from any effects of friendly, neutral, or enemy use of the electromagnetic spectrum that degrade, neutralize, or destroy friendly combat capability (JP 3-85). EP measures eliminate or mitigate the negative impact resulting from friendly, neutral, enemy, or naturally occurring EMI.

Electromagnetic Protection (EP) Tasks

Adversaries are heavily invested in diminishing our effective use of the electromagnetic spectrum. It is crucial we understand the enemy threat and our vulnerabilities to our systems, equipment and personnel. Effective EP measures will minimize natural phenomena and mitigate the enemy's ability to conduct ES and EA actions against friendly forces successfully.

EP tasks include—
- Electromagnetic environmental effects deconfliction.
- Electromagnetic compatibility.
- Electromagnetic hardening.
- Emission control.
- Electromagnetic masking.
- Preemptive countermeasures.
- Electromagnetic security.
- Wartime reserve modes.

Electromagnetic Environmental Effects Deconfliction.

Electromagnetic vulnerability is the characteristics of a system that cause it to suffer a definite degradation (incapability to perform the designated mission) as a result of having been subjected to a certain level of electromagnetic environmental effects (JP 3-85). Any system operating in the EMS is susceptible to electromagnetic environmental effects. Any spectrum-dependent device exposed to or having electromagnetic compatibility issues within an EMOE may result in the increased potential for such electromagnetic vulnerability as safety, interoperability, and reliability issues. Electromagnetic vulnerability manifests when spectrum-dependent devices suffer levels of degradation that render them incapable of performing operations when subjected to electromagnetic environmental effects.

Electromagnetic compatibility, EMS deconfliction, electromagnetic pulse, and EMI mitigation reduce the impact of electromagnetic environmental effects. Recognizing the different types of electromagnetic radiation hazards allows planners to use appropriate measures to counter or mitigate electromagnetic environmental effects. Electromagnetic radiation hazards include— hazards of electromagnetic radiation to personnel, hazards of electromagnetic radiation to ordnance, and hazards of electromagnetic radiation to fuels. Electromagnetic environmental effects can also occur from natural phenomena such as lightning and precipitation static.

Electromagnetic Compatibility.
Electromagnetic compatibility is the ability of systems, equipment, and devices that use the electromagnetic spectrum to operate in their intended environments without causing or suffering unacceptable or unintentional degradation because of electromagnetic radiation or response (JP 3-85). The CEMA spectrum manager assists the G-6 or S-6 spectrum manager with implementing electromagnetic compatibility to mitigate electromagnetic vulnerabilities by applying sound spectrum planning, coordination, and management of the EMS. Operational forces have minimal ability to mitigate

electromagnetic compatibility issues. Instead, they must document identified electromagnetic compatibility issues so that the Service component program management offices may coordinate the required changes necessary to reduce compatibility issues.

Electromagnetic Hardening. Electromagnetic hardening consists of actions taken to protect personnel, facilities, and/or equipment by blanking, filtering, attenuating, grounding, bonding, and/or shielding against undesirable effects of electromagnetic energy (JP 3-85). Electromagnetic hardening can protect friendly spectrum-dependent devices from the impact of EMI or threat EA such as lasers, high-powered microwave, or electromagnetic pulse. An example of electromagnetic hardening includes installing electromagnetic conduit consisting of conductive or magnetic materials to shield against undesirable effects of electromagnetic energy.

Emission Control. Emission control is the selective and controlled use of electromagnetic, acoustic, or other emitters to optimize command and control capabilities while minimizing, for operations security: a. detection by enemy sensors, b. mutual interference among friendly systems, and/or c. enemy interference with the ability to execute a military deception plan (JP 3-85). emission control enables OPSEC by—
- Decreasing detection probability and countering detection range by enemy sensors.
- Identifying and mitigating EMI among friendly spectrum-dependent devices
- Identifying enemy EMI that allows execution of military deception planning.

Emission control enables electromagnetic masking by integrating intelligence, and EW to adjust spectrum management and communications plans. A practical and disciplined emission control plan, in conjunction with other EP measures, is a critical aspect of good OPSEC. *Refer to ATP 3-13.3 for OPSEC techniques at division and below.*

Electromagnetic Masking. Electromagnetic masking is the controlled radiation of electromagnetic energy on friendly frequencies in a manner to protect the emissions of friendly communications and electronic systems against enemy electromagnetic support measures/signals intelligence without significantly degrading the operation of friendly systems (JP 3-85). Electromagnetic masking disguises, distorts, or manipulates friendly electromagnetic radiation to conceal military operations information or present false perceptions to adversary commanders. Electromagnetic masking is an essential component of military deception, OPSEC, and signals security.

Preemptive Countermeasures. Countermeasures consist of that form of military science that, by the employment of devices and/or techniques, has as its objective the impairment of the operational effectiveness of enemy activity (JP 3-85). Countermeasures can be passive (non-radiating or reradiating electromagnetic energy) or active (radiating electromagnetic energy) and deployed preemptively or reactively. Preemptive deployment of passive countermeasures are precautionary procedures to disrupt an enemy attack in the EMS through the use of passive devices such as chaff which reradiates, or the use of radio frequency absorptive material which impedes the return of the radio frequency signal.

Electromagnetic Security. Electromagnetic security is the protection resulting from all measures designed to deny unauthorized persons information of value that might be derived from their interception and study of noncommunications electromagnetic radiation (e.g., radar) (JP 3-85). Changing the modulation and characteristics of electromagnetic frequencies used for radars make it difficult for a threat to intercept and study radar signals.

Wartime Reserve Modes. Wartime reserve modes are characteristics and operating procedures of sensor, communications, navigation aids, threat recognition, weapons, and countermeasure systems that will contribute to military effectiveness if unknown to or misunderstood by opposing commanders before they are used, but could be exploited or neutralized if known in advance (JP 3-85). Wartime reserve modes are held deliberately in reserve for wartime or emergency use.

C. Electromagnetic Support (ES)

Electromagnetic support refers to the division of electromagnetic warfare involving actions tasked by, or under the direct control of, an operational commander to search for, intercept, identify, and locate or localize sources of intentional and unintentional radiated electromagnetic energy for immediate threat recognition, targeting, planning, and conduct of future operations (JP 3-85). In multi-domain operations, commanders work to dominate the EMS and shape the operational environment by detecting, intercepting, analyzing, identifying, locating, and affecting (deny, degrade, disrupt, deceive, destroy, and manipulate) adversary electromagnetic systems that support military operations. Simultaneously, they also work to protect and enable U.S. and Allied forces' freedom of action in and through the EMS.

The purpose of ES is to acquire adversary combat information in support of a commander's maneuver plan. Combat information is unevaluated data, gathered by or provided directly to the tactical commander which, due to its highly perishable nature or the criticality of the situation, cannot be processed into tactical intelligence in time to satisfy the user's tactical intelligence requirements (JP 2-01). Combat information used for planning or conducting combat operations, to include EA missions, is acquired under Command authority; however, partner nation privacy concerns must be taken into account. Decryption of communications is an exclusively SIGINT function and may only be performed by SIGINT personnel operating under Director, National Security Agency and Chief, National Security Service SIGINT operational control (DODI O-3115.07).

ES supports operations by obtaining EMS-derived combat information to enable effects and planning. Combat information is collected for immediate use in support of threat recognition, current operations, targeting for EA or lethal attacks, and support the commander's planning of future operations. Data collected through ES can also support SIGINT processing, exploitation, and dissemination to support the commander's intelligence and targeting requirements and provide situational understanding. Data and information obtained through ES depend on the timely collection, processing, and reporting to alert the commander and staff of potential critical combat information.

Electromagnetic Support (ES) Tasks

When conducting electromagnetic support, commanders employ EW platoons located in the brigade, combat team (BCT) military intelligence company (MICO) to support with information collection efforts, survey of the EMS, integration and multisource analysis by providing indications and warning, radio frequency direction finding and geolocation of threat emissions.

ES tasks include—

- Electromagnetic Reconnaissance.
- Threat Warning.
- Direction finding.

See facing page for an overview and further discussion of electronic support actions.

* Electromagnetic Warfare Reprogramming

Electromagnetic warfare reprogramming is the deliberate alteration or modification of electromagnetic warfare or target sensing systems, or the tactics and procedures that employ them, in response to validated changes in equipment, tactics, or the electromagnetic environment (JP 3-85). The purpose of EW reprogramming is to maintain or enhance the effectiveness of EW and targeting sensing systems. EW reprogramming includes changes to EW and targeting sensing software (TSS) equipment such as self-defense systems, offensive weapons systems, and intelligence collection systems. EW consists of three distinct divisions: EA, EP, and ES, which are supported by EW reprogramming activities.

For more information on EW reprogramming, refer to FM 3-12, app. F.

Electromagnetic Support (ES) Actions

Ref: FM 3-12, Cyberspace Operations and Electromagnetic Warfare (Aug '21), pp. 2-14 to 2-15. See also pp. 3-35 to 3-36 for electronic warfare support techniques from ATP 3-12.3.

Electromagnetic Reconnaissance. Electromagnetic reconnaissance is the detection, location, identification, and evaluation of foreign electromagnetic radiations (energy) (JP 3-85). Electromagnetic reconnaissance is an action used to support information collection and is an element of the tactical task reconnaissance (see Chapter 4). Information obtained through electromagnetic reconnaissance assists the commander with situational understanding and decision making and, can be further processed to support SIGINT activities. Electromagnetic reconnaissance may result in EP modifications or lead to an EA or lethal attack.

Threat Warning. Threat warning enables the commander and staff to quickly identify immediate threats to friendly forces and implement EA or EP countermeasures. EW personnel employ sensors to detect, intercept, identify, and locate adversary electromagnetic signatures and provides an early warning of an imminent or potential threat. EW personnel coordinate with G-2 or S-2 on the long-term impact of detected enemy emitters. Threat warning assists the commander's decision making process in IPB development, updating electromagnetic order of battle, and assisting in the correlation of enemy emitters to communication and weapon systems.

Known electromagnetic signatures should be compared against the electromagnetic order of battle, high-value target, and the high-payoff target list and action taken as warranted by current policy or higher guidance. Unknown radiated electromagnetic signatures detected in the EMS are forwarded to the G-2 or S-2 for analysis. The G-2 or S-2 validates known and unknown systems as part of information collection that feeds the operations process. Staffs analyze and report information to higher and subordinate headquarters, to other Army and joint forces, and to unified action partners in the AO.

Direction Finding. Direction finding is a procedure for obtaining bearings of radio frequency emitters by using a highly directional antenna and a display unit on an intercept receiver or ancillary equipment (JP 3-85). EW personnel leverage various ES platforms with direction finding capabilities to locate enemy forces. Multiple direction finding systems are preferred for a greater confidence level of the enemy location. ES platforms are deployed in various formations to create a baseline and increase the area of coverage. Three or more direction finding systems are considered optimal in triangulating the targeted emitter.

Electromagnetic Support and Signals Intelligence

ES and SIGINT often share the same or similar assets and resources, and personnel conducting ES could be required to collect information that meets both requirements simultaneously. SIGINT consists of communications intelligence, electronic intelligence, and foreign instrumentation SIGINT. Commonalities between ES and SIGINT are similar during the early stages of sensing, collecting, identifying, and locating foreign spectrum emissions. The distinction between ES and SIGINT is determined by who has operational control of assets collecting information, what capabilities those assets must provide, and why they are needed. Information and data become SIGINT when cryptologic processes are applied to a signal to determine its relevance, value, or meaning solely for intelligence. There are also delineating hard lines regarding the systems, signal complexity, and reporting timeliness that divide ES and SIGINT. While both ES and SIGINT report information that meets reporting thresholds directly to the supported unit, SIGINT is obligated further to report acquired information through the U.S. SIGINT system. The added requirement for SIGINT provides accountability and enables the greater intelligence community access to the information for additional intelligence production and dissemination as required.

III. Spectrum Management

Spectrum management is the operational, engineering, and administrative procedures to plan, coordinate, and manage use of the electromagnetic spectrum and enables cyberspace, signal and EW operations. Spectrum management includes frequency management, host nation coordination, and joint spectrum interference resolution. Spectrum management enables spectrum-dependent capabilities and systems to function as designed without causing or suffering unacceptable electromagnetic interference. Spectrum management provides the framework to utilize the electromagnetic spectrum in the most effective and efficient manner through policy and procedure.

See chap. 5, Spectrum Management Operations (SMO/JEMSO).

Electromagnetic Interference (EMI)

Electromagnetic interference is any electromagnetic disturbance, induced intentionally or unintentionally, that interrupts, obstructs, or otherwise degrades or limits the effective performance of electronics and electrical equipment (JP 3-13.1). It can be induced intentionally, as in some forms of EW, or unintentionally, because of spurious emissions and responses, intermodulation products, and other similar products.

See p. 3-36 for related discussion of EMI mitigation, to include an operator EMI troubleshooting checklist. See also p. 3-31.

Frequency Interference Resolution

Interference is the radiation, emission, or indication of electromagnetic energy (either intentionally or unintentionally) causing degradation, disruption, or complete obstruction of the designated function of the electronic equipment affected. The reporting end user is responsible for assisting the spectrum manager in tracking, evaluating, and resolving interference. Interference resolution is performed by the spectrum manager at the echelon receiving the interference. The spectrum manager is the final authority for interference resolution. For interference affecting satellite communications, the Commander, Joint Functional Component Command for Space is the supported commander and final authority of satellite communications interference.

Spectrum Management Operations (SMO)

SMO are the interrelated functions of spectrum management, frequency assignment, host nation coordination, and policy that together enable the planning, management, and execution of operations within the electromagnetic operational environment during all phases of military operations. The SMO functional area is ultimately responsible for coordinating EMS access among civil, joint, and multinational partners throughout the operational environment. The conduct of SMO enables the commander's effective use of the EMS. The spectrum manager at the tactical level of command is the commander's principal advisor on all spectrum related matters.

See chap. 5, Spectrum Management Operations (SMO/JEMSO).

Electromagnetic Warfare Coordination

The spectrum manager should be an integral part of all EW planning. The SMO assists in the planning of EW operations by providing expertise on waveform propagation, signal, and radio frequency theory for the best employment of friendly communication systems to support the commander's objectives. The advent of common user "jammers" has made this awareness and planning critical for the spectrum manager. In addition to jammers, commanders and staffs must consider non-lethal weapons that use electromagnetic radiation. Coordination for EW will normally occur in the CEMA section. It may occur in the EW cell if it is operating under a joint construct or operating at a special echelon.

II. EW Key Personnel

Ref: ATP 3-12.3, Electronic Warfare Techniques (Jul '19), chap. 2.

The conduct of electronic warfare requires highly trained and skilled personnel. This section discusses electronic warfare professionals along with the staff members with roles and responsibilities when planning and conducting electronic warfare operations.

I. Electronic Warfare Personnel

EW personnel on the staff are in the cyberspace electromagnetic activities (CEMA) section at theater army through brigade and consist of a cyber electronic warfare officer (CEWO), electronic warfare technician, electronic warfare noncommissioned officers, and spectrum manager. The CEMA section includes EW trained personnel, personnel trained in electromagnetic spectrum management, and personnel trained in cyberspace operations. Cyberspace electromagnetic activities is the process of planning, integrating, and synchronizing cyberspace and electronic warfare operations in support of unified land operations (ADRP 3-0). EW personnel are responsible to the chief of staff or the assistant chief of staff, operations (G-3) or battalion or brigade operations staff officer (S-3) staff. Battalions have a single EW representative that is a member of the battalion staff.

EW personnel in the CEMA section plan and conduct EW during the full range of military operations. EW personnel conduct CEMA with assistance from, and in coordination with, other members of the CEMA working group. FM 3-12 contains more information on the CEMA working group. EW personnel plan the employment of EA, frequencies for targeting, analyze the probability of frequency fratricide, and collaborate with the assistant chief of staff, signal (G-6) or battalion or brigade signal staff officer (S-6) to mitigate harmful effects from EW to friendly personnel, equipment, and facilities.

The CEWO disseminates key mission status information, such as cancellation of electronic attacks, and coordinates with other staff members within the command post to contribute to situational awareness. The CEWO coordinates with the following sections—

- Assistant chief of staff, intelligence (G-2) or battalion or brigade intelligence staff officer (S-2).
- G-3 (S-3).
- G-5 (S-5).
- G-6 (S-6).
- Fire support coordinator.
- Information operations officer.
- Space support element.
- Special technical operations staff.
- Staff Judge Advocate (SJA) or representative.

II. Theater Army, Corps, Division and Brigade

The Army assigns EW personnel to CEMA sections at theater army, corps, division, and brigade echelons. Each EW professional has specific roles and responsibilities.

A. Cyber Electronic Warfare Officer (CEWO)

The CEWO's EW responsibilities include—
- Integrates, coordinates, and synchronizes EW effects.
- Nominates EW targets for approval from the fire support coordinator and commander.
- Receives, vets, and processes EW targets from subordinate units.
- Develops and prioritizes effects in the EMS.
- Develops and prioritizes targets with the fire support coordinator.
- Monitors and continually assesses measures of performance and measures of effectiveness for EW operations.
- Coordinates targeting and assessment collection with higher, adjacent, and subordinate organizations or units.
- Advises the commander and staff on plan modifications, based on the assessment.
- Advises the commander on how EW effects can impact the operational environment.
- Provides recommendations on commander's critical information requirements.
- Prepares and processes the electronic attack request format (EARF).
- Participates in other cells and working groups, as required, to ensure integration of EW operations.
- Deconflicts EW operations with the spectrum manager.
- Coordinates with the CEMA working group to plan and synchronize EW operations.
- Assists the G-2 (S-2) during intelligence preparation of the battlefield (IPB), as required.
- Provides information requirements to support planning, integration, and synchronization of EW operations.
- Serves as the Jam Control Authority (JCA) for EW operations, as directed by the commander.

B. Electronic Warfare Technician

The electronic warfare technician—
- Serves as the technical subject-matter expert for EW to the CEWO and CEMA working group.
- Plans and coordinates EW across functional and integrating cells.
- Provides input for the integration of threat electronic technical data as part of the IPB process.
- Coordinates target information and synchronizes EA and ES activities with the G-2 (S-2) staff.
- Integrates EW into the targeting process, monitors EW target requests, and conducts battle damage assessment for EW.
- Recommends employment of EW resources.
- Provides technical oversight and supervision for maintenance of EW equipment
- Conducts, maintains, and updates an electromagnetic environment survey.
- Identifies enemy and friendly effects within the EMS.
- Assists in the development and execution of standard operating procedure (SOP) and battle drills.

C. Electronic Warfare Noncommissioned Officer

The electronic warfare noncommissioned officer—
- Plans, manages, and executes EW tasks.
- Manages the availability and employment of EW assets.
- Serves as senior developer and trainer for EW.
- Distributes, maintains, and consolidates EW staff products.
- Collects and maintains data for electromagnetic energy surveys.
- Coordinates and deconflicts EMS resources with the spectrum manager.
- Operates and maintains EW tools.

D. Spectrum Manager

There are spectrum managers in the CEMA section and G-6 (S-6) staff. The G-6 (S-6) staff spectrum manager manages EMS resources that support the friendly use of the EMS. The CEMA section spectrum manager manages EMS resources for EW activities and provides the EW input to the common operational picture. The CEMA section spectrum manager is responsible for—
- Leads, develops, and synchronizes the EW and EP plan by assessing EA effects on friendly force emitters.
- Mitigates harmful impact of EA on friendly forces through coordination with higher and subordinate units.
- Synchronizes with intelligence on the EA effects to support intelligence gain and loss considerations.
- Synchronizes cyberspace operations to protect radio frequency enabled transport layers.
- Coordinates to support protecting radio frequency-enabled information operations.
- Collaborates with staff, subordinate, and senior organizations to identify unit emitters for inclusion on the joint restricted frequency list (JRFL).
- Performs EW-related documentation and investigation of prohibitive electromagnetic interference to support the G-6 (S-6) led joint spectrum interference resolution program.
- Participates in the CEMA working group to deconflict EMS requirements.
- Provides advice and assistance in the planning and execution of EW operations.

Note. The JRFL is a concise list of restricted frequencies and networks categorized as taboo, protected, and guarded.

E. Battalion Electronic Warfare Personnel

Battalions have an EW representative responsible for planning and integrating EW capabilities. The EW representative coordinates with the S-2, S-6 staff, fire support officer, the joint terminal attack controller(JTAC), and other staff sections when assigned. In support of a battalion mission, the battalion EW representative requests effects that require coordination with the brigade CEMA section. Battalion EW representatives' duties and responsibilities include—
- Advising the commander on the employment of EW resources.
- Integrating EW during the military decision-making process (MDMP).
- Recommending and implementing EP activities in close coordination with the S-6.
- Managing the maintenance and employment of EW equipment.

III. Staff Members and Electronic Warfare

Ref: ATP 3-12.3, Electronic Warfare Techniques (Jul '19), pp. 2-4 to 2-6.

The staff contributes to EW by providing unique products and guidance to the CEWO during all phases of an operation. The same staff members participate in the CEMA working group as necessary.

G-2 (S-2) Staff

The G-2 (S-2) staff advises the commander and staff on intelligence aspects of EW operations. The G-2 (S-2) staff—

- Provides threat characteristics to support programming of unit EW systems.
- Maintains appropriate threat EW data.
- Maintains the signals intelligence (SIGINT) priorities of collection and informs the staff for situational awareness.
- Ensures electronic threat characteristics requirements are a part of the information collection plan.
- Determines enemy organizations' network structures, disposition, capabilities, limitations, vulnerabilities, and intentions through collection, analysis, reporting, and dissemination.
- Determines enemy EW vulnerabilities and high-value targets.
- Provides intelligence support to targeting operations.
- Assesses the effects of friendly EW activities on the enemy.
- Conducts intelligence gain or loss analysis for EW targets with intelligence value.
- Helps prepare the intelligence-related portion of the EW running estimate.
- Recommends guarded frequencies to the G-6 (S-6) staff for the JRFL.
- Provides updates to the electronic threat characteristics.
- Participates in the CEMA working group to synchronize information collection with EW requirements and deconflict planned EW activities.
- Deconflicts ES and SIGINT operations with the CEMA section.

G-3 (S-3) Staff

The G-3 (S-3) staff is responsible for the overall planning, coordination, and supervision of EW activities. The G-3 (S-3) staff—

- Plans for and incorporates EW into operation plans and orders, in particular within the fire support plan and the information operations plan (in joint operations).
- Tasks EW activities to assigned and attached units.
- Exercises control over EW, including electromagnetic deception plans.
- Directs EP measures based on recommendations from the G-6 (S-6) staff, the CEWO, and the CEMA working group.
- Coordinates EW training requirements.
- Issues EW support tasks within the information collection plan. These tasks are according to the collection plan and the requirements tools developed by the G-2 (S-2) staff and the requirements manager.
- Ensures, through the CEMA working group, that EW activities support the overall plan.
- Integrates EA within the targeting process.

G-6 (S-6) Staff
The network defense technician, network management technician, information services technician, spectrum manager, and information security manager participate in planning EW. The G-6 (S-6) staff—
- Assists the CEWO with the preparation of the EP policy.
- Reports enemy EA activity detected by friendly communications and electronics elements to the CEMA working group for counteraction.
- Assists the unit CEWO with resolving EW systems maintenance.
- Identifies and deconflicts electromagnetic interference (EMI).
- Issues the signal operating instructions (SOI).
- Ensures network connectivity for all EW computer systems.
- Provides EMS resources to the unit or task force (refer to ATP 6-02.70).
- Coordinates for EMS usage with higher echelon G-6 (S-6), communications system directorate of a joint staff, and applicable host-nation and international agencies as necessary.
- Prepares the restricted frequency list and issuance of emissions control guidance.
- Coordinates frequency allotment, assignment, and use.
- Supports the CEMA working group by assisting in the development of electromagnetic deception plans and activities that include EMS resources.
- Coordinates with higher echelon spectrum managers for EMS interference resolution.
- Assists the CEWO in issuing guidance to the unit, including subordinate elements, regarding deconfliction and resolution of interference problems and processes involving EW systems.
- Participates in the CEMA working group to deconflict friendly EMS requirements with EW activities and information collection efforts.
- Supports all subordinate unit software updates and communications security (COMSEC) requirements.
- Compiles and distributes the JRFL (Spectrum Manager).
- Assists the EW section with computer maintenance and troubleshooting.

Information Operations Officer
The information operations officer is responsible for all information operations. To enable information operations, the CEMA section undertakes deliberate actions designed to gain and maintain advantages in the information environment. Typically, but not solely, these actions occur through cyberspace operations and EW. The information operations officer—
- Ensures synchronization and deconfliction with other information operations.
- Considers second- and third-order effects of EW on information operations and proactively plans to enhance intended effects.

Staff Judge Advocate or Representative
The SJA is responsible for all legal advice. The SJA or representative reviews all EW operations to ensure they comply with existing DOD directives and instructions, rules of engagement (ROE), and applicable domestic and international laws, including the law of war. The SJA may also obtain any necessary authorities that are lacking.

- Establishing battalion EW SOP.
- Submitting EW requests and concept of operations to the brigade CEMA section.
- Coordinating with airborne EW assets to provide the aircraft situational awareness of a ground unit's operational environment including actions on the desired target.
- Establishing and enforcing counter radio-controlled improvised explosive device electronic warfare (CREW) employment. For more information about CREW devices, see paragraph 6-56.
- Conducting all EW related training to battalion and company personnel.
- Managing battalion and company EW resource reprogramming activities more information on reprogramming activities is in chapter 3.
- Conducting operational checks and inspections of EW equipment programs.

F. Company Counter Radio-Controlled Improvised Explosive Device Electronic warfare Specialists

Company CREW specialists, operate, maintain, reprogram, and reconfigure CREW devices for the unit. Commanders rely on CREW specialists to manage the devices within the company. CREW specialists—

- Advise the company commander on the employment of CREW and EW resources.
- Train and assist operators in the use and maintenance of CREW equipment
- Perform pre-combat checks and pre-combat inspections for EW equipment.
- Ensure CREW systems are operational and report deficiencies.

G. Electronic Warfare Control Authority

In some instances, EW personnel in an Army headquarters serve as the EW control authority. The EW control authority establishes guidance for EA on behalf of the joint force commander. If designated as the electronic warfare control authority the senior EW staff officer has the following responsibilities—

- Approve, disapprove, and modify EA requests from within the organization and subordinate units.
- Integrate and synchronizing EA activities.
- Maintain a log containing all approved jamming activity.
- Participate in the development of and ensuring compliance with the JRFL and all other EMS use plans.
- Maintain situational awareness of EA capable systems in the area of operations.
- Deconflict EA and ES, in coordination with the G-2 (S-2), to make recommendations to the combatant commander on intelligence gain or loss.
- Coordinate EA requirements with joint force components.
- Investigate unauthorized EA events and implement corrective measures.
- Approve or deny cease jamming requests.

Note. Joint organizations designate EW professionals as an electronic warfare control authority as needed. For more information about EW control authority, refer to JP 3-13.1.

Chap 3
III. EW Preparation & Execution

Ref: ATP 3-12.3, Electronic Warfare Techniques (Jul '19), chap. 4.

Preparation, execution, and assessment are interdependent parts of electronic warfare. This chapter discusses the techniques and resources to prepare, execute and assess electronic warfare effectively. This chapter provides electromagnetic spectrum resource coordination procedures and techniques to mitigate electromagnetic interference.

See pp. 4-15 to 4-28. for discussion of electronic warfare planning.

I. Electronic Warfare Preparation

Peer threats continue to mature their command and control and EW capabilities. To overcome the adversary, units prepare for the contest to dominate the EMS. Preparation begins before arrival on the battlefield and continues through redeployment.

The EW professionals gain proficiencies in EW activities from a combination of military education, doctrinal references, and experience. EW preparation ensures timely support for the commander's scheme of maneuver. Preparation consists of activities that units perform to improve their ability to execute a mission. Preparation for EW includes—

- EW training that includes actual and simulated resources and environments.
- Maintenance activities to ensure that EW equipment is clean and serviceable.
- Practicing the MDMP with other members of the staff. Practicing MDMP fosters teamwork and establishes expectations regarding what the CEWO provides to, and receives from, the staff.
- Rehearsals that include integration of SIGINT & EW resources and capabilities.
- Planning, initiating, and reporting movement of EW resources.
- Coordinating route clearance and escort requirements to mitigate risk and prevent delays during a maneuver.

During preparation, the CEMA section—

- Updates the EW running estimate in coordination with the SIGINT running estimate.
- Requests changes or exceptions to the JRFL and SOI through the G-2 (S-2) and G-6 (S-6) staff.
- Completes risk assessments and develops a risk mitigation strategy.
- Leads the CEMA working group.
- Develops and rehearses battle drills and staff processes including—
- Staffing the EARF and measuring the effectiveness of EW activities.
- Developing EW ground and airborne control authority procedures.
- Integrating information collection activities [G-2 (S-2)] staff.
- Coordinating for external maintenance and reprogramming support for EW assets.
- Initiating EP procedures to counter EMI and enemy jamming actions.
- Developing SOPs.
- Establishing reporting procedures.
- Executes pre-combat checks and inspections of EW assets.

II. Integration of Electronic Warfare and Signals Intelligence

Integrating EW and SIGINT is a force multiplier for unified land operations. EW and SIGINT have similar capabilities that are mutually beneficial. Integrated EW and SIGINT assets present an efficient, holistic approach that reduces duplication of effort, enables additional information collection, and provides flexibility in the employment of EW and SIGINT resources. EW and SIGINT teams collaboratively use DF techniques to locate transmitters, achieving a higher level of fidelity on the location of emitters. Integration techniques take advantage of similar capabilities and the placement of EW and SIGINT resources to increase operational flexibility, such as co-locating capabilities. SIGINT teams can exploit enemy communications characteristics such as verbal content of a transmission and positively identify an emitter as an approved target. The SIGINT teams can inform the EW team for immediate target engagement.

A. Distinctions Between Electronic Warfare and Signals Intelligence

Though EW and SIGINT are similar, there are important distinctions between them. Legal considerations distinguish EW and SIGINT activities, and the authorization for each to support operations, that if not observed, can complicate and delay the execution of electronic warfare effects and SIGINT operations. Commanders and planners should collaborate closely with the SIGINT enterprise and legal authorities to ensure compliance with SIGINT policy when planning electronic warfare.

B. Sensing Activity Distinctions

Commanders have the option to employ SIGINT sensors to support ES activities. The task and purpose are the main factors to decide to use SIGINT or ES capabilities. SIGINT sensors perform ES activities when used to provide immediate threat information including threat warning, avoidance, targeting, and jamming (refer to CJCSI 3320.01D). However, when the SIGINT sensor intercepts, identifies, and locates or localizes sources of intentional and unintentional radiated electromagnetic energy for intelligence purposes, it is no longer supporting an ES task but is conducting a SIGINT mission to satisfy intelligence requirements. These distinctions are identified when answering questions—

- Who tasks or controls the SIGINT sensors?
- What are the sensors tasked to provide?
- What is the purpose of the task driving the employment of the sensors?

ES and SIGINT employ the same or similar capabilities. ES includes actions tasked by, or under direct control of, an operational commander to search for, intercept, identify, and locate or localize sources of intentional and unintentional radiated electromagnetic energy for the purpose of immediate threat recognition, targeting, planning, and conduct of future operations (JP 3-13.1).

Units retain some data from ES to support immediate threat recognition, targeting, and planning of future operations. Units transfer select data from ES activities to the United States SIGINT System for the production of foreign intelligence. Foreign intelligence is information that relates to the capabilities, intentions, and activities of foreign powers, organizations, or persons (JP 2-0). The CEWO and the G-2(S-2) staff develop a structured procedure within each echelon to facilitate information exchange. Units rehearse this procedure during exercises and pre-deployment planning.

Deconflicting the Electromagnetic Spectrum

Ref: ATP 3-12.3, Electronic Warfare Techniques (Jul '19), pp. 4-1 to 4-2.

Deconflicting the EMS requires an understanding of the SOI, JRFL and mission requirements. The CEWO considers the distance, location and the purpose of equipment that is reliant on friendly or restricted frequencies and recommends exceptions to the SOI or JRFL. The SOI contains call signs, call words, frequency assignments, signs, and countersigns for friendly forces.

For more information regarding the SOI and JRFL, refer to ATP 6-02.70.

Frequency deconfliction is a systematic management procedure to coordinate the use of theelectromagnetic spectrum for operations, communications, and intelligence functions. Frequencydeconfliction is one element of electromagnetic spectrum management (JP 3-13.1).

Mission requirements may drive modifications to the SOI and JRFL. Modifications require staffingand approval through the G-2 (S-2) and G-6 (S-6) staff. For SOI and JRFL deconfliction, the CEWOconsiders the following—
- The purpose of the frequency.
- Waveform characteristics.
- Location and time of use.

Note. When EW activities conflict with the SOI or JRFL, the commander decides which has priority.

Due to security concerns, frequencies employed in intelligence roles may not be included in the SOI.The CEWO maintains awareness of the frequencies used in support of SIGINT activities throughcoordination with the G-2 (S-2) staff.

Electromagnetic Spectrum Resources

The authorization to use EMS resources is not always available. The G-6 (S-6) section spectrummanager uses the EMS certification process to gain the use of previously unallocated EMS resources, whichrequires completing a standard frequency action format.

Host nations have EMS usage plans that assist in the management of frequencies. The spectrum manager assigned to the G-6 (S-6) assists the CEMA section in frequency use authorization for EW activities.The G-6 (S-6) spectrum manager requests frequency resources through an online database. The onlinedatabase enables managers to determine the historical EMS supportability of like systems. The hyperlink tothe DD Form 1494, Application for Equipment Frequency Allocation, to request frequencies is in thereferences section of this publication.

See chap. 5, Spectrum Mananagement Operations.

III. Electronic Warfare Execution

Ref: ATP 3-12.3, Electronic Warfare Techniques (Jul '19), pp. 3-12 to 3-13.

The CEWO addresses targets and employs EW assets in support of an operation. Targeting requires continuous involvement from the CEWO. After planning, the CEWO participates in the targeting board and assesses the effects using measures of effectiveness. During execution the CEWO—

- Prosecutes approved EW targets in support of the operation.
- Evaluates the effectiveness of EW.
- Maintains situational understanding of EW activities and associated effects.
- Oversees the movement and placement of EW assets in support of operational requirements.
- Continues to identify and assess risk.
- Receives information from EW assets and disseminates to the staff:
- Detection and location of targeted and potential enemy emitters, including enemy EW assets.
- Indicators and warnings of enemy activity from EW.
- Maintains direct liaison with the fires cell, G-2 (S-2) and G-6 (S-6) staff to ensure integration and deconfliction of EW activities.
- Coordinates and manages EW missions tasked to subordinate units and requests for nonorganic EW support.
- Continues to assist the targeting working group in target development and to recommend targets for attack and reattack.
- Anticipates EW equipment outages and initiates the capability replacement plan.
- Validates and disseminates cease-jamming requests.
- Coordinates and expedites EMI reports with the G-2 (S-2) and G-6 (S-6) staff for deconfliction.
- Serves as the EW controlling authority when designated.

The CEWO portrays radio wave propagation and EW effects using modeling and simulation techniques with software.

Special Considerations During Execution

EMS resources are congested and contested with friendly and enemy use. EMS resource availability also shifts during an operation. The CEWO updates any changes within the EME and puts them into the common operational picture. During execution, EW planners continually consider—

- The Electromagnetic Order of Battle (EOB) *(See p. 5-15.)*
- The signal operating instructions (SOI)
- The Joint Restricted Frequency List (JRFL) *(See p. 4-22)*
- Anticipated or reported Electromagnetic Interference *(See pp. 3-10 & 3-31.)*

IV(a). Electronic Attack Techniques

Ref: ATP 3-12.3, Electronic Warfare Techniques (Jul '19), chap. 6. See also pp. 3-3 to 3-5.

This section discusses the techniques for conducting electronic attack and describes their characteristics. Electronic attack enables the commander to dominate the electromagnetic and supports the scheme of maneuver during Army operations.

I. Planning Electronic Attack

Commanders use EA to affect threat communications and noncommunications capabilities and for defense. EA is a single action or supplements other lethal or nonlethal attacks. Dynamics in an operational environment require the CEWO to employ different EA techniques based on operational variables. EA techniques include countermeasures and electromagnetic deception. Army operations employ offensive and defensive EA such as—

- Jamming adversary radar or command and control systems.
- Using antiradiation missiles to suppress adversary air defenses.
- Using electronic deception to confuse adversary surveillance and reconnaissance systems.
- Employing self-propelled, towed, or stationary decoys.
- Using self-protection and force protection measures such as use of expendables (e.g., flares and active decoys)
- Employing directed energy or infrared countermeasures systems.

EA includes both offensive and defensive activities. Offensive EA disrupts or destroys threat capability. Defensive EA protects friendly personnel and equipment. When planning EA, the CEMA section, in conjunction with the staff consider—

- Interference of friendly communications.
- Intelligence gain or loss.
- EMS use by locals and non-hostile parties.
- The persistence of effects.
- Electronic signatures.

EA depends on ES and SIGINT to provide targeting information and battle damage assessment. Throughout the MDMP and the targeting process, the CEWO coordinates and deconflicts spectrum requirements with the CEMA working group.

Refer to JP 3-13.1 for more information about EA and defensive EA planning.

A. Electronic Attack Effects

EA denies the enemy or adversary the ability to use the EMS, use equipment, or affects personnel and their decision making or courses of action. The effects that EA creates include denying, destroying, degrading, deceiving, delaying, diverting, neutralizing, or suppressing enemy or adversary EMS capabilities. These effects are mutually exclusive, and these terms are common when describing the desired effects. There may be other terms appropriate to describe desired effects other than those listed. For more information on effects, refer to JP 3-60.

The different EA systems have varying capabilities. The EW personnel planning and employing the variety of systems consider each of the system-specific parameters, the environment, and mission requirements. Each system has specific capabilities and may require ingenuity during planning to ensure mission success.

B. Electronic Attack (EA) Considerations

Ref: ATP 3-12.3, Electronic Warfare Techniques (Jul '19), pp. 6-2 to 6-4.

The CEMA working group plans and rehearses EMS deconfliction procedures. When EA conflicts with the G-2 (S-2) information collection efforts, the commander decides which has priority or the G-3(S-3) decides based on commander's guidance.

The potential for threat intelligence collection also affects EA planning. A well-equipped adversary can detect EA by employing ES techniques to gain intelligence on U.S. force locations and intentions. To develop an understanding of the adversary's intelligence collection capability, the CEWO and the G-2 (S-2) staff develop the enemy EOB. CEWOs protect EA assets through EP and risk mitigation techniques to counter threat ES and EA. For more information about EP, see chapter 7.

A red team provides an independent capability to explore alternatives in plans and operations in the context of an operational environment and from the perspective of enemies, adversaries, and others (JP 2-0). In conjunction with the red team, the CEWO and the G-2 (S-2) staff determine what intelligence the adversary can gain.

Threat Electronic Warfare Persistence

6-9. Aside from antiradiation missiles, the effects of jamming are less persistent than effects achieved by lethal means. The effects of jamming persist as long as the jammer itself is emitting and is in range to affect the targeted receiver. These effects last a matter of seconds or minutes, which makes the timing of such missions critical. Timing is important when units use jamming in direct support of aviation or ground platforms. For example, in a mission that supports suppression of enemy air defense, the time on target and duration of the jamming must account for the speed of attack of the aviation platform. They must also account for the potential reaction time of threat air defensive countermeasures. Because jamming may cause the threat to take unexpected actions or use other means of communications to avoid the intended effect, the CEWO uses ES techniques to sense and validate the persistence of known threat transmissions.

Electronic Attack to Destroy

6-10. An electromagnetic pulse creates a permanent effect and destroys equipment rendering it useless until the threat repairs or reconstitutes the capability. An electromagnetic pulse is the electromagnetic radiation from a strong electronic pulse, most commonly caused by a nuclear explosion that may couple with electrical or electronic systems to produce damaging current and voltage surges (JP 3-13.1). Units at echelons theater army and below seeking to destroy a target using an electronic pulse rely on strategic level decisions and support to achieve this effect.

Countermeasures

The Army uses countermeasure techniques to mitigate threat EW sensing and attack activities. Countermeasures are that form of military science that, by the employment of devices and techniques, by design impairs the operational effectiveness of threat activity (JP 3-13.1). Countermeasures can be active or passive and deployed preemptively or reactively. Countermeasure devices and techniques include flares, chaff, radar jammers, CREW systems, and decoys. Chaff is radar confusion reflectors, consisting of thin, narrow metallic strips of various lengths and frequency responses, which are used to reflect echoes for confusion (JP 3-13.1).

Electromagnetic Deception

Deception mission techniques include misleading transmissions that present false indications of friendly force battle rhythms. Control and coordination are necessary to avoid confusing friendly activities with deception missions. When planning an electromagnetic deception mission, the EW planners consider activities that support the current, friendly operation as well as those that will support the deception mission and perform integration and deconfliction. EW supports all deceptions plans, both military deception andtactical deception, using the electromagnetic deception and scaling appropriately for the desired effect.Electromagnetic deception is the deliberate radiation, reradiation, alteration, suppression, absorption, denial,enhancement, or reflection of electromagnetic energy in a manner intended to convey misleading information to an adversary or to adversary electromagnetic dependent weapons, thereby degrading or neutralizing the enemy's combat capability (JP 3-13.1). Electromagnetic deception can increase or decrease ambiguity affecting the enemy decision maker's understanding. This can prove to an enemy commander the certainty of a course of action or create confusion on their behalf.

The G-3 (S-3) staff plans and supervises deception missions. The information operations officerdevelops deception plans. Integration of electromagnetic deception with information operations is necessarywhen conducting deception missions. EW supports information related capabilities and deception plans usingelectromagnetic deception techniques. The CEWO is responsible for the EW portion of the deception plan.

Simulative Electromagnetic Deception
Simulative electromagnetic deception attempts to represent friendly notional or actual capabilities tomislead threat forces. Simulative electromagnetic techniques require extensive command and staffcollaboration to present a believable deception plan. What the threat detects electronically should beconsistent with other sources of intelligence reports. Simulative electromagnetic deception transmissionsrequire close attention. Electromagnetic deception effects are often of short duration. Simulative electromagnetic deception includes the use of systems that give off emissions indicative aparticular organization. A countermortar or counter-battery radar is organic to an artillery unit. By turningon that type of radar, you can identify the probable location of an artillery unit. Simulative electromagneticdeception also includes using emitters to imply a type or change of activity by a unit, for example, placing surveillance radars in a typical defensive array, when in fact the intent is to attack.

Manipulative Electromagnetic Deception
Manipulative electromagnetic deception uses communication or noncommunication signals to conveyindicators that mislead the enemy. For example, to indicate that a unit is going to attack when it is going towithdraw, the unit might transmit false plans and requests for ammunition. CEWO's use manipulativeelectromagnetic deception to mislead the enemy to misdirect their EA and ES assets, while interfering lesswith friendly communications. Manipulative electromagnetic deception seeks to eliminate, reveal, or conveymisleading indicators of friendly intentions. Success in manipulative electromagnetic deception andsimulative electromagnetic deception depends on understanding how friendly transmitters appear to thethreat. The EW planners consider what is occurring with the friendly transmitters. Then the EW plannersdetermine how to portray the friendly command's transmission infrastructure (JP 3-13.1).

Imitative Electromagnetic Deception
Imitative deception mimics threat emissions with the intent to mislead them. Imitative electromagneticdeception, if recognized by the enemy, can compromise SIGINT efforts. Imitative deception normallyrequires approval from higher echelon commands. An example of imitative electromagnetic deception includes entering the adversary's communicationnets by using their call signs and radio procedures and then giving threat commanders instructions to initiateactions, which are to the advantage of friendly forces. Targets for imitative electromagnetic deception includeany threat receiver and range from cryptographic systems to plain-language tactical nets. Imitativeelectromagnetic deception can cause a unit to be in the wrong place at the right time, to place ordnance onthe wrong target, or to delay attack plans. Imitative deception efforts foster decisions based on falseinformation that, to the enemy, appears to have come from within. Imitative electromagnetic deception canbe decisive on the battlefield. However, to be effective, imitative electromagnetic deception requireselectronic equipment capable of convincingly duplicating the emissions of enemy equipment (JP 3-13.1).

(Electronic Warfare) IV(a). Electronic Attack Techniques 3-23 *

II. Preparing Electronic Attack

In preparation for EA, the CEWO gathers target information from ES sensors and the EOB. The information includes the location of the targeted asset, electronic characteristics, and the frequencies in use. Using location, characteristics and frequency, the CEWO determines which assets are best to conduct EA. The CEWO then completes calculations to determine the power required to jam the targeted receiver. The CEWO gives guidance to subordinate units about EA. The guidance includes information that allows the subordinate unit to prepare for the EA. EA guidance includes—

- Target identification.
- Target location.
- Special coordination requirements and procedures.
- Jamming technique.
- Jamming duration.
- Desired effect.
- Battle damage assessment method of delivery and prescribed format.

Note. The CEWO uses formulas to determine minimum power output requirements used for targeting. Refer to ATP 3-12.3, Appendix A.

Electronic Attack Considerations

The selection of EA assets is a significant factor when preparing to conduct EA. EA considerations include—

- Concealment characteristics.
- Power output capability.
- Availability of physical protection.
- Time available for the mission.
- Route clearance and escort requirements to conduct friendly maneuver.
- Augmented security coordination.
- Airspace considerations for airborne EW assets.

See facing page for discussion of Electronic Attack Requests (EARFs).

III. Executing Electronic Attack

The CEWO has multiple options to choose from when executing EA. The CEWO prosecutes EA from air and ground (including fixed and mobile) platforms and monitors the EA activities during the mission. Mobile platforms consist of vehicular mounted and dismounted configurations. Units conduct EA using the chosen jamming technique and report the results of the jamming efforts to the CEWO.

Close Air Support (CAS)

Close air support (CAS) delivers EA using a variety of air platforms. There are two types of CAS requests: preplanned and immediate. The CEWO reviews the air tasking order (ATO) calendar when resourcing EA requirements. When CAS is available, the CEWO submits a request to use CAS for the EA mission.

The air support operations center provides the ATO calendar, which has detailed information on aircraft, crews, and missions. Preplanned CAS requests occur during planning. The ATO calendar is broken down into 24-hour duty cycles. The specific theater or joint operations area supporting joint air operations command and control center will establish cut-off times to receive preplanned air support requests for inclusion in the ATO. Immediate air support requests arise from situations that develop outside the planning stages of the joint air tasking cycle. It is important to understand that air assets available to satisfy immediate air support requests already exist in the published ATO.

For more information about CAS and the ATO calendar, refer to JP 3-09.3.

Electronic Attack Requests (EARFs)

Ref: ATP 3-12.3, Electronic Warfare Techniques (Jul '19), pp. 6-4 to 6-5. Refer to ATP 3-09.32 and ATP 3-12.3, app D for more information. See also pp. 4-27 to 4-28 (EARF).

The objective of EA is to disrupt or degrade the threat's ability to receive electromagnetic signalsradiating from their transmitters, or process signals from other sources, such as friendly transmissions, withconfidence. CEWOs integrate EA into the tactical plan by coordinating with the targeting board and theCEMA working group. The targeting list is an output from the targeting board and specifies the targets andtimes of attack regardless of the method used. When preparing for EA, the CEWO considers—

- The commander's intent.
- The ROEs.
- The location and identity of the targeted receiver and associated transmitter.
- The electronic threat characteristics of the targeted receiver & associated transmitter.
- The target engagement calculations.
- The associated risk when targeting with EA.

The CEWO makes coordination with the staff to plan EA. The G-2 (S-2) staff provides electronicthreat characteristics to aid in the development of targets. The electronic threat characteristics include thetechnical characteristics of the target. The CEWO maintains electronic threat characteristics for futuretargeting efforts. Threat characteristics regarding targets include—

- Threat's unit or organization.
- Frequencies in use.
- Call signs.
- Location.
- Power of transmitters.
- Bandwidth.
- Equipment nomenclature.
- Modulation type.
- Multiplex capability.
- Pulse duration.
- Pulse repetition frequency.
- Antenna type.
- Antenna height.
- Antenna orientation.
- Antenna gain.

The CEWO determines the minimum power output required to attack the target receivers. ExcessiveEA power makes it easier for the threat to locate and attack the friendly EA asset. Distances between thethreat transmitter and receiver and the friendly EA asset are critical considerations for EA asset placement.

Terrain is a factor because LOS is necessary between the EA asset and the location of the targetedreceiver. The adversary uses terrain to mask transmitted signals from friendly detection and attack. Otherterrain considerations include—

- Urban infrastructure.
- Bodies of water.
- Soil composition.
- Vegetation density.

A. Airborne Electronic Attack

Ref: ATP 3-12.3, Electronic Warfare Techniques (Jul '19), pp. 6-6 to 6-7.

Airborne EA delivers jamming from rotary, fixed-wing and unmanned aircraft systems. Althoughsome of these platforms are organic to the Army, much of the airborne EA capability resides in other Services.Requesting airborne EA often requires coordination with joint forces. Effective airborne EA requiresintegrating procedures and communications between the supported unit and the airborne EA asset owner.The EARF includes the prescribed communications method.

Communications between the aircrew, CEWO, and JTAC throughout the mission is beneficial for maintaining situational understanding and for retasking an asset. Best practices include activecommunications between the CEWO and the aircraft that is delivering the EA.

When the CEWO cannot communicate with the aircrew or the JTAC, the supporting aircraft continueswith the airborne EA mission specified in the EARF. A technique is to note in the EARF regarding what todo in the event of a communication failure (FM 3-12).

Canceling and Retasking Airborne Electronic Attack

Changes within an operational environment and EA missions make it necessary for reprioritization ofassets. Air platforms are in demand for other purposes such as surveillance supporting intelligence missionsor signal missions. The CEWO can request dynamic retasking of airborne EA assets and requests retaskingwith the JTAC and the air operations center.

Joint Tactical Attack Controller

The JTAC conducts air and ground coordination. The JTAC initiates requests and maintainscommunications with the designated airborne EA point of contact for the duration of the mission.

Air Operations Center

The air operations center, which can be joint or allied depending on the task organization, coordinatesall assigned aerospace forces. The air operations center conducts the following activities—

- Coordinates and approves airspace.
- Coordinates aerial refueling.
- Makes ATO changes.
- Issues retasking instructions.

Airborne Electronic Attack Cancellations at the Battalion and Brigade

Sometimes it is necessary to cancel airborne EA missions. CEWOs communicate cancellations to theasset owner and requestor points of contact. Reporting cancellations ensures the most efficient use of EAassets and availability for other missions.

Advanced Cancellation of Preplanned Mission

When a CEWO cancels an airborne EA mission more than six hours before a preplanned mission is aroutine task. The requestor includes the reason for cancellation. The CEWO immediately communicates acancellation of a mission to release the airborne EA asset for other missions. The CEWO also notifies the fire support officer and the air liaison officer. Cancellations made during operations include direct voice communications when possible to ensure someone is available and ready to process the cancellation.

Short Notice Cancellation of Preplanned Mission
Short notice airborne EA cancellations are cancellations that occur less than six hours before apreplanned mission. Short-term cancellations require immediate action to avoid mission launch and theunnecessary employment of an asset. The CEWO informs the designated point of contact that a cancellationis coming by the most expeditious means available. Following the initial notification, the CEWO sends theofficial cancellation joint tactical air strike request (JTASR) to the appropriate point of contact as soon aspossible. Since the cancellation may require communications that bypass normal chain of commandr elationships, CEWOs include the process in the written unit SOPs and battle drills.

Immediate Cancellation of Preplanned Mission
CEWOs use this technique for canceling missions within one-hour of the expected execution time.CEWOs use the fastest communication means possible, such as Internet relay chat or voice communications,to distribute the necessary cancellation information. Immediately following an immediate cancellation,CEWOs contact the prescribed point of contact and provide an official cancellation using the points of contacton the JTASR and EARF to ensure units receive information promptly. Effective units include this processin the unit SOP and battle drills.

Dynamic Retasking
The staff makes every effort to provide immediate EA in response to an urgent request, including theallocation of available airborne EA assets. The retasking of airborne EA assets fulfills requests for on-demand requirements.

The process for retasking airborne EA platforms varies depending on joint command and control andArmy mission command arrangements, task organization, force disposition, and unit boundaries. Therequesting unit submits a request to their supporting EW representative. The retasking format is available in ATP 3-09.32.

If the requesting unit previously submitted a JTASR for EA support, the CEWO modifies the existingJTASR with a numbered change. Some units make the change using red for easier identification. If therequesting unit has not submitted a JTASR for the mission, the CEWO creates a new JTASR. The CEWO provides status updates to the requesting unit. Effective units address the knowledge management processes for maintaining updated JTASRs in their SOPs and battle drills.Due to the dynamic nature of an urgent requirement, there is no way to calculate the amount of time needed for coordinating the airborne EA. The CEWO or JTAC notifies the appropriate EW representativeand air support operations center when it is apparent that the duration of EA will exceed the initiallyanticipated time. The air support operations center notifies the airborne EA asset and coordinates anyadditional fuel requirements or determines the need to re-task another airborne EA asset. The air supportoperations center then informs the CEWO and JTAC of what support to expect. The JTAC or CEWO contacts the air support operations center to release airborne EA assets upon mission completion or cancellation.

Jamming Techniques
CEWOs use jamming techniques to disrupt the threat's ability to effectively receive or processelectromagnetic signals by overcoming the threat receiver with higher power transmissions. Successfuljamming of receivers requires an understanding of available jamming techniques. CEWOs consider whichtechnique is appropriate to support the commander's intent. Jamming techniques include—
- Electromagnetic jamming.
- Electromagnetic intrusion.
- Electromagnetic pulse.
- Electronic probing.

B. Defensive Electronic Attack

Defensive EA degrades the threat's ability to employ weapons that use electromagnetic activated triggers. Defensive EA protects friendly personnel and equipment. Counter radio-controlled improvised explosive device (RCIED) systems implement this EA technique.

Defensive EA uses the EMS to protect personnel, facilities, capabilities, and equipment. Examples include self-protection and other protection measures such as the use of expendables (flares and active decoys), jammers, towed decoys, directed-energy infrared countermeasures, and counter RCIED systems(FM 3-12).

Counter Radio-Controlled Improvised Device (CREW)

A common form of defensive electronic attack is counter radio-controlled improvised explosive device electronic warfare (CREW). CREW systems jam threat radio frequencies to prevent RCIEDs from receiving a triggering signal, thus stopping the RCIED from detonating. Units program CREW systems with threat-specific loadsets based on various sources of intelligence, including the technical exploitation of recovered RCIEDs. The loadset is what the device uses to determine its operational frequency range, change rate, andother attributes of the system. The loadset is essentially what programs the system to operate under predetermined parameters based on an operational environment. The Army employs mounted, dismounted,and fixed CREW systems as electronic countermeasures to RCIED attacks.

Cyber Electronic Warfare Officer Role

The CEWO is the commander's subject matter expert on CREW. The CEWO plans the inclusion of CREW in support of operations, establishes maintenance procedures and ensures reprogramming and configuration of CREW devices.

IV. Electronic Attack Techniques in Large Scale Combat Operations

Peer and near-peer adversaries rely on the EMS for command and control, sensing and targeting, and EW. Units require EA capabilities during large-scale combat operations to counter adversary communications and noncommunications emitters.

When jamming threat communications, the CEWO aligns EW capabilities with targets. The EA does not jam every threat communication. The EA is only disrupting the communication between the enemy battalion and enemy company. The close proximity and transmit power of the radios of the enemy tanks in a company formation allows them to maintain uninterrupted communications. The battalion transmissions to the company have a greater distance to travel and weaker signal at the receiving antenna leaving the communications vulnerable to EA.

Adversaries employ multiple sensors and noncommunications emitters, such as radars, to detect and locate friendly forces during large-scale combat operations. The CEWO uses EW activities, such as electromagnetic deception, combined with EW techniques to disrupt the adversary's ability to target friendly forces. The CEWO also disrupts adversary SIGINT and ES sensors to prevent detection, locating, and exploitation of friendly transmitters.

The CEWO understands that during large-scale combat operations, the threat has EW capabilities that can negatively affect friendly operations. The threat conducts EA to degrade communications and achieve a tactical advantage during operations. Units must incorporate EP techniques to counter threat EA activities.

IV(b). Electronic Protection Techniques

Ref: ATP 3-12.3, Electronic Warfare Techniques (Jul '19), chap. 7.

The greatest threat to mission command information systems at the tactical level is the enemy's use of electronic warfare assets to geolocate and jam friendly communications. This chapter discusses electronic protection and the techniques used to overcome electromagnetic interference. Successful electronic protection requires planning and execution by all members of the unit.

EP is the sum of technology, equipment, and techniques used to counter threat EW activities. EP is not force protection or self-protection. EP is an EMS-dependent system's use of electromagnetic energy or physical properties to preserve itself from direct or environmental effects of friendly and adversary EW, thereby allowing the system to continue operating (JP 3-13.1).

See also pp. 3-6 to 3-7.

Commander's Electronic Protection Responsibilities

EP is a command responsibility. Commanders ensure that all Soldiers in their units' train to apply EP techniques. Commanders rely on the staff to mitigate electronic vulnerabilities. The staff continuously measures the effectiveness of the applied EP techniques. Commanders' EP responsibilities are—

- Read after action reviews and reports about threat jamming or deception efforts and assess the effectiveness of EP.
- Ensure the staff reports and analyzes EMI, deception, or jamming.
- Analyze the impact of threat efforts to affect friendly communications.
- Ensure the unit incorporates appropriate EP techniques such as—
- Changing network call signs and frequencies in accordance with the SOI.
- Using approved COMSEC devices.
- Loading and using prescribed encryption keys.
- Using planned authentication procedures.
- Controlling emissions.

I. Planning Electronic Protection

Electronic protection uses techniques such as limiting transmissions and using natural or manmade objects to mask radiated energy from traveling to undesirable destinations. Electronic protection is essential to prevent the adversary from learning behavior and intentions within the EMS.

The CEWO considers friendly communications asset characteristics, their priorities for protection and their purpose of employment when planning EP. Additionally, the CEWO considers adversarial EW and SIGINT capabilities and their use against friendly systems. The G-6 (S-6) is the primary source for gaining the characteristics of friendly communications resources while the G-2 (S-2) is the CEWO's primary resource to gain electronic threat characteristics.

See following page for an overview of EP considerations.

Electronic Protection Considerations

Ref: ATP 3-12.3, Electronic Warfare Techniques (Jul '19), pp. 7-1 to 7-2.

EP includes physical security, COMSEC measures, system technical capabilities, such as frequencyhopping, shielding of electronics, electromagnetic spectrum management, and emission control procedures.EP is an EMS-dependent system's use of electromagnetic energy and/or physical properties to preserve itself from direct or environmental effects of friendly and adversary EW, thereby allowing the system to continue operating (JP 3-13.1). The CEWO considers the following for EP—

- Vulnerability analysis and assessment of friendly communications assets.
- EP monitoring techniques and feedback procedures.
- EP effects on friendly capabilities.

Vulnerability Analysis and Assessment

Vulnerability analysis and assessment form the basis for developing EP plans. The CEWO reviews theunit EP techniques and procedures to determine weaknesses and develops plans for improvement. The G-6(S-6), United States Cyber Command, and the Defense Information Systems Agency provide a variety ofcybersecurity services, including vulnerability analysis and assessments.

The National Security Agency monitors COMSEC and provides security posture feedback to units. Itsprograms focus on telecommunications systems using wire and electronic communications. Their programscan support and remediate the command's COMSEC procedures.

Electronic Protection Effects on Friendly Capabilities

The CEWO and the G-6 (S-6) consider effects on friendly communications when developing an EPplan. A plan that maximizes EP can overly restrict the friendly use of communications assets. The CEWOmaintains a balance regarding the unit's ability to communicate with the planned level of EP. EP effects onfriendly communications are included in the CEWO's risk assessment. The CEWO and G-6 (S-6) present therisk assessment to the commander during the MDMP. The commander decides what level of risk isacceptable. For EP planning, the CEWO and G-6 (S-6) consider the following—

- Electromagnetic hardening.
- Electronic masking.
- Emission control.
- Electromagnetic spectrum management.
- Wartime reserve modes.
- Electromagnetic compatibility.

II. Electromagnetic Interference (EMI)

EMI prevents successful transmissions. Units must recognize and mitigate EMI to create the conditions required to use the EMS to communicate. Electromagnetic interference is any electromagnetic disturbance, induced intentionally or unintentionally, that interrupts, obstructs, or otherwise degrades or limits the effective performance of electronics and electrical equipment (JP 3-13.1).

A. Recognizing Electromagnetic Jamming

Radio operations require that radio operators can recognize electromagnetic jamming. Recognizing electromagnetic jamming is not always an easy task; the cause of EMI can be internal and external. If the EMI remains after grounding or disconnecting the antenna, the disturbance is most likely internal and caused by a malfunction of the radio. Contact maintenance personnel for repairs or replace the faulty equipment. Eliminate or substantially reduce the EMI or suspected jamming by grounding the radio equipment or disconnecting the receiver antenna. If measures to eliminate the radio as the source of the disturbance are unsuccessful, it is most likely external to the radio. Check external EMI further for threat jamming or unintentional EMI.

Sources, other than jamming, cause EMI. Unintentional EMI is caused by—

- Friendly and threat use of the same frequencies.
- Other electronic or electric and electromechanical equipment.
- Atmospheric conditions.
- Malfunction of the radio.
- A combination of any of the above.

Unintentional EMI normally travels a short distance; a search of the immediate area may reveal its source. Moving the receiving antenna short distances may cause noticeable variations in the strength of the interfering signal. Conversely, little or no variation normally indicates threat jamming. Regardless of the source, take appropriate actions to reduce the effect of EMI on friendly communications.

Signal	Description
Random Noise	It is indiscriminate in amplitude and frequency. It is similar to normal background noise. Random noise degrades all types of signals. Operators often mistake it for receiver or atmospheric noise and fail to take appropriate electronic protection actions.
Stepped Tones	Tones transmitted in increasing and decreasing pitch. They resemble the sound of bagpipes. Single-channel amplitude modulation or frequency modulation use stepped tones for voice circuits.
Spark	Spark is one of the most effective jamming signals. Spark uses short intensity and high intensity; they repeat at a rapid rate. This signal is effective in disrupting all types of radio communications.
Gulls	Generated by a quick rise and slow fall of a variable radio frequency and are similar to the cry of a seagull. It produces a nuisance effect and is very effective against voice radio communications.
Random Pulse	Pulses of varying amplitude, duration, and rate are generated and transmitted. They disrupt teletypewriter, radar, and all types of data transmission systems.
Wobbler	A single frequency modulated by a low and slowly varying tone. The result is a howling sound that causes a nuisance effect on voice radio communications.
Recorded Sounds	Any audible sound, especially of a variable nature, distracts radio operators and disrupts communications. Music, screams, applause, whistles, machinery noise, and laughter are examples of recorded sounds.
Preamble Jamming	A broadcasted tone over the operating frequency of secure radio nets resembles the synchronization preamble of the speech security equipment. Preamble jamming results in all radios being locked in the receive mode. It is especially effective when employed against radio networks using speech security devices.

Ref: ATP 3-12.3, Electronic Warfare Techniques (Jul '19), table 7-2. Common jamming signals.

B. Remedial Electronic Protection Techniques

Remedial EP techniques that help reduce the effectiveness of threat jamming efforts are the—

- Identification of threat jamming signals.
- Determination of the EMI as being obvious or subtle jamming.
- Recognition of jamming causing EMI by—
- Determining whether the EMI is internal or external to the radio.
- Determining whether the EMI is deliberate or unintentional.
- Reporting of jamming and other EMI incidents.
- Overcoming of jamming and EMI by adhering to the following techniques—
- Continue to operate.
- Diagnose the root cause of EMI.
- Improve the signal-to-jamming ratio.
- Adjust the receiver settings.
- Increase the transmitter power output.
- Adjust or change the antenna.
- Establish a retransmission station.
- Relocate the antenna.
- Use an alternate route for communications.
- Change the frequencies.
- Acquire another satellite or retransmission station.
- Installation of firmware and update software.
- Use enhancements to tactical radio ancillary communications electronics equipment and COMSEC devices.

C. Concealment

EP plans include provisions to conceal communications personnel and equipment. Though physical concealment is ineffective in changing the EMS signature, obscuring the physical attributes may prevent positive identification of the equipment as a communications system. Units use camouflage material to cover communications assemblages and their power generators. It is difficult to conceal most communications systems. However, installing antennas as low as possible on the backside of terrain features, and behind manufactured obstacles help conceal communications equipment while still facilitating effective communications.

D. Threat Electronic Attack on Friendly Command Nodes

Adversaries attack or exploit friendly command nodes that support operations. They have developed and equipment and techniques to contest the friendly use of the EMS. Friendly units use EP measures to counter threat EW and exploitation actions against friendly communications nodes.

Adversary attack on friendly command nodes can disrupt or destroy information, intelligence gathering efforts, and communications that support weapons systems. Threat forces expend considerable resources gathering intelligence about U.S. forces. Goals or effects may include—

- Jam friendly communications.
- Enter friendly radio networks.
- Collect information and intelligence about friendly forces.

E. Electromagnetic Interference (EMI) Battle Drill

Ref: ATP 3-12.3, Electronic Warfare Techniques (Jul '19), pp. 7-8 to 7-9.

Some prohibitive EMI has a measurable, operational impact. Units execute battle drills to addressprohibitive EMI. An EMI battle drill helps isolate the cause of interference and dispel erroneous assumptionsabout its cause. For example, knowing that CREW devices are jammers may lead to a hasty assumption thata CREW device impairs the use of combat net radios when operator error or faulty equipment is the cause ofthe EMI. The uninformed assumption that CREW systems are the problem leads to an unnecessary loss ofconfidence in EW equipment. Lack of confidence in equipment can lead to reluctance to prosecute EW andcan have a negative impact on operations. Proper analysis uses sensors and indicators that identify interferingfrequencies, the levels of transmission power and receiver strength.

Note. Watts express the radio transmission output levels, while decibels (dB) express radio receive signal strength. For more information about decibels, refer to ATP 2-22.6-2.

7-42. The lowest element or individual experiencing the EMI should report the interference via the JointSpectrum Interference Resolution Website. If unable to access the website, contact someone to input theinformation into the website at the earliest convenient time. On a staff, normally the G-6 (S-6) staff submitsJSIR reports to resolve interference. When appropriate, the staff disseminates the mitigating steps tosubordinate units as lessons learned and best practices to avoid future interference. A well-constructed EMIbattle drill, guides units to respond to JSIR reports in a consistent, methodical manner. Table 7-3 provides an example of an EMI troubleshooting battle drill.

Signal	Description
1	Follow equipment troubleshooting (verify frequency, cable and antenna connections, communications security). If EMI continues, then follow remaining steps.
2	Determine start and stop times or duration of EMI.
3	Identify EMI effect (interfering voice, noise, static).
4	Identify other emitters in area of operations.
5	Check adjacent and nearby units for similar problems.
6	Prepare and submit a joint spectrum interference resolution report to S-6.
LEGEND	
EMI	electromagnetic interference
S-6	battalion or brigade signal staff officer

Ref: ATP 3-12.3, Electronic Warfare Techniques (Jul '19), table 7-3. Electromagnetic interference troubleshooting battle drill.

III. Staff Electronic Protection Responsibilities

The staff implements the EP plan for the commander. Staff responsibilities are—
- Planning, coordinating, and supporting the execution of EP activities (CEMA working group).
- Advising the commander of threat EMS related capabilities [G-2 (S-2) staff].
- Supervising the CEMA section and include EP scenarios in command post, field training exercises, and evaluates employed EP techniques [G-3 (S-3) staff].
- Work with the CEWO to prepare and conduct the unit EP training program. Ensure there are PACE means of communications to support mission command. Distribute COMSEC. Perform friendly frequency management duties and issues the SOI. Review the JRFL and prepares which includes a restricted frequency list of taboo, protected and guarded frequencies [G-6 (S-6) staff].

Note. The PACE plan compliments EP as it provides multiple means of communications and designates the order in which an element will move through available communications methods until contact can be established with the desired recipient.

Preventive EP techniques include all measures taken to avoid threat detection and threat EA. EP seeks to mitigate threat information collection and intelligence gathering efforts. Electronic communications equipment has built-in features used to mitigate threat EA, ES and SIGINT actions. CEWOs advise the use of built-in features and user tactics, techniques, and procedures for countermeasures against threat actions.

IV. Equipment and Communications Enhancements

Some communications equipment has embedded capabilities used to prevent jamming, locating and listening by threat forces. Operators use the embedded capabilities when supporting operations.

Frequency-Hopping Mode
Some peer and near-peer adversaries with advanced EW equipment can jam radios that use frequency-hopping techniques. Single channel transmissions are vulnerable to jamming by unsophisticated transmitters, so units use frequency-hopping mode but remain vulnerable to threat EA and DF efforts. Frequency hopping is useful in mitigating the effects of threat jamming, and in keeping friendly position location data from threat forces.

Adaptive Antenna Techniques
Adaptive antenna techniques result in more survivable communications. These techniques typically link with spread spectrum waveforms to combine frequency hopping with pseudo-noise coding. Pseudo-noise coding is a technique to make spread spectrum waveforms and frequency-hopping mode appear to be unintelligible noise to an unintended receiver. Spread spectrum is a form of wireless communication in which the frequency of the transmitted signal varies deliberately. This uses more bandwidth than the signal would have otherwise, making it less susceptible to interference.

Frequency Hop Multiplexer (FHMUX)
The frequency hop multiplexer (FHMUX) and vehicular whip antennas that support FHMUX are available for use to enhance very high frequency (VHF) communications. The FHMUX is an antenna multiplexer used with single channel ground and airborne radio system in stationary and mobile operations. This FHMUX allows multiple radios to transmit and receive through one VHF antenna while operating in the frequency-hopping mode, single channel mode, or a combination of both. Using one antenna reduces visual and electronic profiles of command posts and reduces emplacement and displacement times.

IV(c). Electronic Warfare Support Techniques

Ref: ATP 3-12.3, Electronic Warfare Techniques (Jul '19), chap. 5.

I. Planning Electronic Warfare Support

Threat forces use the EMS to give orders, monitor and manage operations, detect aircraft using radar,and conduct DF. Locating threat transmitters aids in the development of situational understanding and assists with targeting. ES uses direction-finding techniques to find threat transmitters. Once located, the commander can direct fires in the form of lethal attack, request offensive cyberspace operations or use EA to gain the desired effects.

See also pp. 3-8 to 3-9.

A. Electronic Reconnaissance

5-2. Electronic warfare personnel conduct electronic reconnaissance to understand the types of threat emissions. Electronic reconnaissance is the detection, location, identification, and evaluation of foreign electromagnetic radiations (JP 3-13.1). The CEWO acquires electronic threat characteristics from the G-2(S-2). The electronic threat characteristics provide technical data including—

- Threat EMS resources in use.
- Antenna orientation and polarization.
- Radio transmit power levels.
- Radio range.

B. Electronic Warfare Support Considerations

The task and purpose of the mission determine whether a SIGINT or EW asset is appropriate for a given mission. ES assets conduct immediate threat recognition, targeting, future operations planning, andother tactical actions such as threat geolocation for avoidance.

The adversary employs electronics security measures to prevent the detection of emitters. Electronics security is the protection resulting from all measures designed to deny unauthorized persons information of value that might be derived from their interception and study of noncommunications electromagnetic radiations, e.g., radar (JP 3-13.1). When the adversary employs electronic security measures, the CEWO may require assistance from SIGINT to understand the nature of the emissions.

II. Preparing Electronic Warfare Support

The CEMA section uses ES assets to scan the EME for transmissions and then illustrates the results in a manner that the commander and staff can understand. Units develop an electromagnetic environment survey using air, ground, and sea platforms. The G-2 (S-2) staff assists the CEMA section by developing and maintaining the electromagnetic environment survey (refer to FM 2-0). The electromagnetic environment survey aids the CEWO to understand the nature, limitations, and sources of EMI in an operational environment and plan the employment of ES equipment. The CEMA section submits requests for information to address information gaps to the G-2 (S-2) staff.

(Electronic Warfare) IV(c). EW Support Techniques 3-35 *

The electromagnetic environment survey provides input, and the CEMA section enters the information into automated tools to maintain a current environment survey.

III. Executing Electronic Warfare Support

The CEWO and G-2 (S-2) mutually develop the SOPs and battle drills for integration of EW support and SIGINT information collection activities. Integration techniques take advantage of similar equipment capabilities and fuse EW and SIGINT resources to increase flexibility. SIGINT teams pass targeting information to EW teams. The SIGINT DF equipment compliments geolocation efforts and transitions a LOB into a cut or a fix for targeting. Integration facilitates immediate sharing of information and reduces delays in targeting.

A. Electromagnetic Environment (EME) Survey

Like weather reports for aircraft pilots, the EME survey informs the CEWO about the activities and conditions of the EME, enabling the CEWO to choose optimal COAs for EW.

EME surveys begin with the enemy EOB. The enemy EOB provides the CEWO with an initial overview of threat EMS capabilities derived from IPB. The enemy EOB assists the CEWO in making EW plans that exploit adversary vulnerabilities while preserving friendly capabilities. The enemy EOB is the baseline for the EME survey.

Electromagnetic Environment Survey

A unit tasks an airborne EW asset to support suppression of enemy air defense missions. During mission planning, the crew receives the EOB for the area of operations. The airborne EW crew identifies threat emitters they will likely encounter during the mission by priority, and de-conflicts friendly and neutral emitters.

As the airborne EW crew enters the target area of operations, they conduct an EME survey that confirms the presence of friendly, neutral, and threat emitters. Conducting and EME survey allows the crew to prioritize their activities against confirmed threat emitters by only targeting systems that are active.

B. Direction Finding (DF)

When conducting DF, the CEWO leverages the arrayed ES assets and coordinates support from the G-2 (S-2) for SIGINT resources to sense transmitters, collect information and triangulate the location of specified emitters of interest. The CEWO provides targeting requirements to the targeting board. Additionally, the CEWO shares the information collected from ES assets during DF activities with the G-2(S-2). The G-2 (S-2) considers information derived from ES when developing intelligence.

DF provides LOBs, cuts, and fixes to locate transmitters. A LOB is a single approximate azimuth from a sensor providing the approximate azimuth to the transmitter. A cut is two approximate azimuths providing the general location of a transmitter by determining where two LOBs intersect. A fix is three or more approximate azimuths providing a location using a triangulation method. A cut or fix may use approximate azimuths from one sensor receiving the signal multiple times from different locations, or from different sensors.

Refer to ATP 3-12.3, Electronic Warfare Techniques (Jul '19), pp. 5-3 to 5-10 for an overview of and discussion of line of bearing, cuts, fixes, establishing a direction finding baseline, and what causes direction finding errors.

Chap 4

IPB Cyberspace Considerations

Ref: ATP 2-01.3, Intelligence Preparation of the Battlefield (Mar '19), app. D.

Intelligence Preparation of the Battlefield (IPB)

Intelligence preparation of the battlefield is the systematic process of analyzing the mission variables of enemy, terrain, weather, and civil considerations in an area of interest to determine their effect on operations (ATP 2-01.3). Led by the intelligence officer, the entire staff participates in IPB to develop and sustain an understanding of the enemy, terrain and weather, and civil considerations. IPB helps identify options available to friendly and threat forces.

IPB consists of four steps. Each step is performed or assessed and refined to ensure that IPB products remain complete and relevant:
- Define the Operational Environment
- Describe Environmental Effects on Operations
- Evaluate the Threat
- Determine Threat Courses of Action

IPB begins in planning and continues throughout the operations process. IPB results in intelligence products used to aid in developing friendly COAs and decision points for the commander. Additionally, the conclusions reached and the products created during IPB are critical to planning information collection and targeting.

Refer to BSS6: The Battle Staff SMARTbook, 6th Ed., pp. 3-3 to 3-52 for complete discussion of Intelligence preparation of the battlefield from ATP 2-01.3.

* Refer also to INFO1: The Information Operations & Capabilities SMARTbook, pp. 4-17 to 4-34 for related discussion of information environment analysis (IO and intelligence preparation of the battlefield) from ATP 3-13.1.

As an essential part of the information environment, there is a massive global dependence on the cyberspace domain for information exchange. With this dependence and the associated inherent vulnerabilities, the cyberspace domain must be considered during each step of the IPB process:

- **Step 1—define the OE:** Visualize cyberspace components and threats through the three layers of cyberspace.
- **Step 2—describe environmental effects on operations**: Use military aspects of terrain.
- **Step 3—evaluate the threat**: Evaluate threats and HVTs in cyberspace against the warfighting functions by performing critical factors analysis (CFA).
- **Step 4—determine threat COAs**:
    - Consider the threat's historical use of cyberspace and incorporate threat COAs.
    - Determine HVT lists within the cyberspace domain.
    - Assist the S-6 staff to identify friendly networks that require protection.

(Planning) IPB Cyberspace Considerations 4-a *

To gain situational understanding, the following staff sections, in addition to assistance and support from the cyber mission force, provide the G-2/S-2 enough information to develop IPB products that include cyberspace considerations. The G-2/S-2 relies on the—

- G-3/S-3 to task operational assets to report items significant to cyberspace (such as satellite dish locations, cyber cafés, cellular network towers), since the G-3/S-3 is typically aware of maneuver and/or reconnaissance elements moving through specific designated AOs that have the potential to interact with the populace and the ability to visually confirm relevant infrastructure.
- G-6/S-6 for the friendly force network design to determine where the threat can possibly access friendly systems.
- G-9/S-9 to assist in identifying and confirming civil considerations that are pertinent to the cyberspace domain. For example, civil affairs teams may assist in ascertaining existing and planned network infrastructure in the AO, as well as identifying key leaders and landowners to determine their internet presence, activity, or cyber-personas.
- Information operations officer to primarily synchronize and deconflict information-related capabilities employed to support unit operations. With information provided by the intelligence, the information operations officer contributes to IPB by analyzing the information environment and developing the combined information overlay. Working with the intelligence staff, the information operations officer develops products that portray the information infrastructure of the AO and aspects of the information environment that can affect operations. These products include information all audiences and other decision makers, key people, and significant groups in the AO. They also address potential strengths and vulnerabilities of adversaries and other groups. The information operations officer will also assist in identifying how the populace communicates within the logical network layer, such as local government websites, heavily used social media sites, any group or individual blog sites. Additionally, the information operations officer can possibly identify threat TTP for deception and denial of information within the logical layer.
- Cyberspace electromagnetic activities section to provide information on enemy cyber forces' doctrine, tactics, and equipment, and for cyber capabilities for information collection. Cyberspace capabilities cross cue with SIGINT capabilities to provide better situational awareness of threat forces operating in the cyberspace domain.

Note. Although the intelligence, operations, and signal staff sections are the primary collaborators regarding gaining situational understanding in cyberspace, all staff sections are valuable, to some degree, and should not be disregarded during the staff integration process.

Step 1 — Define the Operational Environment

When defining the OE, cyberspace includes information and its communications. Although there are other variables in cyberspace that warrant attention (such as individuals, organizations, and systems), they either process, disseminate, or act on information.

A. Step 1 Cyberspace Considerations

When defining the OE, consider the three layers of cyberspace—physical network, logical network, and cyber-persona. When evaluating the OE, staff collaboration and reachback assets are essential.

Physical Network Layer

Depicting the physical network layer within the AO allows the intelligence staff to analyze the physical network layer as it relates to friendly and threat operations. Analysts derive the physical network layer depiction from single-source reporting, all-source intelligence products, cyber mission forces reporting, and other reporting sources. These products assist in developing the physical network layer.

When analyzing the physical network layer, identify—

- Threat C2 systems that traverse the cyberspace domain.
- Critical nodes the threat can use as hop points in the AO and area of influence. Note. Data packets pass through bridges, routers, and gateways as they travel between their sources and destinations. Each time data packets pass to the next network device, a hop occurs.
- Physical network devices in the AO, such as fiber optic cables, internet exchanges, public access points (internet cafés), server farms, and military or government intranets.
- Elements or entities (threat and nonthreat) interested in and possessing the ability to access data and information residing on and moving through the network.
- Physical storage locations with the most critical information and accessibility to that information.
- Critical nodes and entry points the threat is most likely to use to penetrate the network, including mobile tactical communications systems.
- Implemented measures that prevent threat actors from accessing the networks.

Logical Network Layer

Depicting the threat's logical network layer discloses how and where it conducts cyberspace operations. It is also useful to understand how and where the population exists, socializes, and communicates within the logical network layer. Additionally, network maps often depict the logical network layer in relation to the physical network layer. All-source intelligence analysis can enhance this depiction.

Reporting from many sources can provide information about the logical network layer of threat cyberspace, including but not limited to protocols, internet protocol address blocks, and operating systems. The network's key systems can be assessed using the depiction on the logical network layer.

When analyzing the logical network layer, identify—

- Websites or web pages that influence or have a social impact on the AO.
- Friendly logical network configurations and vulnerabilities and the friendly physical network configurations.
- Current activity baselines on friendly networks, if possible.
- Through which uniform resource locaters (known as URLs), internet protocol addresses, and other locations that critical mission data can be accessed on the internet.
- How friendly data is shared and through which software.

Cyber-Persona Layer

Depicting the threat cyber-persona layer begins with understanding the organizational structure. Assessment of the organizational structure is an all-source intelligence task. Understanding the organizational structure leads to assessing the cyber-personas associated with the organization. These include cyber-personas that represent the organization, subordinate elements, and personnel.

(Planning) IPB Cyberspace Considerations 4-c *

When analyzing the cyber-persona layer, identify—
- Threat presence in and usage of the cyberspace domain.
- Data and information consumers in the AO.
- Hacktivists in the AO, specifically with the intent to disrupt.
- Entities capable of penetrating the networks.
- How local actors interrelate with the physical network (mobile phone or internet café) and logical network (websites or software) layers.

A primary objective when analyzing the cyber-persona layer from an all-source perspective is to identify the physical persons that created and/or used cyber-personas of interest. All-source analysts gain valuable insight by using various tools and techniques, such as link diagrams (refer to ATP 2-33.4) informed by internet and social media usage, linking or associating one or more of the following, both suspected and confirmed, but not limited to cyber-personas, people, websites, internet protocol addresses, organizations or groups, buildings or facilities, and activities.

While on the internet, multiple users can use a single cyber-persona and a single user can use multiple cyber-personas. A user may have multiple cyber-personas for various reasons. This is not necessarily an indicator of illicit activity. However, multiple users using a single cyber-persona may indicate a group's activity or common affiliations.

B. Cyber-Centric Activities and Outputs for Step 1

The intelligence staff completes the graphic display of significant characteristics and components of cyberspace in relation to the unit's AO and area of influence, as illustrated in figure D-2. If known, it may be beneficial to label those websites frequently visited by the local populace, including Dark websites. Figure D-2 also exhibits the contrasts between a traditional AO overlay and an AO overlay with cyberspace considerations.

Note. Since cyberspace is a global domain, threats can potentially affect a BCT's battlefield from anywhere in the world. This must be considered when analyzing and establishing the AOI and area of influence.

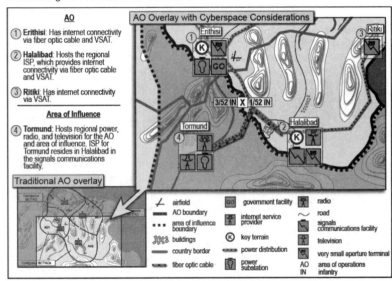

Ref: ATP 2-01.3, fig. D-2. Area of operations and area of influence example.

* 4-d (Planning) IPB Cyberspace Considerations

Step 2 — Describe Environmental Effects on Operations

Although steps 3 and 4 of the IPB process offer a detailed analysis of threats within the OE, the type of threat and their cyberspace capabilities should be defined during step 2. The significance of a cyber force presence should be considered with and weighed against identified variables within the OE.

A. Step 2 Cyberspace Considerations

For environmental effects on operations associated with step 2, describe how the following can affect friendly and threat operations:

- Threats in cyberspace.
- Terrain in cyberspace.
- Weather, light, and illumination data.
- Civil considerations.

Military Aspects of Terrain

Conduct terrain analysis of the cyberspace domain using traditional methods. Examine the five military aspects of terrain (OAKOC) factors, which can be displayed in a MCOO. Analyzing terrain in cyberspace, as in geographic terrain, can favor either friendly or threat forces. Table D-1 presents the military aspects of terrain with corresponding cyberspace considerations. This allows commanders to understand the terrain's impact, both geographically and in cyberspace, on friendly and threat operations.

Military aspects of terrain (OAKOC factors)	Cyberspace considerations
Observation and fields of fire	Ability to see subnets within networks, intrusion detection systems, password protections, and encryptions used in the area of operations. It is essential to understand what portion of the network can be seen and from where it can be seen. This may include the ability to see using physical surveillance. Additionally, closed networks may prevent observation on friendly and threat networks. Intrusion protection systems may eliminate possible threats across the network.
Avenues of approach	Method of network access, such as an access point, threat intrusion, or path to the physical or logical key terrain, such as switches, routers, servers, and vectors. Mobility corridors can be identified and grouped according to network speed, where slow speeds can cause restricted or severely restricted terrain. The volume of network activity may create additional avenues of approach.
Key terrain	Key terrain can be applied to the physical network, logical network, or cyber-persona layer. Key terrain associated with cyberspace can be considered as a physical node or data that is essential for mission accomplishment. Examples include major lines of communications, key waypoints for grouping incoming threats, domain name servers, network operating systems, switches, spectrum-dependent devices, main internet service provider inputs, mission-critical parts of the threat information network. The intelligence staff can determine key terrain in cyberspace by overlapping the threat's critical asset list, mission, and intent. **Note.** In cyberspace, it is possible for friendly and threat forces to occupy the same key terrain, potentially without either knowing of the other's presence.
Obstacles	Network features that can impede cyberspace operations include intrusion detection systems, firewalls, antivirus software, password protections, encryptions, reliability of network connectivity, data limits, and write-protections that prevent data manipulation.
Cover and concealment	The threat electromagnetic signature, cyberspace hygiene, noise awareness, and ability to limit attribution are considered cover and concealment within the cyberspace domain. Intelligence staffs determine collaboration or intelligence reach— ⊠ If threat actors are hiding their true identity using multiple cyber-personas, honeypots, or Dark webs. ⊠ Threat defensive measures (firewalls, software patches, antivirus software, encryption software, nonattributable proxy systems). ⊠ Time and volume of network activity. These may support concealment of activity on the network.

Ref: Table D-1. Terrain analysis and corresponding cyberspace considerations.

Civil Considerations

When analyzing the environment from a cyberspace perspective, apply civil considerations (ASCOPE) by cross-walking with the operational variables (PMESII-PT). When analyzing the cyberspace domain, intelligence staffs consider the informa-

(Planning) IPB Cyberspace Considerations 4-e *

B. Cyber-Centric Activities & Outputs (Step 2)

Ref: ATP 2-01.3, Intelligence Preparation of the Battlefield (Mar '19), pp. D-6 to D-9.

The S-2 ensures the intelligence staff accomplishes the following activities and outputs by the end of step 2, incorporating cyberspace considerations where applicable: threat overlay; threat description table; terrain analysis or MCOO; terrain effects matrix; weather, light, and illumination charts or tables; and civil considerations data files, overlays, and assessments.

Threat Overlay

A threat overlay graphically depicts the threat's current physical location in the AO, AOI, and area of influence, including the threat's identity, size, location, and strength. A cyberspace perspective (see figure D-3) should evaluate—

- Physical and nonphysical AOs and AOIs by identifying the physical network layer, such as media communications infrastructure and server locations, and the logical network layer, such as hosts or the threat's use of social media sites or websites.

- Known or suspected physical or cyber-personas, threats, groups, or disseminating liaisons—size, strength, and physical or logical locations, if known or suspected.

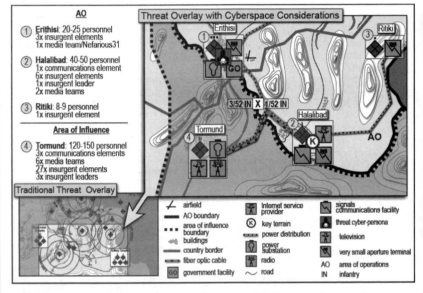

Ref: ATP 2-01.3, fig. D-3. Threat overlay with cyberspace components example.

Threat Description Table

A threat description table describes the broad capabilities of each threat depicted on the threat overlay (see table D-2). A cyberspace perspective should consider—

- Possible interdependencies between the threat's cyber and military capabilities (for example, the reliance on network communications infrastructure).

- Annotating any known or suspected technical capabilities, expertise, or programs.

Identity	Location	Disposition	Description
Nefarious31 (cyber-persona)	Erithisi	Operates from internet café as Nefarious31 using open Wi-Fi (802.11) weekly	• Greatest cyber threat in the area of operations • Capable of offensive cyberspace operations using malware • Likely coordinating with government facility to increase cyber capability • Works closely with media elements to assist in propaganda/recruiting effort
2x squads (16-18 personnel)	Erithisi government facility	Population provides sanctuary to threats	• Armed conventional/irregular forces that protect government officials and secure government network • Government facility capable of distributed denial-of-service attack
1x squads (8-9 personnel)	Erithisi southern boundary	Possible screening operations	Armed conventional/irregular forces that prevent U.S. forces from entering or occupying the area
Media element/ Recruitment	Erithisi	Operates from internet café using open Wi-Fi (802.11)	Disseminates threat propaganda to sympathetic population and actors in and around Erithisi, Ritiki, and Halalibad via social media and email campaigns

Ref: ATP 2-01.3, table D-2. Threat description table with cyberspace considerations example.

Modified Combined Obstacle Overlay

The output from the terrain analysis is used to develop the MCOO, which should reflect the physical network, logical network, or cyber-persona layers of cyberspace when applicable. (See figures D-4 and D-5.) The MCOO traditionally includes natural and man-made OAKOC factors, built-up areas, and civil infrastructure. To add cyberspace considerations into a traditional MCOO, an intelligence staff should include (not all inclusive) public-switched telephone networks, radio stations, media kiosks, internet cafés, electric power, and other supervisory control and data acquisition systems.

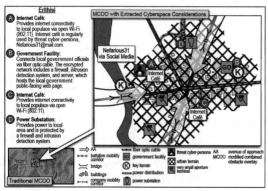

Ref: Figure D-4. MCOO, physical network and cyber-persona layers example

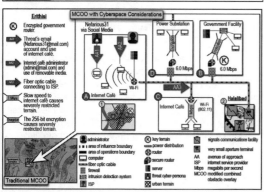

Ref: Figure D-5. MCOO, physical network, logical network, and cyber-persona layers example.

B. Cyber-Centric Activities & Outputs (Cont.)

Ref: ATP 2-01.3, Intelligence Preparation of the Battlefield (Mar '19), pp. D-6 to D-9.

Note. Intelligence staffs, in conjunction with cyber support elements and echelons above corps, develop cyberspace considerations to the MCOO with organic assets. Fiber optic lines, which are physical connections that make it part of the physical network layer in cyberspace, are typically co-located or near existing LOCs, such as roads.

Terrain Effects Matrix

Using the MCOO as a guide, a terrain effects matrix describes OAKOC factor effects on friendly and threat operations. Table D-3 presents a terrain effects matrix for operations in the cyberspace domain.

OAKOC factors (military aspects of terrain)	Terrain effects with cyberspace aspects (As related to figures D-4 and D-5)
Observation and fields of fire	• Internet café networks are wide-open and very accessible, thus allowing ability to see network configurations and the threat's capabilities.
Avenues of approach	• Primary access through unencrypted, open Wi-Fi in internet cafés (Nefarious31 and administrator accounts). • Secondary access through regional internet service provider.
Key terrain	• Regional internet service provider hosts regional power, radio, and television for area of operations. • Internet café router provides internet access to local populace, which is used to spread propaganda throughout the area of operations.
Obstacles	• Intrusion detection systems, firewalls, secure routers, and 256-bit encryptions in both power substation and government facility. • Open Wi-Fi (802.11) in internet cafés with slow download and upload speeds (severely restricted).
Cover and concealment	• Government network defended with intrusion detection systems, firewalls, secure routers, and encryptions. • Power substation also uses intrusion detection systems, firewalls, secure routers, and encryption.

Ref: ATP 2-01.3, table D-3. Terrain effects matrix with cyberspace considerations example. A network component can be associated with more than one military aspect of terrain, such as a firewall that can be both an obstacle and provide cover from fires (on the network).

Weather, Light, and Illumination Charts or Tables

Weather, light, and illumination charts or tables describe weather, light, and illumination effects on friendly and threat operations. Potential cyberspace considerations comprise any weather, including weather in space, that affects data transmissions, such as solar flares, high winds, and extreme weather conditions, such as sand storms, thunderstorms, or blizzards.

Civil Considerations Data Files, Overlays, and Assessments

Civil considerations data files may include raw data such as voting locations, base locations, and organizational hierarchies. These data files support and are supplemented by civil consideration overlays, such as, population and demographic overlays, and civil considerations assessments. Cyberspace considerations may include the use of non-governmental organizations to provide tacit or explicit support, such as proxy media disseminators or internet cafés. Additionally, consider the threat's use of governmental and noncombatant facilities for cyberspace or media activities or propaganda production.

tion and infrastructure variables. However, cyberspace operations affect, to varying degrees, the following civil considerations:
- **Areas:** In cyberspace, intelligence staffs should consider cellular phone coverage, internet service providers, and electricity distribution to industrial, commercial, and residential areas.
- **Structures:** Some cyberspace examples include power plants, moveable bridges and dams, communications/broadcast facilities (internet service providers, server farms, cell towers), internet cafés, and any building with an internet connection relevant to the AO or area of influence.
- **Capabilities:** For capabilities in cyberspace, consider internet access (and the capability to throttle or restrict access), cell phones, Wi-Fi, Bluetooth, fiber optic connections, cable television, modern information technological systems, internet and cellular network types.
- **Organizations:** Nonmilitary groups or institutions that can influence the AO (for example, hacktivists, community organizations, journalists, universities, and schools with a cyber curriculum, commercial and industrial unions, outside influencers or regional sympathizers, and online social media groups).
- **People:** Nonmilitary persons encountered by military personnel (for example, religiously and politically motivated hackers, network administrators, technologically proficient individuals, and commercial and industrial workers).
- **Events:** Routine, cyclical, historic, planned, or spontaneous activities and events that significantly affect organizations, people, and military operations.

Step 3 — Evaluate the Threat

Intelligence staffs determine threat force capabilities, doctrinal principles, and TTP employed by threats in and through the cyberspace domain. The threats' use of cyberspace varies; they use the cyberspace domain differently to accomplish or support objectives. In step 3 of the IPB process, with input from individual intelligence disciplines, the intelligence staff evaluates the threat, creates threat models, develops broad threat COAs (attack, defend, reinforce, and retrograde) or capabilities in a narrative format, and identifies HVTs.

When creating a threat model that incorporates cyberspace considerations, identify how the threat has executed and integrated cyberspace operations independently of and in concert with traditional operations, and what the threat's capabilities are in and through cyberspace. It is also crucial to realize that the physical manifestation of the threat is not at the core of the threat. For example, where the threat appears is not necessarily where the threat is likely to be. Attributing an attack to a specific threat can be very difficult and consequently makes evaluating the threat especially challenging. For example, the use of a proxy allows the threat to conceal its true location. Tapping into intelligence reach assets is necessary to develop threat models that include TTP or signatures of different threats or groups in cyberspace.

A. Step 3 Cyberspace Considerations

When evaluating the threat, understand that threats have varying cyberspace capabilities across all warfighting functions. However, the cyberspace domain likely affects each warfighting function to some degree. Therefore, it is prudent to evaluate how the threat uses the cyberspace domain to support operations by incorporating cyberspace considerations into each warfighting function to increase overall situational understanding. (See table D-4.)

Warfighting function	Cyberspace considerations
Command and Control	Delegation of authority, synchronization, and direction of forces throughout the cyberspace domain (for example, the use of email or websites to administer guidance to subordinate elements).
Movement and maneuver	Movement of forces, physically or logically, to achieve an advantage over a threat in the cyberspace domain (for example, the execution of a distributed denial of service to disrupt the threat's movement of forces).
Intelligence	The information derived through cyberspace, which enables understanding of the threat, terrain, or civil considerations (for example, the collection of threat open-source data).
Fires	The collective or coordinated use of indirect, cyberspace, missile defense, and joint fires through the targeting process (for example, the threat's use of offensive cyberspace operations or a threat's automated fire systems).
Sustainment	Cyberspace-enabled synchronized or coordinated support and services to enable freedom of maneuver, extending reach and endurance (for example, use of databases or cyberspace-enabled order processes of a threat's equipment or mission essential supplies).
Protection	Cyberspace-enabled methods to preserve the force, allowing commanders to apply maximum combat power (for example, the threat's use of defensive cyberspace operations to prevent geolocation or the targeting of its systems or networks).

Ref: ATP 2-01.3, table D-4. Cyberspace considerations for the warfighting functions.

In addition to considering and evaluating traditional threats on the battlefield, it is necessary to evaluate other relevant actors and threats that may conduct operations in cyberspace relevant to the AO:

- **Nation-state actors**. Nations that either conduct operations directly or outsource them to third parties to achieve national goals. They generally have access to domestic resources and personnel not typically available to other actors. They may involve traditional threats as well as traditional allies when conducting espionage.
- **Transnational nonstate actors or terrorists**. Formal and informal organizations not bound by national borders. These actors use cyberspace to raise funds, communicate, recruit, plan operations, destabilize confidence in governments, and conduct terrorist actions within cyberspace.
- **Criminal organizations or multinational cyber syndicate actors**. National or international, these criminal organizations steal information for their use or they sell it to raise capital. Nation states or transnational nonstate actors may use these criminal organizations as surrogates to conduct attacks or espionage through cyberspace.
- **Individual actors, hacktivists, or small groups**. These actors are known to illegally disrupt or gain access to networks or computer systems. Their intentions are as diverse as the number of groups or individual threats in cyberspace. These actors gain access to systems to discover vulnerabilities, sometimes sharing the information with owners. However, they may have a malicious intent. Political motivators often drive their operations, so they use cyberspace to spread their message. These actors can be encouraged or hired by others, such as criminal organizations or nation states, to conceal the attribution of those larger organizations.
- **Insider threats**. Any persons using their access wittingly or unwittingly to harm national security interests through unauthorized disclosure, data modification, espionage, or terrorism.

Note. Friendly elements not practicing proper cybersecurity represent the greatest threat to friendly networks.

B. Cyber-Centric Activities and Outputs for Step 3

In step 3, the intelligence staff ensures the development of threat models—the primary outputs for this step that accurately depict how threat forces typically execute operations, and how they historically have reacted in similar circumstances relative to the specified mission and environment. The compilation of these threat models for each identified threat in the AO guides the development of threat COAs in step 4

of the IPB process. Step 3 may require the following IPB activities and outputs with cyberspace considerations, time permitting:
- Creating and updating threat characteristics files.
- Creating or refining the threat model.
- Creating a threat capability statement.

Note. Upon completing steps 3 and 4 of the IPB process, update the intelligence estimate with current threat model details. Additionally, refine and update any requests for information or requests for collection.

Threat Characteristics
Analyze the threat in cyberspace applying the broad threat characteristics normally considered when analyzing any threat (see chapter 5 and appendix C). Cyberspace considerations may include—
- Attributing electronic devices to specific cyber-personas and/or persons.
- Social networking hierarchy.
- Historical threat TTP or malware signatures.
- C2 nodes.
- Threat intentions towards friendly networks.
- Insider threat potential from host-nation forces operating against friendly forces, or from a foreign intelligence physical threat.

Threat Model
The threat model comprises three parts:
- Threat template.
- Threat tactics, options, and peculiarities.
- HVT identification.

See following pages for further discussion of the threat model.

Threat Capabilities
Identify physical and nonphysical threats' operational patterns and capabilities in cyberspace by considering—
- If threats emit any unique electronic signatures.
- Media's production flow locally, regionally, and globally.
- If threats use any specific malware.
- Threats' or other relevant actors' skill level.
- Networks used to conduct operations and operations security.
- Threats' intent, for example, reconnaissance, espionage, and destructive malware.
- Threats' planning, scanning, and exploitation TTP.
- Threats' exfiltration TTP and their ability to move laterally across networks.
- Threat assets' C2.

Threat capability statements are used to identify threat capabilities, including cyberspace threat capabilities, and the broad options and supporting operations the threat can conduct to influence the accomplishment of friendly missions. This statement is a narrative that addresses an action the threat can complete. Major units may be portrayed on the threat template along with the activities of each warfighting function.

(Planning) IPB Cyberspace Considerations 4-k *

Threat Model (& Cyber Kill Chain)

Ref: ATP 2-01.3, Intelligence Preparation of the Battlefield (Mar '19), pp. D-11 to D-13.

A threat template graphically depicts the threat's preferred deployment patterns, dispositions, and capabilities for a type of operation, when not constrained by OE effects. While there are several analytic programs, figure D-6 provides an example of a traditional threat template with cyberspace considerations using the Cyber Kill Chain methodology.

Cyber Kill Chain

The Cyber Kill Chain is an analytic framework that describes the seven steps or the process the threat follows to achieve some offensive objective against a friendly network in cyberspace. Regarding IPB, it can be used as a cyber equivalency to a traditional threat template. It depicts a generalized, yet systematic approach that the threat takes to gain access to friendly resources in cyberspace when not constrained by OE effects. Understanding how attacks proliferate, the anatomy of cyberspace attacks, and historical pattern analysis of attackers in the AO can enhance the situational understanding of existing threats in cyberspace.

The following describes the seven phases of a Cyber Kill Chain:

- **Phase 1: Reconnaissance**. The threat collects information on the target before the actual attack begins.
- **Phase 2: Weaponization**. The threat exploits and creates or obtains a malicious payload to send to a victim associated with the targeted friendly network.
- **Phase 3: Delivery**. The threat sends the malicious payload to the victim by email or other means. This represents one of many intrusion methods the attacker can use.
- **Phase 4: Exploitation**. The threat exploits a vulnerability to execute code on the victim's system.
- **Phase 5: Installation**. The threat installs malware on the victim's system.
- **Phase 6: C2**. The threat creates a C2 channel to continue communications and operations of installed botnet or manipulation of the victim's system.
- **Phase 7: Actions on objectives**. The threat performs the steps to achieve goals inside the friendly forces' network.

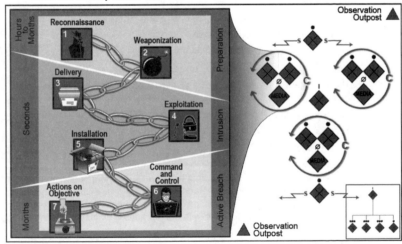

Ref: ATP 2-01.3, fig. D-6. Threat template with cyberspace considerations example.

Although intelligence staffs have little to no capability to identify or detect activity related to the Cyber Kill Chain, this analytical framework provides a platform for them to articulate logically to commanders current and potential threats against friendly networks, as well as an attack's progress on the friendly network. The right half of figure D-6 depicts a generic threat formation for occupying a village or town without OE constraints. The left half of figure D-6 shows the steps and processes the threat's cyber element, which is imbedded with the threat's media element, takes to conduct a nondescript cyberspace attack against a friendly network.

Note. The Cyber Kill Chain provides a common model for identifying and preventing cyber intrusions activity; however, the phases can occur nonsequentially.

Threat Tactics, Options, and Peculiarities

The threat model includes a description of the threat's preferred tactics. To assess threat tactics in cyberspace, identify—

- Similar TTP patterns against comparable networks worldwide.
- Any threats with the intent or capability to penetrate friendly networks, and the specific techniques they use.
- Threats' preferred methods of lateral movement.
- Any common malware used by any threat or threat elements.

High-Value Targets

HVTs can be depicted and described on the threat template. HVTs related to cyberspace are identified and evaluated using the same resources as traditional methodologies-databases, intelligence studies, patrol debriefs, the threat template with supporting narrative, and tactical judgement. The intelligence staff's tactical judgement should be influenced and informed by performing a thorough CFA—normally associated with a center of gravity analysis—of the threat and other relevant actors. A CFA is one of the most useful structured analytic techniques to identify and frame the threat's capabilities in cyberspace. (See JP 5-0.) Additionally, in step 3, regarding general COAs identified in the threat model, a CFA assists in identifying HVTs in cyberspace.

Critical Factors Analysis (CFA) consists of three major areas, which are evaluated and analyzed:

- **Critical capability** is a means that is considered a crucial enabler for a center of gravity to function as such and is essential to the accomplishment of the specified or assumed objective(s) (JP 5-0).
- **Critical requirement** is an essential condition, resource, and means for a critical capability to be fully operational (JP 5-0).
- **Critical vulnerability** is an aspect of a critical requirement which is deficient or vulnerable to direct or indirect attack that will create decisive or significant effects (JP 5-0).

Note. A completed CFA may also act as the catalyst for another analytic tool—the (modified) CARVER criteria tool used in step 4 of the IPB process. (See chapter 5.)

In evaluating HVTs, the intelligence staff should—

- Identify those assets critical to a threat's ability to conduct primary operations, sequels, or branches using cyberspace operations as a main effort or in a supporting role.
- When assessing HVTs in cyberspace, consider them based on the three layers of cyberspace (physical network, logical network, and cyber-persona).
- Identify those threat units explicitly tasked to conduct offensive cyberspace operations and those specifically tasked to conduct defensive cyberspace operations. The initial HVT list can be determined by mentally war-gaming and thinking through any specified operations under consideration.

Step 4 — Determine Threat Courses of Action

In step 4, the final step of the IPB process, intelligence staffs identify and develop the full range of COAs available to the threat and describe threat COAs that can influence friendly operations. They develop the most likely and most dangerous COAs, incorporating cyberspace threats and considerations. The level of detail always depends on the time available.

It is essential to consider how threat COAs are fundamentally affected by the cyberspace domain. For example, upon identifying methods of threat communications, consider secondary and tertiary effects on threat COAs if any or all of those threat communications are denied through degraded, disrupted, destroyed, or manipulated. Identify HVTs for each COA, such as nodes, C2 centers, communications towers, satellites, internet service providers, fiber optic lines, and local power substations. Additionally, develop initial collection requirements for each COA.

A. Step 4 Cyberspace Considerations

When determining threat COAs regarding cyberspace, consider—

- Threats' historical use of cyberspace and possible types of cyberspace operations conducted:
    - Malware—viruses, spyware, worms, network-traveling worms, socially engineered Trojans.
    - Password attacks—brute-force and dictionary attacks.
    - Denial-of-service or distributed denial-of-service attacks.
    - Advanced persistent threat.
    - Phishing attacks.
- Specific units with a task and purpose to produce cyberspace effects in the cyberspace domain.
- Threats' ability and desire to employ cyberspace operations against specific friendly operations.
- If threat forces will be arrayed distinctively based on cyberspace operations or effects.
- Threats that may be located outside of the AO.
- Threat COAs that may use proxies worldwide, which may be outside of the AOI.
- COAs that address the use of the cyberspace domain in completely different ways.

B. Cyber-Centric Activities and Outputs for Step 4

At the end of step 4, the S-2 ensures the intelligence staff accomplished the following IPB activities and outputs, including cyberspace considerations, as time allows:

- Refined threat COA statement.
- Threat situation template.
- Event template and event matrix:
    - Identify potential objectives, decision points, NAIs, and TAIs.
    - Provide input to the information collection plan.
    - HVT list and input to the HPT list.

Refined Threat Course of Action Statement

The refined threat COA statement is a narrative that describes the situation template. It should typically contain—

- The threat situation, mission, objectives and end state, and task organization.
- Capabilities.
- Vulnerabilities.
- Decision points.
- The decisive point.
- Failure options.

Each of these categories should be considered from a cyberspace perspective, either integrating a cyberspace narrative into each category or creating a separate cyberspace narrative at the end of the threat COA statement. Use the technique that best describes the threat's use of cyberspace to the commander. The level of emphasis on cyberspace should be comparable to the threat's use of and effectiveness in cyberspace.

Threat Situation Template

The threat situation template is a graphic overlay that depicts the threat's expected disposition upon the threat's selection of a COA. Typically, the situation template is accomplished by overlapping the threat template with the MCOO, which incorporates environmental effects on operations, and displaying the threat executing a specific COA.

In cyberspace, the situation template can depict a threat that is physically located within the AO and integrated with regular threats, as shown in figure D-7 below. It can also be depicted from the physical network layer perspective, which may also contain logical network elements, as shown in figure D-8 on p. 4-r.

The level of cyberspace detail in the situation template should be proportional to the level of the threat in cyberspace and the friendly unit's mission.

Note. Threats associated with cyberspace may be integrated with larger, regular threats, or they may be independent entities with no known connection to the local threat.

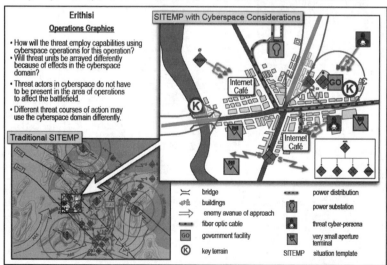

Ref: ATP 2-01.3, fig. D-7. Threat situation template with cyberspace considerations, example 1. See p. 4-r for a second example (fig. D-8).

(Planning) IPB Cyberspace Considerations 4-o *

Event Template and Event Matrix

Ref: ATP 2-01.3, Intelligence Preparation of the Battlefield (Mar '19), pp. D-16 to D-18.

An event template is a graphic overlay that confirms or denies threat COAs. This enables the development of the information collection plan. An event matrix always accompanies the event. The event template traditionally results from overlapping the developed situation templates to identify those areas or indicators that identify a COA as being unique. Prominent differences are marked as NAIs. In contrast, NAIs in cyberspace are likely not determined by overlapping situation templates and can be physical or logical.

In cyberspace, as in the land, air, maritime, and space domains, a historical record of TTP on how the threat fights assists in determining NAIs, showing possible, expected activity at a specified location. Consider that NAIs regarding cyberspace are likely related to locations or activity on a network—possibly indicating a specific type of cyberspace operations. Each NAI is linked to an assigned task and the party responsible for collecting and reporting any illicit activity or items associated with those NAIs.

Note. It is not possible to stop all malicious activity on a network. A determination should be made between which systems are mission-critical and need to be secured, versus systems that just need to be monitored.

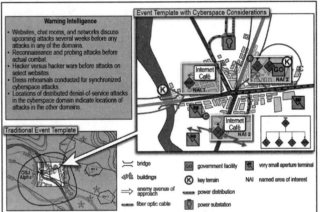

Ref: ATP 2-01.3, fig. D-9. Event template with cyberspace considerations, example 1. Figure D-9 illustrates an event template with developed NAIs for a local threat present in the AO.

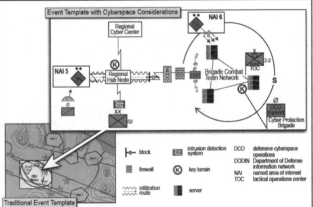

Ref: ATP 2-01.3, fig. D-10. Event template with cyberspace considerations, example 2. Figure D-10 illustrates the same threats attacking a friendly network, primarily focused on the physical network layer aspect.

* 4-p (Planning) IPB Cyberspace Considerations

Event Matrix

An event matrix describes indicators and activity expected to occur in each NAI. Although there is no prescribed format for the event matrix, it normally associates each NAI and threat decision point with indicators and the times they are expected to occur, as well as COAs they confirm or deny. (See table D-5.)

The time that a threat activity may or may not occur in cyberspace is likely influenced more by intangible variables such as the stealth and persistence of the resource being used (for example, the malware designated for an attack):

- Stealth of the resource refers to the probability that if the threat uses the resource, the resource will still be available for use in the future.
- Persistence of the resource refers to the probability that if the threat refrains from using the resource, the resource will still be useable in the future.

The timing of a threat's cyberspace attack is tied less to typical environmental factors (such as increased visibility due to daylight)—which are considered imperative for some traditional operations—and more to the logical aspects of the network. For example, the volume of network activity may spur threat operations because it can mitigate attribution, which increases stealth.

Named area of interest	Indicators	Threat decision point	Time	Threat course of action indicated
1	• Uses email • Targeting is specific • Sophisticated, appears to come from associate, client, or acquaintance • May be contextually relevant to work	1	Time of cyberspace operations may be synchronized with land or other operations.	Spear-phishing attack
2	• Unusually slow network performance • Unavailability of a particular website • Unable to access any website • Stark increase in the number of spam emails received (also known as an email bomb)	2	Cyberspace operations may be planned over a period of months or years	Denial-of-service attack
3	• Social media sites contain an increase in negative messaging • Intelligence assets discover different media in the area of operations containing threat messaging	3	• Timing may be seasonal or synchronized with other threat operations • Timing may be linked to negative effects of friendly operations	Propaganda campaign

Ref: ATP 2-01.3, table D-5. Event matrix with cyberspace considerations example.

(Planning) IPB Cyberspace Considerations 4-q *

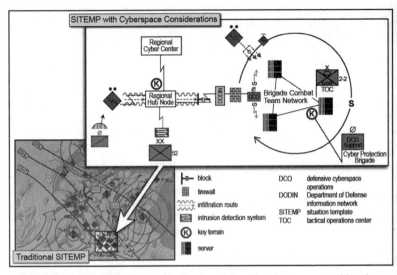

Ref: ATP 2-01.3, fig. D-8. Threat situation template with cyberspace considerations, example 2.

I(a). Cyberspace (CEMA) Operations Planning

Chap 4

Planning (Cyber & EW)

Ref: FM 3-12, Cyberspace and Electronic Warfare Operations (Apr '17), chap. 3, app. B, and app C.

The commander and staff include the cyberspace planner during the MDMP for operations. The cyberspace planner is the subject matter expert to create effects in cyberspace and the EMS, with considerations from the CEMA section. Involving the cyberspace planner early in development of the commander's vision and planning allows for synchronization and integration with missions, functions, and tasks. A consideration of cyberspace operations is the lead time required for effects support. Early involvement, inclusion in operations orders preparation, and effects approval early in the process enhance the possibility of effects in cyberspace and the EMS supporting an operation. The two primary methodologies commanders and staffs use for planning cyberspace and EW operations are the Army design methodology and the MDMP.

Planning is the art and science of understanding a situation, envisioning a desired future, and laying out effective ways of bringing that future about (ADP 5-0). Planning is one of the four major activities of mission command that occurs during operations process (plan, prepare, execute, and assess). Commanders apply the art of command and the science of control to ensure cyberspace and EW operations support the concept of operations.

The full scope of planning for cyberspace and EW operations is not addressed by the Army design methodology or the MDMP. These methodologies will allow Army forces to determine where and when effects in cyberspace and EW can be integrated to support the concept of operations. Army forces plan, prepare, execute, and assess cyberspace and EW operations in collaboration with the joint staff and other joint, interorganizational, and multinational partners as required. Whether cyberspace and EW operations are planned and directed from higher headquarters or requested from tactical units, timely staff actions and commander's involvement coupled with continued situational awareness of cyberspace and the EMS are critical for mission success.

Army commanders and staffs will likely coordinate or interact with joint forces to facilitate cyberspace operations. For this reason, commanders and staffs must have an awareness of joint planning systems and processes that facilitate cyberspace operations. Some of these processes and systems include the—

- Joint Planning Process *(see pp. 4-41 to 4-44)*
- Adaptive Planning and Execution System
- Review and Approval Process Cyberspace Operations *(refer to CJCS Manual 3139.01 and appendixes A and C).*

Refer to BSS6: The Battle Staff SMARTbook, 6th Ed. for further discussion. BSS6 covers the operations process (ADP 5-0); commander's activities; Army planning methodologies; the military decisionmaking process and troop leading procedures (FM 7-0 w/Chg 2); integrating processes (IPB, information collection, targeting, risk management, and knowledge management); plans and orders; mission command, C2 warfighting function tasks, command posts, liaison (ADP 6-0); rehearsals & after action reviews; and operational terms and military symbols (ADP 1-02).

(Planning) I(a). Cyberspace/CEMA Planning 4-1

I. Army Design Methodology (Including Cyberspace and Electronic Warfare Operations)

The Army design methodology is a methodology for applying critical and creative thinking to understand, visualize, and describe unfamiliar problems and approaches to solving them. Given the unique and complex nature of cyberspace, commanders and staffs benefit from implementing the Army design methodology to guide more detailed planning during the MDMP.

Framing an operational environment involves critical and creative thinking by a group to build models that represent the current conditions of the operational environment (current state) and models that represent what the operational environment should resemble at the conclusion of an operation (desired end state). A planning team designated by the commander will define, analyze, and synthesize characteristics of the operational and mission variables and develop desired future end states. Cyberspace should be considered within this framing effort for opportunities as they envision desired end states.

Framing a problem involves understanding and isolating the root causes of conflict discussed and depicted in the operational environment frame. Actors may represent obstacles for commanders as they seek to achieve desired end states. Creating and employing cyberspace capabilities shapes conditions in the operational environment supporting the commander's objectives.

II. The Military Decision-Making Process (with Cyberspace and Electronic Warfare Operations)

Cyberspace and EW operations planning is integrated into MDMP, an iterative planning methodology to understand the situation and mission, develop a course of action (COA), and produce an operation plan or order (ADP 5-0). The commander and staff integrate cyberspace and EW operations throughout the MDMP. They ensure courses of action are supported by the scheme of cyberspace operations and meet requirements for suitability, feasibility, and acceptability. Staff members responsible for planning and integrating cyberspace operations participate in the MDMP events and CEMA working groups. The MDMP consists of the following seven steps—

Step 1: Receipt of Mission

Commanders initiate the MDMP upon receipt or in anticipation of a mission. Staff members responsible for planning and integrating cyberspace and EW operations initiate coordination with higher headquarters staff counterparts to obtain information on current and future cyberspace and EW operations, running estimates, and other cyberspace and EW operations planning products.

MDMP Step 1: Receipt of Mission

Key Inputs	Process	Key Outputs
Higher headquarters plan or order Planning products from higher headquarters including the cyberspace effects running estimate	Begin updating the cyberspace effects and electronic warfare running estimates Gather the tools to prepare for mission analysis specific to cyberspace operations Provide cyberspace and electronic warfare operations input for formulation of the commander's initial guidance and the initial warning order	Updated cyberspace effects and electronic warfare running estimate

Ref: FM 3-12, *Cyberspace and Electronic Warfare Operations* (Apr '17), table 3-2.

Operational (PMESII-PT) & Mission Variables (METT-TC)

Commanders and staffs use the operational and mission variables to help build their situational understanding. They analyze and describe an operational environment in terms of eight interrelated operational variables: political, military, economic, social, information, infrastructure, physical environment, and time (PMESII-PT). Upon receipt of a mission, commanders filter information categorized by the operational variables into relevant information with respect to the mission. They use the mission variables, in combination with the operational variables, to refine their understanding of the situation and to visualize, describe, and direct operations. The mission variables are mission, enemy, terrain and weather, troops and support available, time available, and civil considerations (METT-TC).

See pp. 2-20 to 2-21 for related discussion of operational and mission variables as related to cyberspace operations.

Step 2: Mission Analysis

Commanders and staffs perform mission analysis to better understand the situation and problem, identify what the command must accomplish, when and where it must be done, and why (the purpose of the operation). Staff members responsible for planning and integrating cyberspace and EW operations gather, analyze, and synthesize information on current conditions of the operational environment with an emphasis on cyberspace, the EMS, and the information environment.

MDMP Step 2: Mission Analysis

Key Inputs	Process	Key Outputs
Commander's initial guidance Army design methodology product Higher headquarters' plans, orders, or knowledge products	Analyze inputs and develop information requirements Participate in the intelligence preparation of the battlefield process Identify and develop high-value targets Identify vulnerabilities of friendly, enemy, adversary, and neutral actors Determine cyberspace and electronic warfare operations specified, implied, and essential tasks Determine cyberspace operations limitations and constraints Identify cyberspace critical facts and assumptions Identify and nominate cyberspace related commander's critical information requirements Identify and nominate cyberspace critical information Provide input to the combined information overlay Provide input for the development of the mission analysis brief and warning order Participate in the mission analysis brief	List of cyberspace information requirements Intelligence preparation of the battlefield products to support cyberspace and electronic warfare operations Most likely and most dangerous enemy courses of action List of cyberspace operations specific and implied tasks List of cyberspace limitations and constraints List of cyberspace assumptions Updated cyberspace operations running estimate

Ref: FM 3-12, Cyberspace and Electronic Warfare Operations (Apr '17), table 3-3.

The Army design methodology may have been performed before the MDMP, scheduled to occur in parallel with the MDMP, or it may not be fulfilled at all. Army design products, if and when available, should be reviewed by the commander and staff to enhance situational understanding and for integration into the MDMP.

Intelligence preparation of the battlefield is the systematic process of analyzing the mission variables of enemy, terrain, weather, and civil considerations in an area of interest to determine their effect on operations (ATP 2-01.3). Intelligence support to cyberspace and EW operations begins with the IPB and continues throughout the operations process. Staff members responsible for planning cyberspace operations will coordinate with the intelligence staff to identify enemy and adversary capabilities and their use of cyberspace and the EMS to assist in the development of models, situation templates, event templates, high-value targets, named areas of interest, and other outputs from the intelligence process, which include enemy and adversary cyberspace information.

Refer to ATP 2-01.3 for additional information on the IPB.

Step 3: Course of Action Development

COA development generates options for subsequent analysis and comparison that satisfy the commander's intent and planning guidance. Staff members responsible for planning and integrating cyberspace operations apply knowledge gained from the mission analysis step to help with overall COA development. During COA development, staff members responsible for planning cyberspace and EW operations develop an initial scheme of cyberspace and EW operations consisting of cyberspace support tasks. The scheme of cyberspace and EW operations describes how the commander intends to use cyberspace operations to support the concept of operations with an emphasis on the scheme of maneuver.

MDMP Step 3: Course of Action Development

Key Inputs	Process	Key Outputs
Initial commander's planning guidance, mission, and intent	Develop information requirements for the information collection plan	Updated cyberspace operations and electronic warfare information requirements
Initial commander's critical information requirements	Integrate and synchronize cyberspace operations into the scheme of maneuver and concept of operations	Cyberspace operations initial input for high-payoff target list and target folders
Updated intelligence preparation of the battlefield products	Analyze high-value targets and develop a list of tentative high-payoff targets	Draft scheme of cyberspace operations including objectives and effects
Updated cyberspace effects and electronic warfare running estimates	Provide cyberspace input for the combined information overlay	Updated cyberspace operations running estimate
Higher headquarters' plans, orders, or knowledge products	Develop initial scheme of cyberspace and electronic warfare operations	
	Provide cyberspace operations and electronic warfare input for the development of the course of action development brief	
	Begin development of cyber effects request format	
	Begin development of electronic attack request format	
	Submit cyber effects request format (if sufficient guidance on COAs exist)	

Ref: FM 3-12, Cyberspace and Electronic Warfare Operations (Apr '17), table 3-4.

Upon completion of COA development, many outputs from the mission analysis should be updated such as the cyberspace and EW operations-related input for the commander's critical information requirements and essential elements of friendly information. The staff updates the portions of the operations orders, including annexes and appendixes that contain cyberspace and EW operations information.

Step 4: Course of Action Analysis

COA analysis enables commanders and staffs to identify difficulties or coordination problems as well as probable consequences of planned actions for each COA under consideration. Staff members responsible for planning and integrating cyberspace and EW operations use the draft products from COA development to participate in COA analysis. During COA analysis, they refine their scheme of cyberspace and EW operations, ensuring that it nests with the scheme of maneuver.

Upon completion of COA analysis, operational planning continues with drafting and submitting the CERF and then updating these requests when the COA is later refined. Development and submission of the CERF is one method by which Army forces request, coordinate, and integrate effects to support cyberspace and EW operations. The CERF contains baseline information for coordinating and integrating effects in cyberspace and EW.

MDMP Step 4: Course of Action Analysis

Key Inputs	Process	Key Outputs
Revised commanders planning guidance	Provide cyberspace operations and input and participate in the war-game briefing as required	Refined cyberspace and electronic warfare input to commander's critical information requirements
Cyberspace operations and electronic warfare initial requirements for high-payoff target list and supporting target folders	Develop cyber effects request format	Refined cyberspace operations and electronic warfare input to the high-payoff targets list
	Develop input for electronic attack request format	
Draft scheme of cyberspace operations	Continue development of scheme of cyberspace operations	Refined scheme of cyberspace operations
Updated cyberspace effects and electronic warfare running estimates	Provide cyberspace operations and electronic warfare input for the development of the decision support matrix and decision support template	Updated cyberspace effects and electronic warfare running estimate
Higher headquarters' plans, orders, or knowledge products		
Feedback from submitted cyber effects request formats	Provide refined cyberspace operations and electronic warfare input to the combined information overlay	

Ref: FM 3-12, Cyberspace and Electronic Warfare Operations (Apr '17), table 3-5.

(Planning) I(a). Cyberspace/CEMA Planning 4-5

Step 5: Course of Action Comparison

COA comparison is an objective process to evaluate each COA independently and against set evaluation criteria approved by the commander and staff. Staff members responsible for cyberspace and EW operations may not be directly involved in this process, but will provide recommendations for consideration during the process. Upon completion of the COA comparison, output products and the base operation order, become final draft.

MDMP Step 5: Course of Action Comparison

Key Inputs	Process	Key Outputs
War-game results	Conduct an analysis of advantages and disadvantages for each course of action	Recommended course of action
Refined cyberspace operations and electronic warfare input to commander's critical information requirements	Provide cyberspace operations input to the decision matrix tool as required	Updated cyberspace effects running estimate
Refined cyberspace operations and electronic warfare input to the high-payoff target list	Provide cyberspace operations input for the risk assessment (collateral effects evaluations)	
Refined scheme of cyberspace operations	Develop recommendation for the most supportable course of action from a cyberspace operations perspective	
Updated cyberspace effects and electronic warfare running estimate		
Higher headquarters' plans, orders, or knowledge products	Provide cyberspace and electronic warfare operations input for the development of the course of action decision brief as required	
Feedback from submitted cyber effects request formats		

Ref: FM 3-12, Cyberspace and Electronic Warfare Operations (Apr '17), table 3-6.

Step 6: Course of Action Approval

During COA approval the commander selects the COA to best accomplish the mission. The best COA must first be ethical, and then the most effective and efficient possible. The commander will issue final planning guidance including refined commander's intent, commander's critical information requirements, and any additional guidance on priorities for the warfighting functions.

MDMP Step 6: Course of Action Approval

Key Inputs	Process	Key Outputs
Updated cyberspace effects running estimate including refined products for each course of action Evaluated courses of action Recommended course of action Higher headquarters' plans, orders, or knowledge products Feedback from submitted cyber effects request formats	Receive and respond to final planning guidance from the commander Assess implications and take actions to revise operation order products Finalize and submit cyber effects request formats Finalize and submit input for electronic attack request format Finalize scheme of cyberspace operations	Commander approved course of action Final draft Tab A (Offensive Cyberspace Operations) to Appendix 12 (Cyberspace Electromagnetic Activities) to Annex C (Operations) Final draft Tab B (Defensive Cyberspace Operations) to Appendix 12 (Cyberspace Electromagnetic Activities) to Annex C (Operations) Final draft Tab C (Electronic Attack) to Appendix 12 (Cyberspace Electromagnetic Activities) to Annex C (Operations) Final draft Tab D (Electronic Protect) to Appendix 12 (Cyberspace Electromagnetic Activities) to Annex C (Operations) Final draft Tab E (Electronic Warfare Support) to Appendix 12 (Cyberspace Electromagnetic Activities) to Annex C (Operations) Final draft cyberspace operations input to Appendix 1 (Defensive Cyberspace Operations) to Annex H (Signal) Final draft cyberspace operations input to Appendix 2 (DODIN Operations) to Annex H (Signal) Final cyberspace operations input to Annex B (Intelligence) and Annex L (Information Collection) Nominated targets in cyberspace and the EMS Updated intelligence preparation of the battlefield products to support cyberspace and electronic warfare operations

Ref: FM 3-12, Cyberspace and Electronic Warfare Operations (Apr '17), table 3-7.

Step 7: Orders Production, Dissemination, and Transition

The final step of the MDMP is orders production, dissemination, and transition. All planning products are finalized including the cyberspace and EW operations running estimate and CERF. As time permits, the staff may conduct a more detailed war game of the selected COA. Outputs are internally reconciled and approved by the commander. Table 3-8 details cyberspace and EW operations planning inputs, actions, and outputs for step 7.

MDMP Step 7: Orders Production, Dissemination, Transition

Key Inputs	Process	Key Outputs
Commander-approved course of action and any modifications	Participate in the staff plans and orders reconciliation as required	Final Tab A (Offensive Cyberspace Operations) to Appendix 12 (Cyberspace Electromagnetic Activities) to Annex C (Operations)
Final draft operations order products	Participate in the staff plans and orders crosswalk as required	
Higher headquarters' plans, orders, or knowledge products	Provide final input to the risk assessment specific to cyberspace operations	Final Tab B (Defensive Cyberspace Operations) to Appendix 12 (Cyberspace Electromagnetic Activities) to Annex C (Operations)
Feedback from submitted cyber effects request formats	Finalize and submit cyber effects request formats	
	Finalize and submit input for evaluation request messages as required	Final Tab C (Electronic Attack) to Appendix 12 (Cyberspace Electromagnetic Activities) to Annex C (Operations)
	Produce operations order products	Final Tab D (Electronic Protect) to Appendix 12 (Cyberspace Electromagnetic Activities) to Annex C (Operations)
	Participate in the operations order brief and confirmation brief as required	Final Tab E (Electronic Warfare Support) to Appendix 12 (Cyberspace Electromagnetic Activities) to Annex C (Operations)
		Final cyberspace operations input to Appendix 1 (Defensive Cyberspace Operations) to Annex H (Signal)
		Final cyberspace operations input to Appendix 2 (DODIN Operations) to Annex H (Signal)
		Final cyberspace operations input to Annex B (Intelligence) and Annex L (Information Collection)

Ref: FM 3-12, Cyberspace and Electronic Warfare Operations (Apr '17), table 3-8.

Operations Orders, fragmentary orders, and warning orders include cyberspace operations information. The information is throughout the orders in different attachments found in Annex C and Annex H. The sections in the base operations orders and fragmentary orders with cyberspace operations information are paragraph 3, c; 3, g; and paragraph 5, c. Warning orders have cyberspace operations information in paragraph 5, c.

See pp. 4-35 to 4-39 for further discussion and a sample format.

I(b). Requesting Cyberspace Effects (CERF)

Ref: FM 3-12, Cyberspace Operations and Electromagnetic Warfare (Aug '21), app. E.

In conjunction with the necessary legal and operational authorities, commanders select organic EW capabilities to create desired effects on targets identified for EAs. If a unit's organic EW capabilities do not fulfill the targeting requirements to support the commander's intent, or if the commander does not have the authority to employ a particular EW capability, the CEMA section requests support from the next higher echelon. To request EA that will be administered by aircraft, the CEMA section uses the Joint Tactical Air Strike Request and the support request tool.

As requests pass from echelon to echelon, each echelon processes the Joint Tactical Air Strike Request to assess their ability to provide the support that meets the requesting unit's requirements. The requirement elevates either until it reaches an echelon that can support the requesting unit or until the highest echelon denies the request. Supporting a requesting unit may not be possible due to prioritization, timing, capabilities, authorization, or conflict with other EW capability requirements. Commanders ultimately have the responsibility for denying resource requests and may delegate that authority to their staff. The joint force commander may refuse a request for joint air resources, but not the joint force air component commander.

See pp. 4-27 to 4-28 for discussion of electromagnetic attack requests to include DD Form 1972 (Joint Tactical Air Strike Request).

Corps and below units do not have organic cyberspace capabilities to conduct DCO-IDM, DCO-RA, or OCO missions. The G-3 or S-3 requests support through higher headquarters. The G-6 or S-6 and the CEMA section coordinate to request DCO-IDM after determining that a threat in friendly cyberspace is beyond the scope of cyberspace security. DCO-IDM is an enabler for DCO-RA. Cyber mission forces performing DCO-IDM request DCO-RA upon deciding that a cyberspace threat requires a defensive attack beyond friendly cyberspace. OCO is used to create desired effects on targets identified for cyberspace attacks on the integrated target list. DCO-RA and OCO are similar except that DCO-RA is only used to deter a threat, whereas OCO is used to project power.

I. Requesting Cyberspace Effects (CERF)

Cyber Effects Request Format (CERF) is the format corps and below units use to request cyberspace support. Support in response to a CERF may come from joint cyberspace forces such as the combat mission teams, from other joint or Service capabilities, or Service-retained cyberspace forces.

Effects Approval at Echelons Corps and Below

During the operations process at echelons corps and below, the commander and staff identify the effects desired in and through cyberspace to support operations against specific targets. If the requesting and higher echelons determine that a current capability is insufficient, the commander and staff approve and processes the CERF. The routing process continues to each echelon until the CERF reaches the joint force land component command it is converted to an RFS, and forwarded to the JTF headquarters. The CERF approval process at echelons corps and below follow the below steps—

- Identify targets of cyberspace effects.
- Verify if organic capabilities can create desired effects.

- Approve target for cyberspace effects.
- Forward to next higher Army echelon for deconfliction and synchronization.
- Verify if other organic capabilities can create desired effects if organic cyberspace capabilities do not exist.
- If current capabilities fulfill the requirement, synchronize operations.
- If current capabilities do not fulfill the requirement, approve target for cyberspace effects.
- Forward to next higher Army echelon for approval until CERF enters the joint process.
- Synchronize operation with cyberspace effect (if possible).

Note. The joint force land component command may require the requesting corps to convert the CERF to an RFS format before submitting it into the joint process.

Effects Approval at Echelons Above Corps

Cyberspace operations provide a means by which Army forces can achieve periods or instances of cyberspace superiority to create effects to support the commander's objectives. Cyberspace attack capabilities are tailored to create specific effects and must be planned, prepared, and executed using existing processes and procedures. Commander and staff at all echelons apply additional measures for determining where, when, and how to use cyberspace effects.

Commanders and staff at each echelon will coordinate and collaborate regardless of whether the cyberspace operation is directed from higher headquarters or requested from subordinate units. The Army intelligence process, informed by the joint intelligence process, provides the necessary analysis and products from which targets are vetted and validated, and aim points are derived. As a result of the IPB process, and in collaboration with the joint intelligence preparation of the operational environment process, intelligence personnel develop network topologies for enemy, adversary, and host nation technical networks.

Targets determined during the planning process are described broadly as physical and logical entities in cyberspace consisting of one or more networked devices used by enemy and adversary actors. The G-2 may label these targets as named areas of interest and target areas of interest. Additionally, an analysis of friendly force networks will inform the development of critical information and provide a basis for establishing key terrain in cyberspace. Critical network nodes are key terrain in cyberspace. They include those physical and logical entities in friendly force technical networks of such extraordinary importance that any disruption in their operation would have debilitating effects on accomplishing the mission.

As part of CEMA, the staff will perform a key role in target network node analysis. While determining cyberspace attack effect-types for targets and defensive measures for critical network nodes, the CEMA section will prepare, submit, and track the CERF. This request will elevate above the corps echelon and integrate into the joint targeting cycle for follow-on processing and approval.

Cyber Effects Request Format (CERF)

Ref: FM 3-12, Cyberspace and Electronic Warfare Operations (Apr '17), fig. C-2.

Format 26. Cyber Effects Request Format (CERF)		
SECTION 1 REQUESTING UNIT INFORMATION		
SUPPORTED MAJOR COMMAND:	DATE:	TIME SENT:
REQUESTED UNIT:	BY:	
POINT OF CONTACT::	CLASSIFICATION (Unclassified Until Filled in)	
	USCYBERCOM J3 USE ONLY:	
SUPPORTED OPLAN/COMPLAN/ORDER:	RECEIVED BY JOC	
	DATE:	TIME:
SUPPORTED MISSION STATEMENT:	NAME/RANK:	
SUPPORTED COMMANDER'S INTENT:	CERF TRACKING NUMBER:	
	ASSIGNED TO:	
SUPPORTED COMMANDER'S ENDSTATE	STAFF SECTION:	
SUPPORTED CONCEPT OF OPERATION:	DATE:	
	TIME:	
SUPPORTED OBJECTIVE (STRAP/OP/TACT):	POC:	
SUPPORTED TACTICAL OBJECTIVE/TASK	REMARKS:	

SECTION 3 - COMPUTER NETWORK OPERATIONS (CNO) SPECIFIC INFORMAITON			
TYPE OF TARGET: SCHEDULED ON-CALL	**TARGET PRIORITY:** EMERGENCY PRIORITY ROUTINE		
TARGET NAME:	TARGET LOCATOR		
TARGET DESCRIPTION:	DESIRED EFFECT:		
TARGET FUNCTION:	TARGET SIGNIFICANCE:		

TARGET DETAILS:
Include any relevant device information such as type; operating system version and patch level, software, number of users, activity, friendly actors in the area of operations, surrounding/adjacent/parallel devices, etc.

CONCEPT OF CYBER OPERATION:
Include Task, Purpose, Method and Endstate. Also specify intelligence collection plan for battle damage assessment (BOA), to include allocated resources, measures of performance (MOPs), measure of effectiveness (MREs), and Measures of Effectiveness indicators (MOEs).

TARGET EXPECTATION STATEMENT:

REMARKS: If any of the following information is available, please provide
1.) Time on target/Duration of Effect
2.) No Earlier Than/No Later Than Need Time
3.) Trigger Event, or Conditions for Execution
4.) Persistence Requirement (ie., effect must persist through a restart of the target, trigger event))
5.) Command and Control Requirement (ie., effect must be able to be turned on/off remotely)
6.) Self-Destruct / Auto Delete Requirement (ie., effect must stop itself if C2 is lost after X amount of time)
7.) Level of Attribution Requirement (ie., unattributable to CONUS or USG, misattributed, attributed to USG, etc)
8.) Level Desectability allowed (ie., should not be detected by (a) administrator, (b) user © forensic analyst, etc)
9.) Level Co-optability allowed (ie., low, medium, high)
10.) Remote Monitoring Requirement (ie., effect should be able to be monitored by (a) operator, (b) JOC, etc)
11.) Infrastucture Requirement (ie., effect should be launched from (a) National Security Agency (NSA) Tailored Access Operations (TAO) (b) naval vessel, etc)
12.) Reversability Requirement (ie., effect should be reversible/not reversibility)

II. Cyber Effects Request Format Preparation

Although the requesting unit may not have the specific target network topology information it should provide current target information. The approval process for cyberspace effects may take longer than other targeting capabilities.

Figure C-2 (previous page) shows an example the format and instructions required to complete the CERF. The requesting unit will complete all sections except the USCYBERCOM operations directorate of a joint staff (J-3) portion of the CERF as described below.

A. Cyber Effects Request Format Section 1 Requesting Unit Information

Section 1 of the CERF requests the following unit information—

- Supported Major Command. Enter the major command authorized to validate and prioritize the CERF. For Army units at corps level and below this entry will commonly include the geographic or functional combatant command.
- Date. Enter the date the completed CERF(s) are submitted to higher headquarters.
- Time Sent. Enter the time the CERF is submitted to higher headquarters.
- Requesting Unit. Enter the name of the unit originating the requirement for the creation of effect(s) or conduct of specific activities.
- By. Enter the rank, last, and first name of the unit point of contact that time stamped and processed the CERF.
- Point of Contact. Enter the rank, last, and first name of the unit point of contact from the requesting unit. Also, enter phone number and e-mail.
- Classification. Enter the overall classification of the document. Ensure classification markings are applied to each section and supporting documentation.

B. Cyber Effects Request Format Section 2 Supported Operation Information

Section 2 of the CERF requests the following supported operation information—

- Supported OPLAN/CONPLAN/Order. Describe key information within the plan that the requested effect(s) will support.
- Supported Mission Statement. Describe the unit's essential task(s) and purpose that the requested effect(s) will support.
- Supported Commander's Intent. Describe key information within the commander's intent that the requested effect(s) will support.
- Supported Commander's End State. Describe key information within the commander's end state that the requested effect(s) will support.
- Supported Concept of Operations. Describe key information within the concept of operations that the requested effect(s) will support.
- Supported Objective (strategic, operational, and tactical). Describe the supported objective(s) that the requested effect(s) will directly support.
- Supported Tactical Objective/Task. Describe the tactical objectives and tasks that the requested effect(s) will directly or indirectly support.

The remaining portion of Section 2 is completed by the USCYBERCOM J3.

C. Cyber Effects Request Format Section 3 Computer Network Operations

Section 3 of the CERF requests the following computer network operations and specific information—

Type of Target
- Indicate "scheduled" if specific dates, times, and or supporting conditions are known.
- Indicate "on-call" if trigger events or supporting conditions are known.

Target Priority
- Indicate "emergency" if target requires immediate action. Indicate "priority" if target requires a degree of urgency.
- Indicate "routine" if target does not require immediate action or a degree of urgency beyond standard processing.

Target Name
Enter name of target as codified in the Modernized Integrated Database.

Target Location
- Provide target location according to CJCSI 3370.01, Enclosure D.
- Disregard if the request is for specific activities to support DODIN operations or DCO.

Target Description
- Provide target(s) description according to CJCSI 3370.01, Enclosure D.
- Provide description of network node(s) wherein specific activities are to support DODIN operations or DCO.

Desired Effect
- Enter deny, degrade, disrupt, destroy, or manipulate for OCO.
- Provide timing as "less than 96 hours", "96 hours to 90 days", or "greater than 90 days".

Target Function
Enter target(s) primary function and additional functions if known.

Target Significance
Describe why the target(s) is important to the enemy's or adversary's target system(s) and/or value in addition to its functions and expectations.

Target Details
Describe additional information about the target(s) if known. This information should include any relevant device information such as type; number of users; activity; friendly actors in the area of operations; and surrounding/adjacent/parallel devices.

Concept of Cyberspace Operations
Describe how the requested effect(s) would contribute to the commander's objectives and overall concept of operations.

- Include task, purpose, method, and end state.
- Describe the intelligence collection plan and specific assessment plan if known.
- Provide reference to key directives and orders.

Target Expectation Statement

According to CJCSI 3370.01, Enclosure D, describe how the requested effect(s) will impact the target system(s). This description must address the following questions.

- How will the target system be affected if the target's function is neutralized, delayed, disrupted, or degraded? (Two examples are operational impact and psychological impact.)
- What is the estimated degree of impact on the target system(s)?
- What is the functional recuperation time estimated for the target system(s) if the target's function is neutralized, delayed, disrupted, or degraded?
- What distinct short-term and/or long-term military or political advantage/disadvantage do we expect if the target's function is neutralized, delayed, disrupted, or degraded?
- What is the expected enemy or adversary reaction to affecting the target's function?

II(a). Electronic Warfare Planning

Ref: ATP 3-12.3, Electronic Warfare Techniques (Jul '19), chap. 3.

I. Electronic Warfare Contributions to the Military Decision-Making Process

EW planners follow the MDMP. In a time-constrained environment, they follow the abbreviated MDMP appropriately. The CEWO ensures planned EW activities contribute to the operation. Staff planners with the necessary expertise, and in some cases access to sensitive compartmented information facilities, are essential for planning EW and related capabilities. Integrating EW into operations requires placing planners at the brigade combat team level with experience in capabilities, such as special technical operations and special access program effects. Throughout the MDMP, the CEWO continuously identifies risks and appropriate risk mitigation techniques.

The CEWO participates in the MDMP by planning and synchronizing EW and cyberspace operations actions. During planning, the CEWO considers joint, interorganizational, and multinational dependencies and interdependencies of EW resources.

The members of the CEMA section assist the CEWO during the MDMP by conducting terrain and radio wave propagation analysis relevant to friendly and threat forces within an operational environment. The results of the analysis contribute to staff products, such as map overlays depicting EW assets and their associated range of effectiveness. The staff uses the products to refine the EW portions of the plan. The CEMA section builds and staffs operations order appendices and annexes and submits them to the G-3 (S-3)staff for dissemination. The CEWO provides EA information to the fires staff for inclusion in Annex D of the operations order (FM 6-0).

The CEMA section considers policies, laws, and ROE that affect EW operations when participating in the MDMP process. The SJA and the CEMA working group develop the ROE for commander review. Planners and the SJA clarify the ROE or develop supplemental ROE when necessary.

See pp. 4-2 to 4-8 for discussion of EW operations (and cyberspace operations) planning and the MDMP.

II. Electronic Warfare Planning Considerations

Several considerations are important to planning EW operations to include equipment type,configurations, logistics, availability, and risks. The running estimate is a tool to assist with planning and maintaining awareness of EW capabilities, current missions, and future mission requirements.

See following pages (pp. 4-16 to 4-17) for discussion of the EW running estimate.

A. Planning Factors

The CEWO visualizes an operational environment and EME using maps and simulation programs that predict the behavior of radio waves used during unified land operations. The course of action proposed by the CEWO require analysis to determine the capabilities and limitations of the systems. For example, man-pack EW systems are lightweight and highly mobile but also have limited transmit power for EA. Vehicle mounted systems allow for higher power output but have line of sight (LOS) limitation in dense terrain.

B. Electronic Warfare Running Estimate

Ref: ATP 3-12.3, Electronic Warfare Techniques (Jul '19), pp. 3-3 to 3-5.

The CEWO prepares and continually updates the running estimate. A running estimate is the continuous assessment of the current situation used to determine if the current operation is proceeding according to the commander's intent and if planned future operations are supportable (ADP 5-0). Information in the running estimate are committed and reserved assets, maintenance status of EW equipment and training proficiency of EW personnel. Resources that are useful in developing a running estimate are the maintenance report and the commanders' training assessments. Threat information is available from online databases, unit intelligence assets, and national intelligence sources.

The purpose of the CEMA section running estimate is to provide a consolidated list of information about cyberspace and the electromagnetic spectrum to assist the CEMA section in planning, preparing, and executing operations. The information serves as a foundation for the Appendix 12 to Annex C and tabs, and is dependent on information requirements with other staff such as Operations, Fires, Intelligence, and Signal as sources of information. Some of this information will be redundant with other staff section planning products. Table 3-1 below is an example of an EW running estimate.

1. Friendly electronic warfare systems.
    a. System nomenclature and disposition by echelon.
        i. Planning, modeling, and simulation tools.
        ii. Organic systems.
        iii. Echelons above corps and joint assets.
    b. System capabilities.
        i. Frequency range.
        ii. Modulation type(s).
        iii. Maximum power output.
        iv. Antenna configuration and characteristics.
        v. Command and control details (mesh network parameters, data paths, and bandwidth requirements).
    c. Modeling and simulation of each system based on differing parameters and area of operations
        i. Differing power ratios.
        ii. Antenna configuration.
        iii. Terrain.
    d. Constraints and limitations associated with each system.
2. Friendly spectrum-dependent systems.
    a. System nomenclature and disposition by echelon.
        i. VHF radios
        ii. Satellite communications terminals.
        iii. Radar sets.
        iv. Unmanned aircraft systems
    b. System characteristics.
        i. Frequency ranges.
        ii. Bandwidth requirements.

iii. Power.
   iv. Modulation.
  c. Modeling and simulation of each system, based on differing parameters and area of operations.
  d. Constraints and limitations associated with each system.
3. Friendly electronic warfare systems.
  a. System nomenclature and disposition by echelon.
  b. System capabilities.
    i. Frequency range.
    ii. Modulation type(s).
    iii. Maximum power output.
    iv. Antenna configuration and characteristics.
    v. Command and control details (mesh network parameters, data paths, and bandwidth requirements.
  c. Threat electronic warfare tactics, techniques, and procedures.
  d. Modeling and simulation of each system, based on differing parameters and area of operations.
  e. Critical capabilities and vulnerabilities by system.
4. Threat spectrum-dependent systems.
  a. System nomenclature and disposition by echelon.
  b. System characteristics.
    i. Frequency ranges.
    ii. Bandwidth requirements.
    iii. Power.
    iv. Modulation.
  c. Tactics, techniques, and procedures.
  d. Frequency allocations.
  e. Cueing cycles (radar sets)
  f. Modeling and simulation of each system, based on differing parameters and area of operations.
  g. Critical capabilities and vulnerabilities by system.
5. Civil infrastructure considerations.
  a. Networks in the area of operations.
    i. SCADA.
    ii. Internet service providers.
    iii. Fiber (regional, national, and international).
  b. Spectrum resources and allocations (with characteristics of each).
    i. Wi-Fi.
    ii. Broadcast television.
    iii. Broadcast radio.
    iv. Satellite ground stations.
  c. Physical access to structures and equipment.

Ref: ATP 3-12.3, Electronic Warfare Techniques (Jul '19), table 3-1. Example of an electronic warfare running estimate.

Airborne platforms offer the best LOS of EW systems, but are vulnerable to enemy air defense systems and have limited dwell time on target.

Additional Factors for Airborne Planning

Maintenance activities and other missions reduce the availability of aircraft to support EW requirements. Airborne platform unavailability for EW is attributed to—
- Poor weather and visibility that restrict flight.
- Planned and unplanned maintenance.
- Transport missions.
- Intelligence, surveillance, and reconnaissance missions.
- Communications missions.

Logistical Considerations

Units conduct scheduled and unscheduled maintenance on EW equipment. Maintenance ensures readiness for current and future operations. The CEWO, with assistance from logistics staff, develops an SOP that includes maintenance procedures. The CEWO or representative prioritizes maintenance efforts ensuring a unity of effort, as maintainers are a limited resource.

The planner considers—
- An EW capability replacement plan for potential coverage gaps and unexpected outages.
- Parts availability for maintenance to prevent non-mission capable equipment.
- Power resources including:
- Batteries.
- Generators and fuel.
- Shore power.
- Vehicle or transport power sources

Commanders allocate EW resources to support various units. When EW resources support another unit, the supported unit—
- Identifies EW requirements.
- Protects and defends EW assets.
- Provides logistical support.

Risk Management

EW can cause unwanted radio frequency (RF) exposure to personnel. High levels of RF exposure can damage external and internal human tissue. The CEWO identifies risks associated with EW activities and develops mitigating steps to reduce the risk to friendly personnel and equipment. The CEWO then coordinates with the staff to refine the risk mitigating recommendations and presents them to the commander. For more information about risk management, refer to ATP 5-19.

Planners synchronize EW with lethal and nonlethal capabilities to achieve desired effects. The CEWO uses predetermined formulas to calculate EA and ES.

For additional information on predetermined formulas and jamming calculations, refer to ATP 3-12.3, Electronic Warfare Techniques (Jul '19), appendix B.

EW actions can mitigate operational risk, though using EA, both offensively and defensively, has inherent risk associated with the systems due to emissions. The risks include hazards of electromagnetic radiation to personnel, hazards of electromagnetic radiation to fuels, and hazards of electromagnetic radiation to ordnance.

Hazards of electromagnetic radiation to personnel is the danger to personnel from the absorption of electromagnetic energy by the human body. Personnel hazards are associated with the absorption of RF energy above certain power levels in certain frequency bands for certain lengths of time. DODI 6055.11 specifies the allowable amounts of radiation and personnel exposure time to RF fields at particular intensities and frequencies.

Hazards of electromagnetic radiation to fuels is the hazard associated with the possibility of igniting fuel or other volatile materials through RF energy-induced arcs or sparks. It takes a certain amount of arc energy to ignite a fuel, and modern fuels are much safer than older fuels. This is a major concern when there is limited separation between EW capabilities and fuel, such as airfields, forward armament and refueling point, and refueling on-the-move locations. Fortunately, there are many operational safeguards against this problem.

Hazards of electromagnetic radiation to ordnance refers to the susceptibility of electro-explosive devices to RF energy. Electro-explosive or electrically-initiated devices are the control devices to detonate explosives, launch ejection seats, cut tow cables, and other similar functions. Modern communications and radar transmitters can produce high levels of electromagnetic energy that are potentially hazardous to ordnance. These environments can cause premature actuation of sensitive electro-explosive and electrically initiated devices.

III. Staff Contributions to EW Planning

EW personnel are dependent on the staff for a variety of products to understand an operational environment, targeting, and EP requirements. The EW personnel can plan EW activities once they have sufficient situational awareness of an operational environment.

See following pages (pp. 4-20 to 4-21) for an overview and discussion of EW Contributions to the staff.

A. G-2 (S-2) Staff

EW planners rely on the G-2 (S-2) staff for threat characteristics identified during IPB. The CEWO submits requests for information to address gaps identified during IPB.

In most cases, the CEWO relies on SIGINT-derived enemy electronic technical data to plan and conduct EW targeting operations. Therefore, the G-2 (S-2) staff supports the CEWO during the alignment of EW and SIGINT assets against the commander's priorities of effort to achieve the best possible outcomes. SIGINT and EW resources, synchronized with the commander's scheme of maneuver significantly, enhances situational awareness while increasing the precision of the targeting process. For more information about a line of bearing (LOB), a cut, and a fix, see paragraph 5-8.

Useful products G-2 (S-2) creates or assists in creating include—

- High-value target list (HVTL) during IPB.
- High-payoff target list (HPTL) during MDMP.
- Enemy electronic order of battle (EOB).

B. G-6 (S-6) Staff

The CEWO uses the joint restricted frequency list (JRFL) and friendly network architecture to plan EW and avoid EMI. The CEWO and the G-6 (S-6) use this information to develop the unit EP plan.

See p. 4-22 for an overview and discussion of the joint restricted frequency list (JRFL).

EW Contributions to the Staff

Ref: ATP 3-12.3, Electronic Warfare Techniques (Jul '19), pp. 3-9 to 3-11.

The CEWO provides information to other staff sections to aid in planning. This information answers requests for information and aids in refining staff products.

Contributions to G-2 (S-2) Staff

The CEWO contributes to the IPB and throughout the MDMP by providing input related to EW activities. IPB involves systematically and continuously analyzing the threat and certain mission variables (terrain, weather, and civil considerations) in the geographical area of a specific mission. Commanders and staffs use IPB to gain information that supports understanding. Some of the CEWO's input to the IPB includes the following:

- Information regarding how the EME affects operational environments.
- Input to likely threat COAs by providing information on threat EMS capabilities, tactics, techniques, and procedures.

When evaluating how the EME affects an operational environment, the CEWO—

- Analyzes the EME and identifies known or suspected threat emitters of interest and neutral emitters in the area of operations.
- Identifies facilities, which may support, operate, or house enemy EW capabilities.
- Contributes to the G-2 (S-2) understanding of the enemy's use of the EMS.

When describing the effects of an operational environment on EW activities, the CEWO—

- Conducts terrain analysis of both the land and air domains using the factors of observation and fields of fire, avenues of approach, key and decisive terrain, obstacles, and cover and concealment.
- Identifies terrain that protects communications and target acquisition systems from activities. Terrain masking reduces friendly vulnerabilities to threat EW actions.
- Identifies how terrain affects LOS, including effects on both communications and noncommunications transmitters. Line of sight is the unobstructed path from a Soldier's weapon, weapon sight, electronic sending and receiving antennas, or piece of reconnaissance equipment from one point to another (ATP 2-01.3).
- Evaluates how vegetation affects radio wave absorption and antenna height requirements.
- Locates power lines and their potential to interfere with radio waves.
- Assesses the likely air and ground avenues of approach, their dangers, and potential support that EW activities could provide for them.
- Determines how weather (including visibility, cloud cover, rain, and wind) may affect ground-based and airborne EW activities and capabilities (for example, when poor weather conditions prevent airborne EW launch and recovery).
- Assists the G-2 (S-2) staff with the development of the modified combined obstacle overlay.
- Considers all other relevant aspects of an operational environment that affect EW activities, using the operational variables (political, military, economic, social, information, infrastructure, physical environment, and time) and mission variables (mission, enemy, terrain and weather, troops and support available, time available, and civil considerations).

The CEWO contributes to the G-2 (S-2) staff's understanding during enemy course of action development by providing—

- Subject-matter-expert input on enemy EW tactics, techniques, and procedures for situation template development.
- A review of named areas of interest and target areas of interest to confirm EW considerations.
- EW options to support decision points.
- EW input to the event template and event matrix.

Contributions to Other Staff

During planning, the CEWO provides information to other members of the staff including—

- EW input to IPB [G-2 (S-2)] staff.
- Input to the HPTL (Fires).
- Input to the commander's critical information requirements including essential elements of friendly information and priority intelligence requirements [G-2 (S-2) and G-3 (S-3)] staff.

Contribution to Fires (Targeting Working Group)

The targeting working group recommends priorities for the targets according to its judgment and the advice of the fires cell, targeting officer and the field artillery intelligence officer. Targeting is the process of selecting and prioritizing targets and matching the appropriate response to them, considering operational requirements and capabilities (JP 3-0). Targeting working groups maintain a HPTL and inform the commander of targets that do not support the commander's guidance. The HPTL includes the recommended priority of targets and target engagement sequence. The HPTL includes the target category, a name, or a number.

The CEWO recommends to the G-3 (S-3) staff and the fire support element whether to engage a target with EW. The fires support element uses decide, detect, deliver, and assess methodology to direct friendly forces to attack the right target with the right asset at the right time. The targeting working group provides the HPTL to the operations, intelligence, and fires support element. The staff employs the HPTL to understand and determine attack guidance and to refine the collection plan. This list may also indicate the commander's operational need for battle damage assessment of the specific target and the time window for collecting and reporting it (ATP 3-60).

The CEWO integrates EW into the targeting process. After the targeting board has approved an EW target, the CEWO deconflicts the EW activity with the friendly use of the EMS. To support targeting, the CEWO—

- Matches EW resources with specific high-payoff targets and high-value targets.
- Ensures EW activities meet targeting objectives.
- Synchronizes EA with friendly use of the EMS.
- Coordinates with the SIGINT staff to gain targeting information that supports ES and EA.
- Provides EW mission management through the command post or joint operations center and the tactical air control party for airborne EA.
- Provides EW mission management as the EW control authority for ground or airborne EA when designated.
- Requests theater EW support.

See pp. 4-29 to 4-34 for further discussion of targeting.

Joint Restricted Frequency List (JRFL)

The JRFL includes—

Taboo Frequencies

Taboo frequencies are friendly frequencies of such importance that must never be deliberately jammed or interfered with by friendly forces. Normally these include international distress, safety, and controller frequencies. They are generally long-standing frequencies, taboo frequencies may be time-oriented, and the restrictions may be removed as the combat or exercise situation changes. During crisis or hostilities, short duration EA may be authorized on taboo frequencies for self-protection to provide coverage from unknown threats or threats operating outside their known frequency ranges, or for other reasons. *For more information about guarded, protected and taboo frequencies, refer to JP 3-13.1.*

Protected Frequencies

Protected frequencies are friendly frequencies used for a particular operation, identified and protected to prevent them from inadvertent jamming by friendly forces while executing active EW operations against hostile forces. These frequencies are of such critical importance that jamming should be restricted unless absolutely necessary or until coordination with the engaged unit is made. They are generally time-oriented and may change with the tactical situation. It is important to update protected frequencies periodically.

Guarded Frequencies

Guarded frequencies are adversary frequencies currently being exploited for combat information and intelligence. A guarded frequency is time-oriented in that the list changes as the adversary assumes different combat postures. These frequencies may be jammed after the commander has weighed the potential operational gain against the loss of the technical information.

C. Staff Judge Advocate

Conducting EW requires an understanding of the ROE and legal authorities. The CEWO consults the SJA for the standing ROE and interpretation. The SJA or representative reviews EW activities to ensure compliance with existing DOD directives and instructions, ROE, and applicable domestic and international laws, including the law of armed conflict.

When considering EA or ES, the SJA will assist in the planning of operations and will review past operations. As part of the assistance, the SJA considers what impacts operations may have on host nation communications and legal implications related to the impacts.

III. Electronic Warfare Configurations
Ref: ATP 3-12.3, Electronic Warfare Techniques (Jul '19), pp. 3-6 to 3-8.

EW equipment requires configuration for successful deployment. Units use EW equipment in man-pack, vehicle, fixed-site, and airborne configurations. Equipment configuration includes—
- Choosing omnidirectional or directional antennas.
- The physical placement of equipment.
- Selecting power resources for EW equipment.
- Primary, alternate, contingency, and emergency (PACE) plan for tasking and reporting.

Power sources for EW equipment include—
- Power generators such as gasoline or diesel powered engines.
- Batteries for man packs and vehicle-mounted configurations.
- Shore power for fixed EW assets.

Manpack Configuration
Manpack configurations include EA and ES capabilities. For manpack configurations, the CEWO considers the following—
- Limited available transmit power for EA.
- Weight of antennas and batteries carried by the Soldier.

Vehicle-Mounted Configuration
Vehicle-mounted EW equipment supports units with EA and ES capabilities. Units use vehicle-mounted EW equipment during maneuver or at the halt. Vehicle-mounted configurations include—
- Mounted and dismounted configurations.
- Jamming capabilities.
- Direction finding (DF) capabilities for locating and targeting threat transmitters.
- PACE plan for tasking and reporting

Fixed-Site Configuration
Fixed-site EW configurations have more available transmitting power than manpack and vehicle EW configurations. Fixed EW configurations have multiple transmitters, receivers, and antennas that enable multiple EW activities to occur simultaneously. A fixed site may include transportable systems that require configuration and operation only at the halt requiring personnel to install or construct the system.

Airborne Configuration
Airborne EW is the coupling of EW assets to airborne platforms such as unmanned aerial systems, tethered balloons, and rotary and fixed-wing aircraft. They provide an extended range over ground-based assets and greater mobility than ground-based assets. In addition, they support ground-based units.

The synchronization of airborne EW missions requires detailed planning. The time on target for airborne EW assets coupled to rotor and fixed-wing platforms is normally brief. Time on target for airborne EW is limited due to the high rate of speed of the aircraft. The short time on target is also a technique used to minimize the threat's abilities to detect the platforms using visual, DF and radar detection techniques.

V. EW Employment Considerations

Ref: ATP 3-12.3, Electronic Warfare Techniques (Jul '19), pp. 3-5 to 3-6.

The CEWO analyzes the operation and EW employment considerations early in the MDMP. These considerations include—
- Survivability of personnel and equipment.
- The time required to build or improve the unit's EP posture and position EA and ES capabilities.
- Ability of EW resources to achieve the desired effects.
- Reprogramming of EW assets.
- Capabilities, limitations, advantages, and disadvantages of available EW and SIGINT assetsequipped with ES capability.
- Intelligence available for targeting.

Note. The G-2 (S-2) manages SIGINT resources that contribute to EW targeting.

Survivability

Survivability of personnel and equipment rely on force protection and EP techniques. EP enhancesforce protection efforts as another method to mitigate environmental and adversarial effects. The CEMAsection plans the mitigation actions, and the commander decides what risk is acceptable for an EW mission.Force protection risk mitigating techniques include coordinating ground or air escort and configuring EWequipment with organic EP capabilities. EP is not force protection or self-protection. EP is an EMS-dependentsystem's use of electromagnetic energy and/or physical properties to preserve itself from direct orenvironmental effects of friendly and adversary EW, thereby allowing the system to continue operating(JP 3-13.1).

EP contributes to survivability. Antennas erected to minimum heights, while maintainingcommunications, prevent visual observation by the threat. This technique contributes to survivability.Survivability is a useful criterion for course of action analysis during the MDMP. For more information aboutEP, see chapter 7.

Time

The CEWO uses available time to configure and position EW assets for optimal performance. Timealso affects the selection of movement techniques for a mission. The CEWO synchronizes EA operationswith maneuver and fire to maximize effects at the appropriate time. The CEWO also plans duration of EAeffects based on target analysis to support survivability of EW assets.

Efficacy

The CEWO considers which EW asset has the appropriate level of efficacy for an EW mission.Efficacy is the likelihood that an EW mission will achieve the desired effect. For example, EA has a minimumtransmission power threshold. Transmission power settings below the threshold have reduced levels ofefficacy to achieve the desired effect, whereas transmission power settings above the threshold have increasedlevels of efficacy to achieve the desired effect.

Electronic Warfare Reprogramming

Electronic warfare reprogramming is the deliberate alteration or modification of electronic warfare ortarget sensing systems, or the tactics and procedures that employ them, in response to validated changes inequipment, tactics, or the electromagnetic environment (JP 3-13.1). When information reveals that the adversary changes frequencies for communications or there are other changes in the EME, the CEWO ensures the reprogramming of EW systems or target sensing systems, to include the employment technique. Reprogramming includes changes to defensive systems, software, firmware, hardware, and information collection systems (JP 3-13.1). The change in the EME may affect friendly communication systems also. The CEWO informs the spectrum manager of the changes to EW requirements to coordinate the adjustment in mission parameters and may recommend friendly communications frequency changes to the G-6 (S-6). The responsibility to reprogram EW equipment is the responsibility of the unit; however, units should be aware of reprogramming efforts when operating with multi-national forces. Reprogramming is a national responsibility due to the effect on the EME. Refer to JP 3-13.1 for more information about reprogramming. EW reprogramming examples include—

- Changing target frequencies for jamming as well as updating restricted frequencies.
- Changes location of sensors due to environmental changes or interference.
- Installing the latest available software, firmware, and hardware for EW and SIGINT equipment.

Electronic Warfare Visualization

The CEMA section visualizes and simulates the EMS, manmade effects, and environmental impacts. The information the section gains informs friendly actions and may provide insight to possible enemy COAs. There are automated tools to assist the CEMA section with the following tasks:

- Providing input to the common operational picture.
- Displaying sensor information from EW and SIGINT assets including—
- Detecting emitters and plotting lines of bearing.
- Analyzing circular error probable ellipse.
- Conducting mission planning and rehearsals.
- Managing EW assets.
- Modeling and visualizing how the EME responds to friendly and enemy EW activities.

The CEMA section analyzes the EME using—

- EMS sensors.
- Threat system databases.
- Intelligence information.
- Operational environment factors.

EW personnel require updates as the situation changes. The tools combined with staff interaction and the command and control system provide the updates.

VI. Electronic Warfare Assessment

EW assessment is continuously monitoring and evaluating the impact of EW on the current situationand the progress of an operation. CEWOs continually assess the current situation and progress of theoperation and compare it with the concept of operations, mission, and commander's intent. Assessment occursthroughout planning, preparation, and execution; it includes three major tasks:

- Continuously identifying threat vulnerabilities and reactions to friendly EW activities.
- Continuously monitoring EW activities to ensure alignment with the commander's desired endstate.
- Evaluating the operation against measures of effectiveness and measures of performance andmaking necessary adjustments.

The targeting working group synchronizes EW effects with other effects. The CEWO coordinates andsynchronizes joint and multinational air and ground EW capabilities. The CEWO also manages the organic EW activities within the main command post.

Measures of Performance and Effectiveness (MOPs/MOEs)

The CEWO develops the measures of performance and measures of effectiveness for evaluating EWactivities during execution. Measures of effectiveness measure the degree to which an EW capability achieved the desired result. Normally, the CEWO measures this by analyzing data collected by both activeand passive means.

Measures of effectiveness help define whether a unit is creating the desired effect(s) or conditions inan operational environment. Example questions to measure EW effectiveness include—

- Did the EA disrupt enemy radar assets?
- Is the enemy radar retuning?
- Is there increased radio traffic on the radar command and control network?

Measures of performance help evaluate whether a unit is accomplishing tasks to standard. In thecontext of EW, example questions of measures of performance include—

- Is the EA asset transmitting at the necessary power?
- Is the EA asset transmitting in the required bandwidth?
- Is the EA asset transmitting using the correct polarization?
- Are all assets for a given mission operating in proper synchronization?

CEWOs use caution when selecting measures of effectiveness to avoid flaws in an analysis of the EWmission. For example, the lack of enemy electronic activity, such as communications or improvised explosivedevice initiation, does not necessarily mean it was the result of the EW mission; other factors may be thecause. Another example of a flawed measure of effectiveness is the premature conclusion that an EAdegraded or disrupted a radio communication that resulted in an enemy commander not being able to directthe maneuver of subordinate forces using a specific frequency during a battle engagement. The enemycommander may have an alternate means of communication.

Effective EW Planning continues during all phases of an operation. The planning of EW requiressignificant preparation to achieve successful execution of EW tasks. The CEWO uses assessment techniquesto measure success.

II(b). Electromagnetic Attack Request

Ref: FM 3-12, Cyberspace Operations and Electromagnetic Warfare (Aug '21), pp. E-9 to E-12. See also p. 3-25.

Typically, Army units at corps and below have the organic capabilities to conduct EW within their assigned AO. The joint force commander typically delegates electromagnetic attack control authority to subordinate commanders conducting EW missions within their assigned AO. Commanders must ensure EW has been integrated and synchronized across the staff and according to the higher commander's guidance parameters.

I. Electromagnetic Attack Request

Dynamic targeting is targeting that prosecutes targets identified too late or not selected for action in time to be included in deliberate targeting. Dynamic targeting is normally employed in current operations planning because the nature and timeframe associated with current operations typically requires more immediate responsiveness than is achieved in deliberate targeting (JP 3-60). Dynamic targeting is used for targets of opportunity that includes unscheduled targets and unanticipated targets. When immediate airborne EA is required for deliberate targeting, for example, when a ground maneuver unit requires jamming enemy communications before engagement, a unit can request support using an EA request. Units also submit an EA request for EA support when a mission cannot pre-plan due to some operations' immediate nature. The EA request prepares the aircrew providing EA support (see ATP 3-09.32). The JTF headquarters, the joint force land component command, the joint force air component command, and the air operations center must collaboratively plan airborne EA before an operation. This planning and coordination provides the joint force air component command the necessary time to identify and prepare an electronic combat squadron that will remain on standby throughout the mission.

Electromagnetic Attack Request

Do not transmit line numbers. Units of measure are standard unless briefed.
Lines 1,2 and 4 are mandatory readback (*). Jam Control Authority (JTAC) may request additional readback.

JCA; " ___Foxfire 06___ . this is ___Forward 09___ ,"
       (aircraft call sign)   (JTAC call sign)

1. Target/ or Effect Description:" ___Disrupt___ ,"
    a. Rapper or Target Name   radio transmitter
    b. Frequency (if known)   107.1 MHZ
    c. Modulation   FM

2. Target Location; " ___N 46° 41' 33.28" W 120° 947.2322"___ ,"
                        (latitude and longitude or MGRS)

3. Remarks:" ___No current remarks or special instructions___ ,"

Legend
JTAC joint terminal attack controller   N North
MGRS military grid reference system   W West

Ref: FM 3-12 (Aug '21), fig. E-5. Electromagnetic attack request.

The electromagnetic attack request is the common format for requesting airborne EA support. The CAOC Non-Kinetic Operations Center may require the use of a different requesting procedure for immediate EA support. The JTF headquarters and its subordinate units must become familiar with the standardized EA requesting format established by the CAOC Non-Kinetic Operations Center or the CCMD before requesting EA support. Additionally, a tactical air control party or joint terminal attack controller may handle electromagnetic attack requests because of its uniqueness and necessity for the particular tactical experience required.

II. Airborne Electromagnetic Attack Support

In a joint environment, Army units can request an airborne cyberspace attack, EA or ES. Targeting using airborne assets for cyberspace and EA can be used for both deliberate and dynamic targets. In this instance, the unit submits DD Form 1972 (Joint Tactical Air Strike Request), accompanied by an airborne cyberspace attack, electromagnetic attack, or electromagnetic support request tool. Each of these requests have unique information requirements and request flows. The requesting unit's fires support element adds the target to its target nomination list. The target is also added to the joint integrated prioritized target list at the JTF headquarters and forwarded to the joint force air component command with its assigned DD Form 1972 (Joint Tactical Air Strike Request) number (see ATP 3-09.32). Once the joint force air component command approves the Joint Tactical Air Strike Request, the request is forwarded to the air operations center to execute the attack.

Table E-1, below, is an example of the request tool used to request an airborne EA or ES and accompanies a DD Form 1972 (Joint Tactical Air Strike Request).

Nonkinetic Effect Discipline (Space, Airborne Electromagnetic Attack, Information Operations, or Cyber)	Applicable phase or Find, Fix Track, Target, Engage, Assess	Nonkinetic Effect	Risk Technical Gain/ Loss	Intel Gain/ Loss	Commander's Acceptable Level or Risk	Aproval Timeline (idea to Approved Execution Order)	Employment Timeline (initial Access to Effect Ready to Fire)	Authority Level (for use in Area of Responsibility)	Execution Authority (Tactical Employment)
----------	----------	----------	----------	----------	----------	----------	----------	----------	----------
----------	----------	----------	----------	----------	----------	----------	----------	----------	----------
Airborne EA	Target	Jam	Gain	Gain	Acceptable	180 mins	180 mins	Local	Local
----------	----------	----------	----------	----------	----------	----------	----------	----------	----------
----------	----------	----------	----------	----------	----------	----------	----------	----------	----------

Ref: FM 3-12 (Aug '21), table E-1. Airborne cyberspace attack, electromagnetic attack, or electromagnetic support request tool.

Ref: FM 3-12 (Aug '21), fig. E-4. DD Form 1972 (Joint Tactical Air Strike Request).

* 4-28 (Planning) II(b). Electromagnetic Attack Request

III. Targeting (D3A)

Ref: FM 3-12, Cyberspace Operations and Electromagnetic Warfare (Aug '21), pp. 4-11 to 4-17.

Targeting is the process of selecting and prioritizing targets and matching the appropriate response to them, considering operational requirements and capabilities (JP 3-0). A target is an entity or object that performs a function for the adversary considered for possible engagement or other actions. (JP 3-60).

When targeting for cyberspace effects, the physical network layer is the medium through which all digital data travels. The physical network layer includes wired (land and undersea cable), and wireless (radio, radio-relay, cellular, satellite) transmission means. The physical network layer is a point of reference used during targeting to determine the geographic location of an enemy's cyberspace and EMS capabilities.

When targeting, planners may know the logical location of some targets without knowing their physical location. The same is true when defending against threats in cyberspace. Defenders may know the logical point of origin for a threat without necessarily knowing the physical location of that threat. Engagement of logical network layer targets can only occur with a cyberspace capability.

The logical network layer provides target planners with an alternate view of the target that is different from the physical network layer. A target's position in the logical layer is identified by its IP addresses. Targets located by their IP address depict how nodes in the physical layer correlate to form networks in cyberspace. Targeting in the logical layer requires the IP address and access to the logical network to deliver cyberspace effects. The ability of adversaries to change logical layer network configurations can complicate fires and effects against both logical and cyber-persona layer targets, but the operational benefit of affecting those targets often outweigh targeting challenges.

The inability to target a cyber-persona in a distinct area or form in the physical and logical network layers presents unique complexities. Because of these complexities, target positioning at the cyber-persona layer often requires multiple intelligence collection methods and an extensive analysis to develop insight and situational understanding to identify actionable targets. Like the logical network layer, cyber-personas can change quickly compared to changes in the physical network layer.

Electromagnetic Attack (EA) is exceptionally well suited to attack spectrum-dependent targets that are difficult to locate physically, cannot be accurately targeted for lethal fires, or require only temporary disruption. The fires support element plans, prepares, executes, and assesses fires supporting current and future operations by integrating coordinated lethal and nonlethal effects through the targeting process. Lethal and nonlethal effects include indirect fires, air and missile defense, joint fires, cyberspace attacks, and EA.

Targeting is a multidiscipline effort that requires coordinated interaction among the commander, the fires support element, and several staff sections that form the targeting working group. The commander prioritizes fires to the targeting working group and provides clear and concise guidance on effects expected from all fires, including cyberspace attacks and EA. Priority of fires is the commander's guidance to the staff, subordinate commanders, fires planners, and supporting agencies to employ fires in accordance with the relative importance of the unit's mission (FM 3-09). The targeting working group determines which targets to engage and how, where, and when to engage them based on the targeting guidance and priorities of the commander.

The targeting working group assigns lethal and nonlethal capabilities, including cyberspace attack and EA capabilities, to produce the desired effect on each target, ensuring compliance with the rules of engagement. The CEMA section participates in the targeting working group and provides recommendations for the employment of cyberspace and EMS-related actions against targets to meet the commander's intent and inclusion in the scheme of fires. Scheme of fires is the detailed, logical sequence of targets and fire support events to find and engage targets to accomplish the supported commander's objectives (JP 3-09).

The CEMA section works closely with the fires support element to coordinate and manage cyberspace and EW assets as part of the fire support plan. This process is called fire support coordination and is the planning and executing of fire so that targets are adequately covered by a suitable weapon or group of weapons (JP 3-09).

Targeting Functions

The G-2 or S-2, in collaboration with the CEMA section and the fires support element, detects, identifies, and locates targets through target acquisition. Effective employment of weapons, including EA and cyberspace attacks, require sufficient intelligence gained through target acquisition. The G-2 or S-2 conducts information collection to provide the fires support element, members of the targeting working group, and members of the targeting board with intelligence information used for targeting. This information includes threat cyberspace and EMS-enabled capabilities that require an individual or combined effect from lethal or nonlethal attacks.

Targeting Methodology

I Decide	II Detect	III Deliver	IV Assess
• Target Development • TVA • HPT and HVT • TSS • Attack Options • Attack Guidance	• Target Deception Means • Detection Procedures • Target Tracking	• Attack • Planned Targets • Targets of Opportunity • Desired Effects • Attack Systems	• Tactical Level • Operational Level • Restrike • Feedback

Ref: ADP 3-19, Fires (Jul '19) and ATP 3-60, Targeting (May '15).

Targeting occurs continuously throughout operations. Army targeting methodology consists of four functions: decide, detect, deliver, and assess (D3A). These targeting functions occur throughout the operations process. Commanders and staff should also be conversant with joint targeting methodology and understand how each of these processes and methodologies relate, because cyberspace operations and EW are usually coordinated by a joint force commander. Table 4-1, page 4-13, illustrates a crosswalk between the operations process, the joint targeting cycle, D3A, and military decision-making process.

Targeting Crosswalk

Ref: FM 3-12, Cyberspace Operations and Electromagnetic Warfare (Aug '21), table 4-1.

Operations Process	Joint Targeting Cycle	D3A	Military Decision-Making process	Targeting Tasks
Continuous Assessment	Plan — 1. Commander's Objectives, Targeting Guidance, and Intent.	Decide	Mission Analysis	• Perform target value analysis to develop fire support (including cyberspace, electromagnetic warfare, and information related capabilities) high-value targets. • Provide fire support, information-related capabilities, cyberspace, and electromagnetic warfare related input to the commander's targeting guidance and desired effects.
	Plan — 2. Target Development and Prioritization.	Decide	Course of Action Development	• Designate potential high-payoff targets. • Deconflict and coordinate potential high-payoff targets. • Develop a high-payoff target list. • Establish target selection standards. • Develop an attack guidance matrix. • Develop fire support, cyberspace, and electromagnetic warfare related tasks. • Develop associated measures of performance and measures of effectiveness.
	Plan — 3. Capabilities Analysis.	Decide	Course of Action Analysis	• Refine the high-payoff target list. • Refine the target selection standard. • Refine the attack guidance matrix. • Refine fire support tasks. • Refine associated measures of performance and measures of effectiveness.
	Plan — 4. Commander's Decision and Force Assignment.	Decide	Orders Production	• Finalize the high-payoff target list. • Finalize target selection standards. • Finalize the attack guidance matrix. • Finalize the targeting synchronization matrix. • Finalize fire support tasks. • Finalize associated measures of performance and measures of effectiveness. • Submit information requirements to battalion or brigade G-2/S-2.
	Prepare — 5. Mission Planning and Force Execution.	Detect		• Execute Information Collection Plan. • Update information requirements as they are answered. • Update the high-payoff target list, attack guidance matrix, and targeting synchronization matrix. • Update fire support, cyberspace, and electromagnetic warfare related tasks. • Update associated measures of performance and measures of effectiveness
	Execute — 6. Assessment	Deliver		• Execute fire support, cyberspace attacks, and electromagnetic attacks according to the attack guidance matrix and the targeting synchronization matrix.
	Assess — 6. Assessment	Assess		• Assess task accomplishment (as determined by measures of performance). • Assess effects (as determined by measures of effectiveness). • Refine fire support tasks and associated measures and reengage target if required

Legend:
D3A   decide, detect, deliver, and assess

I. Decide

The decide function is the first step of the targeting process. It begins with the military decision-making process and continues throughout an operation. The CEMA section conducts the following actions during the decide function of targeting—

- Threat cyberspace and EW-related capabilities and characteristics during target value analysis to identify high-value targets. A high-value target is a target the enemy commander requires for the successful completion of the mission (JP 3-60).
- Identifying potential cyberspace and EW-related HPTs. A high-payoff target is a target whose loss to the enemy will significantly contribute to the success of the friendly course of action (JP 3-60). A high-payoff target is a high-value target that must be acquired and successfully engaged for the success of the commander's mission.
- Specific targets that should be acquired and engaged using a cyberspace attack or EA capability and established target selection standards.
- Location and time that targets are likely to be found through intelligence operations and how long the target will remain fixed.
- Surveillance, reconnaissance, and target acquisition objectives for targets receiving cyberspace attacks or EA and determining if the unit has the necessary cyberspace attack or EA capabilities to deliver appropriate effects.
- Cyberspace and EMS-related IRs essential to the targeting effort.
- When, where, and with what priority should the targets be engaged, and what cyberspace attack or EA capability to employ for effects.
- The level of effectiveness that constitutes a successful cyberspace attack or EA and if the engagement achieved the commander's objective.
- If a cyberspace attack or EA can affect a target, and how and what type of cyberspace attack or EA can create the desired effect.
- How to obtain the information needed to assess a cyberspace attack or EA to determine success or failure, and who will receive and process it.
- Who will be the decision-making authority to determine the success or failure of a cyberspace attack or EA?
- What contingency action will occur if a cyberspace attack or EA is unsuccessful, and who has the authority to direct those actions?
- Identifying the unit's EW assets available for tasking and begin drafting FRAGOS.
- Drafting the RFS for OCO support to meet targeting requirements.
- Collaborating with units at higher, lower, and adjacent echelons for EW support to satisfy identified gaps in EW capabilities.
- Drafting the Joint Tactical Air Request for airborne EA and other necessary EW requesting forms, if required.
- Open communications with the higher command to receive updates on whether anticipated cyberspace attack and EA-related targets have been validated and added to the JTF headquarters' joint target list.
- Discussing cyberspace and EW-related risk that the commander will use to make risk determinations.
- Determining the level of authorities for the engaging targets using cyberspace and electromagnetic attacks.

During the decide function, the targeting working group identifies target restrictions that prohibit or restrict cyberspace attacks or EA on specified targets without approval from higher authorities. The sources of these restrictions include military risk, the law of war, rules of engagement, or other considerations. The JTF annotates entities within the AO prohibited from attack on the no-strike list and targets with restrictions on the restricted target list.

II. Detect

The detect function of the targeting process is the second step of the targeting process; during this step ES capabilities or other target acquisition assets locate and track a specified target to the required level of accuracy in time and space. During the detect function, the G-2 or S-2 coordinates with the targeting working group in developing the information collection plan. Before conducting the deliver function, the targeting team must establish measures of performance and measures of effectiveness for cyberspace and electromagnetic attacks to ensure they meet the commander's objectives.

The targeting working group focuses on the surveillance effort by identifying named areas of interest and target areas of interest integrated into the information collection plan. Named areas of interest are typically selected to capture indications of adversary courses of action but may be related to conditions of the OE.

The targeting working group identifies HPTs during planning and war-gaming. Target areas of interest that require specific engagements using cyberspace attack or EA capabilities differ from engagement areas. An engagement area is an area of concentration where a commander employs all available weapons to engage a target. In contrast, a target area of interest engagement uses a specific weapons system to engage a target. During the detect function, the CEMA section conducts the following—

- Provides cyberspace and EW-related IRs to determine HPTs that, when validated by the commander, are added to the priority intelligence requirement.
- Tasks EW assets, when required, to conduct electromagnetic reconnaissance to support information collection.
- Updates cyberspace attack and EA-related HVTs and HPTs.
- Determines if identified targets can be affected using OCO or EA (or both), and what type of EA capability can create the desired effect

    Note. The CEMA section alone cannot determine the type of cyberspace attack capability to use on targets. The CEMA section must coordinate with higher headquarters CEMA staff and appropriate joint cyberspace entities to develop an understanding of availability, feasibility, and suitability of specific cyberspace capabilities.

- Advocating for the nomination of cyberspace attack and EA-related targets to the JTF headquarters' joint integrated prioritized target list and the joint targeting cycle.
- Developing the RFS for OCO support.

III. Deliver

The deliver function of the targeting process executes the target engagement guidance and supports the commander's battle plan upon confirmation of the location and identity of HPTs. Close coordination between the CEMA section, intelligence, and fires support element is critical when detecting targets and delivering cyberspace attacks and EA. The fire support coordinator or fire support officer details fires coordination in the OPLAN or OPORD or target synchronization matrix.

IV. Assess

The assess function occurs throughout the operations process. During the assess function, targets are continuously refined and adjusted by the commander and staff in response to new or unforeseen situations presented during operations. Combat assessment measures the effectiveness of cyberspace attack and EA capabilities on the target and concludes with recommendations for reattack, continued attack, or to cease an attack. Recommendations for reattack, continued attack, and ceasing EA are combined G-3 or S-3 and intelligence functions approved by the commander.

For more information on the targeting cycle and target development process, refer to ATP 3-60.

Considerations When Targeting

Ref: FM 3-12, Cyberspace Operations and Electromagnetic Warfare (Aug '21), pp. 4-16 to 4-17.

The fires support element, in collaboration with the G-3 or S-3 and G-2 or S-2, uses targeting cycles and target development processes to select, prioritize, determine the type of effects, and duration of effects on targets. CEMA's planning, integrating, synchronizing, and assessing cyberspace operations and EW becomes apparent during the targeting process.

Characteristics of Cyberspace and EW Capabilities

Cyberspace capabilities are developed based on gathered intelligence and from operational and mission variables attained regarding an OE. In cyberspace operations, cyberspace forces consider such conditions as the type of computer operating system used by an enemy or adversary, the make and model of the hardware, the version of software installed on an enemy or adversary's computer, and the availability of cyberspace attack resources before creating effects on a target. EW capabilities are also developed based on gathered intelligence on operational and mission variables attained regarding an EMOE. In EW, targeting planners compare the types and capabilities of known spectrum-dependent devices that enemies use to the availability of EW resources before creating EW effects on a target. Targets include enemy spectrum-dependent devices carried by personnel and spectrum dependent systems used with or in weapons systems, sensory systems, facilities, and cyberspace capabilities that require the use of the EMS.

Cascading, Compounding, and Collateral Effects

The CEMA section should understand the overlaps amongst the military, other government, corporations, and private sectors in cyberspace. These overlaps are particularly important for estimates of possible cascading, compounding, or collateral effects when targeting enemy and adversary cyberspace capabilities. The same level of consideration is required when targeting enemy and adversary spectrum-dependent devices in the EMS.

Cyberspace capabilities can create effects beyond the geographic boundaries of an AO and a commander's area of interest. Employing cyberspace capabilities for attack or manipulation purposes within an area of interest require additional authorities beyond those given to a corps and below commander. Effects resulting from cyberspace attack operations can cause cascading effects beyond the targeted system that were not evident to the targeting planners. Cascading effects can sometimes travel through subordinate systems to attain access to the targeted system. Cascading effects can also travel through lateral or high-level systems to access a targeted system. Compounding effects are a gathering of various cyberspace effects that have interacted in ways that may have been either intended or unforeseen. Effects resulting from EA can cause cascading effects in the EMS beyond enemy or adversary's spectrum-dependent devices, disrupting or denying friendly forces access to the EMS throughout the EMOE. Collateral effects, including collateral damage, are the accidental cyberspace or EW effects of military operations on non-combatant and civilian cyberspace or EW capabilities that were not the intended target when implementing fires.

Reversibility of Effects

Targeting planners must consider the level of control that they can exercise throughout each cyberspace and electromagnetic attack. Categorization of reversibility of effects are—

- **Operator reversible effects.** These effects can be recalled, recovered, or terminated by friendly forces. Operator reversible effects typically represent a lower risk of undesired consequences, including discovery or retaliation.

- **Non-operator reversible effects.** These are effects that targeting planners cannot recall, recover, or terminate after execution. Non-operator reversible effects typically represent a higher risk of response from the threat or undesired consequences.

IV. Cyberspace (CEMA) in Operations Orders

Ref: FM 3-12, Cyberspace Operations and Electromagnetic Warfare (Aug '21), pp. A-14 to A-15.

OPLANs, OPORDs, FRAGORDs, and WARNORDs include cyberspace operations and EW information in various paragraphs and Annex C and Annex H. In OPLANs, OPORDs, and FRAGORDs, the scheme of CEMA is discussed in paragraphs 3.g. (Cyberspace Electromagnetic Activities); and 5.g. (Signal). In WARNORDS, cyberspace operations and EW information are in paragraph 5.g. (Signal).

Note. Paragraph 5g (Signal) has information regarding DODIN operations and spectrum management operations-related information.

Paragraph 3.g. (Cyberspace Electromagnetic Activities) describes how CEMA supports the concept of operations and refers the reader to Appendix 12 (Cyber Electromagnetic Activities) of Annex C (Operations) and Annex H (Signal) as required. Subdivision of Appendix 12 of Annex C and Annex H into the following cyberspace operations and EW-related information is as follows:

ANNEX C–OPERATIONS (G-5 OR G-3 [S-3])
- Appendix 12–Cyberspace Electromagnetic Activities (Electronic Warfare Officer)
- Tab A–Offensive Cyberspace Operations
- Tab B–Defensive Cyberspace Operations (RA & IDM)
- Tab C–Electromagnetic Attack
- Tab D–Electromagnetic Protection
- Tab E–Electronic Support

ANNEX H–SIGNAL (G-6 [S-6])
- Appendix 1—DODIN operations.
- Appendix 2—Voice, Video, and Data Network Diagrams.
- Appendix 3—Satellite Communications.
- Appendix 4—Foreign Data Exchanges.
- Appendix 5—Spectrum Management Operations (CEMA assisted).
- Appendix 6—Information Services.

Appendix 12 (Cyberspace Electromagnetic Activities) to Annex C (Operations) to Operations Plans and Orders

Appendix 12 to Annex C of OPLANs or OPORDs describes the cyberspace operations and EW divisions (EA, EP, and ES) supporting the commander's concept of operations. The CEWO is overall responsible for publishing Appendix 12 of Annex C and oversees the CEMA section in assisting the G-6 or S-2 with the development of Appendixes 1 and 6 of Annex H. Appendix 12 of Annex C describes the scheme of cyberspace operations and EW and CEMA integration and synchronization processes. It also includes cyberspace operations and EW-related constraints from higher headquarters.

See following pages (pp. 4-36 to 4-40) for a sample format for App. 12 to Annex C.

Appendix 12 to Annex C (Sample Format)

Ref: FM 3-12, Cyberspace Operations and Electromagnetic Warfare (Aug '21), p. A-16 to A-20.

[CLASSIFICATION]

Place the classification at the top and bottom of every page of the OPLAN or OPORD. Place the classification marking at the front of each paragraph and subparagraph in parentheses. Refer to AR 380-5 for classification and release marking instructions.

Copy ## of ## copies
Issuing headquarters
Place of issue
Date-time group of signature
Message reference number

Include the full heading if attachment is distributed separately from the base order or higher-level attachment.

APPENDIX 12 (CYBERSPACE ELECTROMAGNETIC ACTIVITIES) TO ANNEX C (OPERATIONS) TO OPERATION PLAN/ORDER [number] [(code name)]—[issuing headquarter] [(classification of title)]

(U) **References**: *Add any specific references to cyberspace electromagnetic activities, if needed.*

1. (U) Situation. *Include information affecting cyberspace and electronic warfare (EW) operations that paragraph 1 of Annex C (Operations) does not cover or that needs expansion.*

    a. (U) <u>Area of Interest</u>. *Include information affecting cyberspace and the electromagnetic spectrum (EMS); cyberspace may expand the area of local interest to a worldwide interest.*

    b. (U) <u>Area of Operations</u>. *Include information affecting cyberspace and the EMS; cyberspace may expand the area of operations outside the physical maneuver space.*

    c. (U) <u>Enemy Forces</u>. *List known and templated locations and cyberspace and EW unit activities for one echelon above and two echelons below the order. Identify the vulnerabilities of enemy information systems and cyberspace and EW systems. List enemy cyberspace and EW operations that will impact friendly operations. State probable enemy courses of action and employment of enemy cyberspace and EW assets. See Annex B (Intelligence) as required.*

    d. (U) <u>Friendly Forces</u>. *Outline the higher headquarters' cyberspace electromagnetic activities (CEMA) plan. List plan designation, location and outline of higher, adjacent, and other cyberspace and EW operations assets that support or impact the issuing headquarters or require coordination and additional support. Identify friendly cyberspace and EW operations assets and resources that affect the subordinate commander. Identify friendly forces cyberspace and EMS vulnerabilities. Identify friendly foreign forces with which subordinate commanders may operate. Identify potential conflicts within the EMS, especially for joint or multinational operations. Deconflict and prioritize spectrum distribution.*

    e. (U) <u>Interagency, Intergovernmental, and Nongovernmental Organizations</u>. *Identify and describe other organizations in the area of operations that may impact cyberspace and EW operations or implementation of cyberspace and EW operations specific equipment and tactics. See Annex V (Interagency) as required.*

[page number]
[CLASSIFICATION]

* 4-36 (Planning) IV. Cyberspace/CEMA in OPORDs

[CLASSIFICATION]

 f. (U) <u>Third Party</u>. *Identify and describe other organizations, both local and external to the area of operations that have the ability to influence cyberspace and EW operations or the implementation of cyberspace and EW operations specific equipment and tactics. This category includes criminal and non-state sponsored rogue elements.*

 g. (U) <u>Civil Considerations</u>. *Describe the aspects of the civil situation that impact cyberspace and EW operations. See Tab C (Civil Considerations) to Appendix 1 (Intelligence Estimate) to Annex B (Intelligence) and Annex K (Civil Affairs Operations) as required.*

 h. (U) <u>Attachments and Detachments</u>. *List units attached or detached only as necessary to clarify task organization. List any cyberspace and EW operations assets attached or detached, and resources available from higher headquarters. See Annex A (Task Organization) as required.*

 i. (U) <u>Assumptions</u>. *List any CEMA specific assumptions.*

1. (U) Mission. *State the commander's mission and describe cyberspace and EW operations to support the base plan or order.*

2. (U) Execution.

 a. <u>Scheme of Cyberspace Electromagnetic Activities</u>. *Describe how cyberspace and EW operations support the commander's intent and concept of operations. Establish the priorities of support to units for each phase of the operation. State how cyberspace and EW effects will degrade, disrupt, deny, and deceive the enemy. State the defensive and offensive cyberspace and EW measures. Identify target sets and effects, by priority. Describe the general concept for the integration of cyberspace and EW operations. List the staff sections, elements, and working groups responsible for aspects of CEMA. Include the cyberspace and EW collection methods for information developed in staff section, elements, and working groups outside the CEMA section and working group. Describe the plan for the integration of unified action and nongovernmental partners and organizations. See Annex C (Operations) as required. This section is designed to provide insight and understanding of the components of cyberspace and EW and how these activities are integrated across the operational plan. It is recommended that this appendix include an understanding of technical requirements. This appendix concentrates on the integration requirements for cyberspace and EW operations and references appropriate annexes and appendixes as needed to reduce duplication.*

  (1) (U) <u>Organization for Combat</u>. *Provide direction for the proper organization for combat, including the unit designation, nomenclature, and tactical task.*

  (2) (U) <u>Miscellaneous</u>. *Provide any other information necessary for planning not already mentioned.*

 b. (U) <u>Scheme of Cyberspace Operations</u>. *Describe how cyberspace operations support the commander's intent and concept of operations. Describe the general concept for the implementation of planned cyberspace operations measures. Describe the process to integrate unified action partners and nongovernmental organizations into operations, including cyberspace requirements and constraints. Identify risks associated with cyberspace operations. Include collateral damage, discovery, attribution, fratricide (to U.S. or allied or multinational networks or information), and possible conflicts. Describe actions that will prevent enemy and adversary action(s) to critically degrade the unified command's ability to effectively conduct military operations in its area of operations. Identify countermeasures and the responsible agency. List the warnings, and how they will be monitored. State how the cyberspace operations tasks will destroy, degrade, disrupt, and deny enemy computer networks. Identify and prioritize target sets and effect(s) in cyberspace. If appropriate, state how cyberspace operations support the accomplishment of*

Continued on next page

[CLASSIFICATION]

Continued from previous page

the operation. Identify plans to detect or assign attribution of enemy and adversary actions in the physical domains and cyberspace. Ensure subordinate units are conducting defensive cyberspace operations (DCO). Synchronize the CEMA section with the IO officer. Pass requests for offensive cyberspace operations (OCO) to higher headquarters for approval and implementation. Describe how DOD information network operations support the commander's intent and concept of operations. Synchronize DODIN operations with the G-6 (S-6). Prioritize the allocation of applications utilizing cyberspace. Ensure the employment of cyberspace capabilities where the primary purpose is to achieve objectives in or through cyberspace. Considerations should be made for degraded network operations. (Reference appropriate annexes and appendixes as needed to reduce duplication).

(1) (U) <u>DODIN Operations</u>. Describe how information operations are coordinated, synchronized, and support operations integrated with the G-6 (S-6) to design, build, configure, secure, operate, maintain, and sustain networks. See Annex H (Signal) as required.

(2) (U) <u>Defensive Cyberspace Operations</u>. Describe how DCO are conducted, coordinated, integrated, synchronized, and support operations to defend the DODIN-A and preserve the ability to utilize friendly cyberspace capabilities.

(3) (U) <u>Offensive Cyberspace Operations</u>. Describe how OCO are coordinated, integrated, synchronized, and support operations to achieve real time awareness and direct dynamic actions and response actions. Include target identification and operational pattern information, exploit and attack functions, and maintain intelligence information. Describe the authorities required to conduct OCO.

c. (U) <u>Scheme of Electromagnetic Warfare</u>. Describe how EW supports the commander's intent and concept of operations. Establish the priorities of support to units for each phase of the operation. State how the EW tasks will degrade, disrupt, deny, and deceive the enemy. Describe the process to integrate and coordinate unified action partner EW capabilities which support the commander's intent and concept of operations. State the electromagnetic attack, electromagnetic protection, and electromagnetic warfare support measures and plan for integration. Identify target sets and effects, by priority, for EW operations. Synchronize with IO officer. See the following attachments as required: Tab C, D, E (Electromagnetic Warfare) to Appendix 12 (Cyberspace Electromagnetic Activities); Appendix 15 (Information Operations of Annex C).

(1) (U) <u>Electromagnetic Attack</u>. Describe how offensive EW activities are coordinated, integrated, synchronized, and support operations. See Tab C (Electromagnetic Attack) to Appendix 12 (Cyberspace Electromagnetic Activities).

(2) (U) <u>Electromagnetic Protection</u>. Describe how defensive EW activities are coordinated, synchronized, and support operations. See Tab D (Electromagnetic Protection) to Appendix 12 (Cyberspace Electromagnetic Activities).

(3) (U) <u>Electromagnetic Warfare Support</u>. Describe how EW support activities are coordinated, synchronized, and support operations. See Tab E (Electromagnetic Warfare Support) to Appendix 12 (Cyberspace Electromagnetic Activities).

d. (U) <u>Scheme of Spectrum Management Operations</u>. Describe how spectrum management operations support the commander's intent and concept of operations. Outline the effects the commander wants to achieve while prioritizing spectrum management operations tasks. List the objectives and primary tasks to achieve those objectives. State the spectrum management, frequency assignment, host nation coordination, and policy implementation plan. Describe the plan for the integration of unified action partners' spectrum management operations capabilities. See Annex H (Signal) as required.

e. (U) <u>Tasks to Subordinate Units</u>. List cyberspace and EW operations tasks assigned to each subordinate unit not contained in the base order.

Continued from previous page

[Classification]

f. (U) <u>Coordinating Instructions</u>. List cyberspace and EW operations instructions applicable to two or more subordinate units not covered in the base order. Identify and highlight any cyberspace and EW operations specific rules of engagement, risk reduction control measures, environmental considerations, coordination requirements between units, and commander's critical information requirements and critical information that pertain to CEMA.

4. (U) <u>Sustainment</u>. Identify priorities of sustainment for cyberspace and EW operations key tasks and specify additional instructions as required. See Annex F (Sustainment) as required.

   a. (U) <u>Logistics</u>. Use subparagraphs to identify priorities and specific instruction for logistics pertaining to cyberspace and EW operations. See Appendix 1 (Logistics) to Annex F (Sustainment) and Annex P (Host Nation Support) as required.

   b. (U) <u>Personnel</u>. Use subparagraphs to identify priorities and specific instruction for human resources support pertaining to cyberspace and EW operations. See Appendix 2 (Personnel Services Support) to Annex F (Sustainment) as required.

   c. (U) <u>Health System Support</u>. See Appendix 3 (Army Health System Support) to Annex F (Sustainment) as required.

5. (U) <u>Command and Signal</u>.

   a. (U) <u>Command</u>.

      (1) (U) <u>Location of Commander.</u> State the location of key cyberspace and EW operations leaders.

      (2) (U) <u>Liaison Requirements</u>. State the cyberspace and EW operations liaison requirements not covered in the unit's SOPs.

   b. (U) <u>Control</u>.

      (1) (U) <u>Command Posts</u>. Describe the employment of cyberspace and EW operations specific command posts (CPs), including the location of each CP and its time of opening and closing.

      (2) (U) <u>Reports</u>. List cyberspace and EW operations specific reports not covered in SOPs. See Annex R (Reports) as required.

   c. (U) <u>Signal</u>. Address any cyberspace and EW operations specific communications requirements. See Annex H (Signal) as required.

ACKNOWLEDGE: Include only if attachment is distributed separately from the base order.

[Commander's last name]
[Commander's rank]

The commander or authorized representative signs the original copy of the attachment. If the representative signs the original, add the phrase "For the Commander." The signed copy is the historical copy and remains in the headquarters' files.

OFFICIAL:
[Authenticator's name]
[Authenticator's position]

Use only if the commander does not sign the original attachment. If the commander signs the original, no further authentication is required. If the commander does not sign, the signature of the preparing staff officer requires authentication and only the last name and rank of the commander appear in the signature block.

[page number]
[CLASSIFICATION]

[Classification]

ATTACHMENTS: *List lower level attachment (tabs and exhibits). If a particular attachment is not used, place "not used" beside the attachment number. Unit standard operating procedures will dictate attachment development and format. Common attachments include the following:*

APPENDIX 12 (CYBERSPACE ELECTROMAGNETIC ACTIVITIES) TO ANNEX C (OPERATIONS) TO OPERATION PLAN/ORDER [number] [(code name)]-[issuing headquarter] [(classification of title)]

ATTACHMENT: *List lower-level attachment (tabs and exhibits)*

Tab A -Offensive Cyberspace Operations
Tab B -Defensive Cyberspace Operations-Response Actions
Tab C -Electromagnetic Attack
Tab D -Electromagnetic Protection
Tab E -Electromagnetic Support

DISTRIBUTION: *Show only if distributed separately from the base order or higher-level attachments.*

[page number]
[CLASSIFICATION]

* 4-40 (Planning) IV. Cyberspace/CEMA in OPORDs

Chap 4

V. Cyberspace Integration into Joint Planning (JPP)

Ref: U.S. Army War College Strategic Cyberspace Operations Guide (Jun '16), chap. 3 and JP 5-0, Joint Planning (Dec '20), chap. III.

Joint planning is the deliberate process of determining how (the ways) to use military capabilities (the means) in time and space to achieve objectives (the ends) while considering the associated risks.

Joint Planning Process (JPP)

The joint planning process (JPP) is an orderly, analytical set of logical steps to frame a problem; examine a mission; develop, analyze, and compare alternative COAs; select the best COpA; and produce a plan or order. JPP helps commanders and their staffs organize their planning activities, share a common understanding of the mission and commander's intent, and develop effective plans and orders.

Joint Planning Process (JPP) Steps

  Planning Initiation

  Mission Analysis

  COA Development

  COA Analysis and Wargaming

  COA Comparison

  COA Approval

  Plan or Order Development

Ref: JFODS5-1: The Joint Forces Operations & Doctrine SMARTbook and JP 5-0.

Refer to JFODS5-1: The Joint Forces Operations & Doctrine SMARTbook (Guide to Joint, Multinational & Interorganizational Operations). Updated for 2019, topics include joint doctrine fundamentals (JP 1), joint operations (JP 3-0 w/Chg 1), an expanded discussion of joint functions, joint planning (JP 5-0), joint logistics (JP 4-0), joint task forces (JP 3-33), joint force operations (JPs 3-30, 3-31, 3-32 & 3-05), multinational operations (JP 3-16), interorganizational cooperation (JP 3-08), & more!

I. Cyberspace Planning Integration

Cyberspace Operations (CO) encompass more than just the network connections upon which the joint force relies. Cyberspace effects are created through the integration of cyberspace capabilities with air, land, maritime, and space capabilities. The boundaries within which CO are executed and the priorities and restrictions on its use should be identified in coordination between the commander, non-DOD government departments and agencies, and national leadership. Effects in cyberspace may have the potential to impact intelligence, diplomatic, and law enforcement (LE) efforts and therefore will often require coordination across the interagency. CO planners are presented the same considerations and challenges that are present in planning for other joint capabilities and functions, as well as some unique considerations. Targeting, deconfliction, commander's intent, political/military assessment, and collateral effects considerations all play into the calculations of the CO planner's efforts. In a similar fashion, all of the principles of joint operations, such as maneuver and surprise, are germane to CO.

However, second and higher order effects in and through cyberspace can be more difficult to predict, necessitating more branches and sequels in plans. Further, while many elements of cyberspace can be mapped geographically in the physical domains, a full understanding of an adversary's posture and capabilities in cyberspace involves understanding the underlying network infrastructure, a clear understanding of what friendly forces or capabilities might be targeted and how, and an understanding of applicable domestic, foreign, and international laws and policy. Adversaries in cyberspace may be nation states, groups, or individuals, and the parts of cyberspace they control are not necessarily either within the geographic borders associated with the actor's nationality, or proportional to the actor's geopolitical influence. A criminal element, a politically motivated group, or even an individual may have a greater presence and capability in cyberspace than many nations do today. Regardless of what operational phase may be underway, it is always important to determine what authorities are required to execute CO. Cyberspace planners must account for the lead time to acquire the authorities needed to implement the desired cyberspace capabilities. This does not change the commander's planning fundamentals, but does emphasize the importance of coordination with interagency partners, who may have authorities that are different from DOD. Despite the additional considerations and challenges of integrating CO in commander planning, planners can use many elements of the traditional processes to implement the commander's intent and guidance.

II. Cyberspace Planning and the JPP

Cyberspace operations capability considerations and options are integrated into JPP, just like all other joint capabilities and functions.

A. Initiation

During the receipt of mission, cyber planners participate in the commander's initial assessment actions and gathers the resources required for mission analysis. Unique to cyberspace, part of the initial assessment determines whether resources can be brought to bear on the mission at hand within a reasonable timeframe or context through the reachback and support processes.

B. Mission Analysis

Cyberspace planners contribute to mission analysis in order to help commanders understand the operational environment and frame the problem. An effective mission analysis considers the potential impact cyberspace on an operational environment. Cyberspace planners do this by participating in planning actions that help form the problem statement, mission statement, commander's intent, planning guidance, initial commander's critical information requirements, essential elements of friendly information, and updated running estimates. Cyberspace planners coordinate with the intelligence directorate (J-2), operations directorate (J-3), communications system directorate (J-6), and other staff elements in reference to mission critical systems, risk assessments, current defense posture, and overall operational requirements. When utilized as an information-related capability the cyberspace planners work closely with the information operations (IO) staff to identify the desired effects for the information environment.

Cyberspace planners further contribute to overall mission analysis by participating in the intelligence preparation of the environment and closely coordinate with the intelligence directorate (J-2) by providing information, advice, and assistance. This ensures the intelligence staff understands what cyberspace products are needed in order for to tailor intelligence preparation of the battlefield products. Threats and vulnerabilities are identified in accordance with adversary offensive cyberspace capabilities. A friendly center of gravity analysis is conducted to ensure thorough planning. A key portion of this analysis is to assess the potential impact of cyberspace operations on friendly assets.

Cyberspace planners then analyze the commander's intent and mission from a cyberspace perspective and determine if cyberspace capabilities are available to accomplish the identified tasks. If organic assets are insufficient, planners draft cyberspace effects requests using the cyberspace effects request form (CERF). A cyberspace support element may be required to support the organic cyberspace planning team.

C. Course of Action (COA) Development

The cyberspace planning team contributes to COA development by determining possible friendly and enemy operations and which friendly Cyberspace capabilities are available to support the operations. Cyberspace planners focus their efforts on achieving an operational advantage at the decision point of each COA. By the conclusion of the COA development, the Cyberspace planners generate a list of cyberspace actions that will accomplish the commander's objectives and desired effects. The team also generates a list of capabilities, information, and intelligence required to perform the tasks for each COA.

D. COA Analysis, Comparison, and Approval

During COA analysis the cyberspace planning team coordinates with each of the warfighting function staff members to integrate and synchronize CO into each COA, thereby identifying which COA best accomplishes the mission. The cyberspace planners address how CO capabilities support each COA and apply them to timelines, critical events, and decision points. During COA comparison all staff members evaluate the advantages and disadvantages of each COA from their perspectives. The cyberspace planner present their findings for the others' consideration. At the conclusion of the COA comparison, the cyberspace planning team generates a list of pros and cons for each COA relative to cyberspace. They also develop a prioritized list of the COAs from a cyberspace perspective. The commander's final guidance provides the cyberspace planners with the commander's intent, any new critical information requirements, risk acceptance, and guidance on the priorities for the elements of combat power, orders preparation, rehearsal, and preparation.

E. Plan or Order Development

Cyberspace planners provide the appropriate input for several sections of the operation order or plan and associated annexes or appendixes as required. This may include input to other functional area annexes such as intelligence, fire support, signal, and civil affairs operations as required.

IV. Cyberspace-Related Intelligence Requirements (IRs)

During mission analysis, the joint force staff identifies significant gaps in what is known about the adversary and other relevant aspects of the operational environment (OE) and formulates IRs. IRs are general or specific subjects upon which there is a need for the collection of information or the production of intelligence. Based on the command's IRs, the intelligence staff develops more specific questions known as information requirements (those items of information that must be collected and processed to develop the intelligence required by the commander). Information requirements related to cyberspace may include: network infrastructures, personnel status and readiness of adversaries' equipment, and unique cyberspace signature identifiers such as software/firmware versions, configuration files, etc.

V. Information Operations (IO)

Cyberspace Operations are one of several information-related capabilities (IRCs) available to the commander. Cyberspace capabilities, when in support of IO, deny or manipulate adversary or potential adversary decision making, through targeting an information medium (such as a wireless access point in the physical dimension), the message itself (an encrypted message in the information dimension), or a cyber-persona (an online identity that facilitates communication, decision making, and the influencing of audiences in the cognitive dimension). When employed in support of IO, CO generally focus on the integration of offensive and defensive capabilities exercised in and through cyberspace, in concert with other IRCs, and coordination across multiple lines of operation and lines of effort.

See pp. 4-45 to 4-50 for discussion of the integrating/coordinating functions of information operations (IO).

VI. Planning Insights

Gaining insight and understanding of available cyberspace capabilities, from the experts listed above, enables planners to merge these capabilities with the other domains.

- Avoid symmetric thinking. Merely because the adversary attacks through cyberspace, does not restrict us to solely cyberspace response options. Commanders and staffs should consider attaching the Cyberspace physical layer as well as conducting operations 'in' cyberspace.

- Identify potential cyberspace needs early Cyberspace capabilities require long approval chains and, sometimes, long development timelines.

- Tailor requests for cyberspace operations. Given cyberspace operations' global nature and potential for cascading effects, authorities rarely grant broad permissions. Planners should craft requirements which are specific (used only in certain situations, limited in duration, and limited networks affected).

- Conducting cyberspace damage assessment is often difficult. A friendly cyberspace operator may report mission accomplishment. However, unlike physical munitions, there will not be a blast crater to verify results. Planners must use other ways to the measure success of a cyberspace operation. One approach is to layer assessments.

- All cyberspace operations require branch plans to accomplish similar effects. Because offensive cyberspace operations (OCO) are often disapproved and susceptible to failure, planners must understand the intent of those cyberspace operations and develop a branch plan to accomplish that intent through other domains.

- Many cyberspace capabilities are classified to avoid exposing vulnerabilities. Lack of sufficient security clearances will hinder a planner's ability to integrate cyberspace capabilities.

Chap 4
VI. Integrating / Coordinating Functions of IO

Ref: JP 3-13 w/change 1, Information Operations (Nov '14), chap. II.

I. Information Operations (IO) and the Information-Influence Relational Framework

Influence is at the heart of diplomacy and military operations, with integration of information-related capabilities (IRCs) providing a powerful means for influence. This section addresses how the integrating and coordinating functions of information operations (IO) help achieve a JFC's objectives. Through the integrated application of IRCs, the relationships that exist between IO and the various IRCs should be understood in order to achieve an objective.

See pp. 0-10 to 0-15 for related discussion of the information environment, information as a joint function, and information operations (IO).

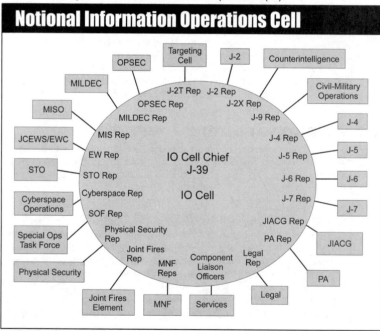

Ref: JP 3-13 (with change 1), Information Operations, fig. II-3, p. II-6.

Refer to INFO1: The Information Operations & Capabilities SMARTbook (Guide to Information Operations & the IRCs). INFO1 chapters and topics include information operations (IO defined and described), information in joint operations (joint IO), information-related capabilities (PA, CA, MILDEC, MISO, OPSEC, CO, EW, Space, STO), information planning (information environment analysis, IPB, MDMP, JPP), information preparation, information execution (IO working group, IO weighted efforts and enabling activities, intel support), fires & targeting, and information assessment.

(Planning) VI. Integrating/Coordinating IO Functions 4-45

II. The Information Operations Staff and Information Operations Cell

Within the joint community, the integration of IRCs to achieve the commander's objectives is managed through an IO staff or IO cell. JFCs may establish an IO staff to provide command-level oversight and collaborate with all staff directorates and supporting organizations on all aspects of IO. Most CCMDs include an IO staff to serve as the focal point for IO. Faced with an ongoing or emerging crisis within a geographic combatant commander's (GCC's) area of responsibility (AOR), a JFC can establish an IO cell to provide additional expertise and coordination across the staff and interagency.

IO Staff

In order to provide planning support, the IO staff includes IO planners and a complement of IRCs specialists to facilitate seamless integration of IRCs to support the JFC's concept of operations (CONOPS). IRC specialists can include, but are not limited to, personnel from the EW, cyberspace operations (CO), military information support operations (MISO), civil-military operations (CMO), military deception (MILDEC), intelligence, and public affairs (PA) communities. They provide valuable linkage between the planners within an IO staff and those communities that provide IRCs to facilitate seamless integration with the JFC's objectives.

IO Cell

The IO cell integrates and synchronizes IRCs, to achieve national or combatant commander (CCDR) level objectives. Normally, the chief of the CCMD's IO staff will serve as the IO cell chief; however, at the joint task force level, someone else may serve as the IO cell chief. The IO cell comprises representatives from a wide variety of organizations to coordinate and integrate additional activities in support of a JFC. It may include representatives from organizations outside DOD, even allied or multinational partners.

III. Relationships and Integration

IO is not about ownership of individual capabilities but rather the use of those capabilities as force multipliers to create a desired effect. There are many military capabilities that contribute to IO and should be taken into consideration during the planning process.

> **Commander's Communications Synchronization (CCS)**
>
> Commander's Communication Synchronization (CCS) entails focused efforts to create, strengthen, or preserve conditions favorable for the advancement of national interests, policies, and objectives by understanding and communicating with key audiences through the use of coordinated information, themes, messages, plans, programs, products and actions, synchronized with the other instruments of national power.
>
> *Refer to Joint Doctrine Note 2-13, CCS (Dec '13) for more information.*

A. Strategic Communication (SC)

The SC process consists of focused United States Government (USG) efforts to create, strengthen, or preserve conditions favorable for the advancement of national interests, policies, and objectives by understanding and engaging key audiences through the use of coordinated programs, plans, themes, messages, and products synchronized with the actions of all instruments of national power. SC is a whole-of-government approach, driven by interagency processes and integration that are focused upon effectively communicating national strategy.

B. Joint Interagency Coordination Group (JIACG)

Ref: JP 3-13 w/change 1, Information Operations (Nov '14), fig. II-4, p. II-8.

Interagency coordination occurs between DOD and other USG departments and agencies, as well as with private-sector entities, nongovernmental organizations, and critical infrastructure activities, for the purpose of accomplishing national objectives. Many of these objectives require the combined and coordinated use of the diplomatic, informational, military, and economic instruments of national power. Due to their forward presence, the CCMDs are well situated to coordinate activities with elements of the USG, regional organizations, foreign forces, and host nations. In order to accomplish this function, the GCCs have established JIACGs as part of their normal staff structures.

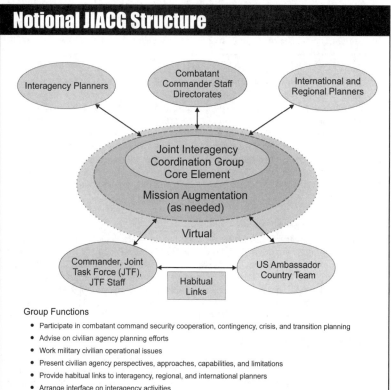

Notional JIACG Structure

Group Functions
- Participate in combatant command security cooperation, contingency, crisis, and transition planning
- Advise on civilian agency planning efforts
- Work military civilian operational issues
- Present civilian agency perspectives, approaches, capabilities, and limitations
- Provide habitual links to interagency, regional, and international planners
- Arrange interface on interagency activities
- Coordinate with regional players

Ref: JP 3-13 (with change 1), Information Operations, fig. II-4, p. II-8.

The JIACG is well suited to help the IO cell with interagency coordination. Although IO is not the primary function of the JIACG, the group's linkage to the IO cell and the rest of the interagency is an important enabler for synchronization of guidance and IO.

(Planning) VI. Integrating/Coordinating IO Functions 4-47

Editor's note: See pp. 0-14 to 0-15 for related discussion of these joint force capabilities, operations, and activities for leveraging information from JP 3-0 with Change 1 (Oct '18).

C. Public Affairs (PA)

PA comprises public information, command information, and public engagement activities directed toward both the internal and external publics with interest in DOD. External publics include allies, neutrals, adversaries, and potential adversaries. When addressing external publics, opportunities for overlap exist between PA and IO.

By maintaining situational awareness between IO and PA the potential for information conflict can be minimized. The IO cell provides an excellent place to coordinate IO and PA activities that may affect the adversary or potential adversary. Because there will be situations, such as counterpropaganda, in which the TA for both IO and PA converge, close cooperation and deconfliction are extremely important. Such coordination and deconfliction efforts can begin in the IO cell. However, since it involves more than just IO equities, final coordination should occur within the joint planning group (JPG).

D. Civil-Military Operations (CMO)

CMO is another area that can directly affect and be affected by IO. CMO activities establish, maintain, influence, or exploit relations between military forces, governmental and nongovernmental civilian organizations and authorities, and the civilian populace in a friendly, neutral, or hostile operational area in order to achieve US objectives. These activities may occur prior to, during, or subsequent to other military operations. In CMO, personnel perform functions normally provided by the local, regional, or national government, placing them into direct contact with civilian populations. This level of interaction results in CMO having a significant effect on the perceptions of the local populace. Since this populace may include potential adversaries, their perceptions are of great interest to the IO community. For this reason, CMO representation in the IO cell can assist in identifying TAs; synchronizing communications media, assets, and messages; and providing news and information to the local population.

Although CMO and IO have much in common, they are distinct disciplines. The TA for much of IO is the adversary; however, the effects of IRCs often reach supporting friendly and neutral populations as well. In a similar vein, CMO seeks to affect friendly and neutral populations, although adversary and potential adversary audiences may also be affected. This being the case, effective integration of CMO with other IRCs is important, and a CMO representative on the IO staff is critical. The regular presence of a CMO representative in the IO cell will greatly promote this level of coordination.

E. Cyberspace Operations

Cyberspace is a global domain within the information environment consisting of the interdependent network of information technology infrastructures and resident data, including the Internet, telecommunications networks, computer systems, and embedded processors and controllers. CO are the employment of cyberspace capabilities where the primary purpose is to achieve objectives in or through cyberspace. Cyberspace capabilities, when in support of IO, deny or manipulate adversary or potential adversary decision making, through targeting an information medium (such as a wireless access point in the physical dimension), the message itself (an encrypted message in the information dimension), or a cyber-persona (an online identity that facilitates communication, decision making, and the influencing of audiences in the cognitive dimension). When employed in support of IO, CO generally focus on the integration of offensive and defensive capabilities exercised in and through cyberspace, in concert with other IRCs, and coordination across multiple lines of operation and lines of effort.

F. Information Assurance (IA)

IA is necessary to gain and maintain information superiority. The JFC relies on IA to protect infrastructure to ensure its availability, to position information for influence, and for delivery of information to the adversary. Furthermore, IA and CO are interrelated and rely on each other to support IO.

G. Space Operations

Space capabilities are a significant force multiplier when integrated with joint operations. Space operations support IO through the space force enhancement functions of intelligence, surveillance, and reconnaissance; missile warning; environmental monitoring; satellite communications; and space-based positioning, navigation, and timing. The IO cell is a key place for coordinating and deconflicting the space force enhancement functions with other IRCs.

H. Military Information Support Operations (MISO)

MISO are planned operations to convey selected information and indicators to foreign audiences to influence their emotions, motives, objective reasoning, and ultimately the behavior of foreign governments, organizations, groups, and individuals. MISO focuses on the cognitive dimension of the information environment where its TA includes not just potential and actual adversaries, but also friendly and neutral populations. MISO are applicable to a wide range of military operations such as stability operations, security cooperation, maritime interdiction, noncombatant evacuation, foreign humanitarian operations, counterdrug, force protection, and counter-trafficking. Given the wide range of activities in which MISO are employed, the military information support representative within the IO cell should consistently interact with the PA, CMO, JIACG, and IO planners.

I. Intelligence

Intelligence is a vital military capability that supports IO. The utilization of information operations intelligence integration (IOII) greatly facilitates understanding the interrelationship between the physical, informational, and cognitive dimensions of the information environment.

By providing population-centric socio-cultural intelligence and physical network lay downs, including the information transmitted via those networks, intelligence can greatly assist IRC planners and IO integrators in determining the proper effect to elicit the specific response desired. Intelligence is an integrated process, fusing collection, analysis, and dissemination to provide products that will expose a TA's potential capabilities or vulnerabilities. Intelligence uses a variety of technical and nontechnical tools to assess the information environment, thereby providing insight into a TA.

A joint intelligence support element (JISE) may establish an IO support office to provide IOII. This is due to the long lead time needed to establish information baseline characterizations, provide timely intelligence during IO planning and execution efforts, and to properly assess effects in the information environment. In addition to generating intelligence products to support the IO cell, the JISE IO support office can also work with the JISE collection management office to facilitate development of collection requirements in support of IO assessment efforts.

J. Military Deception (MILDEC)

One of the oldest IRCs used to influence an adversary's perceptions is MILDEC. MILDEC can be characterized as actions executed to deliberately mislead adversary decision makers, creating conditions that will contribute to the accomplishment of the friendly mission. While MILDEC requires a thorough knowledge of an adversary or potential adversary's decision-making processes, it is important to remember that it is focused on desired behavior. It is not enough to simply mislead the adversary or potential adversary; MILDEC is designed to cause them to behave in a manner advantageous to the friendly mission, such as misallocation of resources, attacking at a time and place advantageous to friendly forces, or avoid taking action at all.

Refer to The Battle Staff SMARTbook, 6th Edition (BSS6) for further discussion.

K. Operations Security (OPSEC)

OPSEC is a standardized process designed to meet operational needs by mitigating risks associated with specific vulnerabilities in order to deny adversaries critical information and observable indicators. OPSEC identifies critical information and actions attendant to friendly military operations to deny observables to adversary intelligence systems. Once vulnerabilities are identified, other IRCs (e.g., MILDEC, CO) can be used to satisfy OPSEC requirements. OPSEC practices must balance the responsibility to account to the American public with the need to protect critical information. The need to practice OPSEC should not be used as an excuse to deny noncritical information to the public.

The effective application, coordination, and synchronization of other IRCs are critical components in the execution of OPSEC. Because a specified IO task is "to protect our own" decision makers, OPSEC planners require complete situational awareness, regarding friendly activities to facilitate the safeguarding of critical information.

L. Special Technical Operations (STO)

IO need to be deconflicted and synchronized with STO. Detailed information related to STO and its contribution to IO can be obtained from the STO planners at CCMD or Service component headquarters. IO and STO are separate, but have potential crossover, and for this reason an STO planner is a valuable member of the IO cell.

M. Joint Electromagnetic Spectrum Operations (JEMSO)

All information-related mission areas increasingly depend on the electromagnetic spectrum (EMS). JEMSO, consisting of EW and joint EMS management operations, enable EMS-dependent systems to function in their intended operational environment. EW is the mission area ultimately responsible for securing and maintaining freedom of action in the EMS for friendly forces while exploiting or denying it to adversaries. JEMSO therefore supports IO by enabling successful mission area operations.

See chap. 3, Electronic Warfare, and chap. 5, Spectrum Management Operations (SMO/JEMSO).

N. Key Leader Engagement (KLE)

KLEs are deliberate, planned engagements between US military leaders and the leaders of foreign audiences that have defined objectives, such as a change in policy or supporting the JFC's objectives. These engagements can be used to shape and influence foreign leaders at the strategic, operational, and tactical,levels, and may also be directed toward specific groups such as religious leaders, academic leaders, and tribal leaders; e.g., to solidify trust and confidence in US forces.

Refer to INFO1: The Information Operations & Capabilities SMARTbook (Guide to Information Operations & the IRCs). INFO1 chapters and topics include information operations (IO defined and described), information in joint operations (joint IO), information-related capabilities (PA, CA, MILDEC, MISO, OPSEC, CO, EW, Space, STO), information planning (information environment analysis, IPB, MDMP, JPP), information preparation, information execution (IO working group, IO weighted efforts and enabling activities, intel support), fires & targeting, and information assessment.

I. Spectrum Management Operations (SMO/JEMSO)

Ref: FM 3-12, Cyberspace & Electronic Warfare Operations (Apr '17), pp. 1-25 & 1-34; JP 3-85, Joint Electromagnetic Spectrum Management Operations (May '20); and ATP 6-02.70, Techniques for Spectrum Management Operations (Oct '19).

I. Electromagnetic Spectrum Operations (EMSO)

Electromagnetic Spectrum Operations (EMSO) are comprised of electronic warfare (EW) and spectrum management operations (SMO). The importance of the EMS and its relationship to the operational capabilities of the Army is the focus of EMSO. EMSO include all activities in military operations to successfully control the EMS. Figure 1-8 illustrates EMSO and how they relate to SMO and EW.

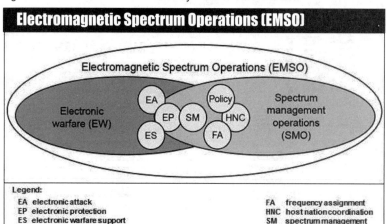

Ref: FM 3-12, Cyberspace & Electronic Warfare Operations (Apr '17), figure 1-8. Electromagnetic spectrum operations. See also chap. 3, Electronic Warfare.

Spectrum Management Operations (Army) *See p. 5-4.*
Spectrum Management Operations (SMO) consists of the interrelated functions of spectrum management, frequency assignment, host nation coordination, and policy that together enable the planning, management, and execution of operations within the electromagnetic operational environment (EMOE), during all phases of military operations (FM 6-02).

Joint Electromagnetic Spectrum Operations (JEMSO) *See p. 5-5.*
JEMSO are military actions undertaken by a joint force to exploit, attack, protect, and manage the EMOE. These actions include/impact all joint force transmissions and receptions of electromagnetic (EM) energy. JEMSO are offensively and defensively employed to achieve unity of effort and the commander's objectives. JEMSO integrate and synchronize electromagnetic warfare (EW), EMS management, and intelligence, as well as other mission areas, to achieve EMS superiority.

See following pages (pp. 5-2 to 5-3) for an overview and discussion of the electromagnetic operational environment (EMOE).

(SMO/JEMSO) I. Overview 5-1 *

Electromagnetic Operational Environment (EMOE)

Ref: JP 3-85, Joint Electromagnetic Spectrum Management Operations (May '20); pp. I-2 to 1-3.

Electromagnetic Spectrum (EMS)

The EMS is a maneuver space consisting of all frequencies of EM radiation (oscillating electric and magnetic fields characterized by frequency and wavelength). The EMS is often organized by frequency bands, based on certain physical characteristics. The EMS includes radio waves, microwaves, infrared (IR) radiation, visible light, ultraviolet radiation, x-rays, and gamma rays.

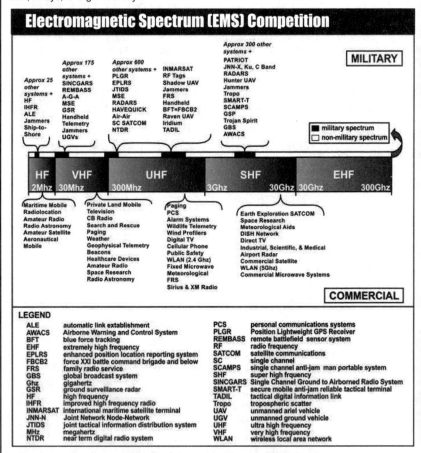

Ref: ATP 6-02.70, Techniques for Spectrum Management Operations (Oct '19), fig. 1-3. Electromagnetic spectrum competition.

Electromagnetic Environment (EME)

The EME is the actual EM radiation encountered in a particular operational area (OA). The EME is the resulting product of the power and time distribution, in various frequency ranges, of the radiated or conducted EM emission levels encountered by a military force, system, or platform when performing its mission in its intended OE. It is important to note that not all EM radiation encountered by joint forces will impact operations.

Electromagnetic Operational Environment (EMOE)

The EMOE is a composite of the actual and potential EM radiation, conditions, circumstances, and influences that affect the employment of capabilities and the decisions of the commander. It includes the existing background radiation (i.e., EME) as well as the friendly, neutral, adversary, and enemy EM systems able to radiate within the EM area of influence. This includes systems currently radiating or receiving, or those that may radiate, that can potentially affect joint operations.

The EMOE has the following attributes:

Physical. The EMOE is part of the physical environment. EM radiation is a physical phenomenon. Both natural and manmade factors (e.g., terrain, weather, atmospheric conditions, sea state, transmitters, power lines, static electricity) influence EM radiation and the organizations and systems that employ it. Military forces maneuver through all environments, including the EMOE, to gain positions of advantage over adversaries and enemies. EMOE maneuver requires effective management of spectrum occupancy.

Pervasive. The EMS permeates all parts of the OE. Military forces use the EMOE to integrate, synchronize, and otherwise enhance their operations. The critical dependencies of modern military operations on EMS activities, coupled with the wide range of effects that can be created through electromagnetic spectrum operations (EMSO), are a potent force multiplier.

Constrained. Although the EMS contains an unlimited number of frequencies, its use for military purposes is limited by physics, policy, and current technology. EM radiation has unique physical properties that dictate its use (e.g., short- or long-range communications, sensing). Additionally, use of the EMS is subject to international treaties and laws, as well as nation-state laws and regulations. Technology bounds those portions of the EMS that are accessible and exploitable (i.e., advances in technological capabilities will result in expanded use of the EMS).

Congested. The EMOE encountered by joint forces is congested due to military and nonmilitary use, resulting in a commensurate increase in the number and density of EM emitters. As a result of physical characteristics and technology, civilian and military organizations increasingly seek to transmit and receive EM energy in the same or adjacent spectral bands. For instance, myriad stakeholders (e.g., cell phone and wireless Internet providers, media) continue to expand their EMS bandwidth requirements, reducing the open EM areas conducive to joint force maneuver. This congestion leads to electromagnetic interference (EMI) to a receiver. EMI is any EM disturbance, induced intentionally or unintentionally, that interrupts, obstructs, or otherwise degrades or limits the effective performance of EMS-dependent systems, electronics, and electrical equipment.

Contested. Since modern military operations are critically dependent on the EMS, a key goal of our adversaries and enemies is to deny our ability to use it successfully. For example: antiradiation missiles and other destructive weapons are used to degrade or destroy transmitters and receivers, while EM energy can be used to disrupt or degrade a receiver's operation.

Dynamic. The EMOE experienced by the joint force is continuously changing, as existing systems are modified, new systems are deployed, units change locations, threats transmit, or natural phenomena occur. Since EM energy travels at the speed of light, military activities in the EMS may provide a decisive advantage by enabling commanders to make decisions, conduct operations, and create effects more rapidly than the threat. Agility in spectrum operations provides joint force operations the flexibility and adaptability to achieve mission success in dynamic EMOEs.

II. Spectrum Management Operations (SMO)

Spectrum management operations (SMO) consists of the interrelated functions of spectrum management, frequency assignment, host nation coordination, and policy that together enable the planning, management, and execution of operations within the electromagnetic operational environment (EMOE), during all phases of military operations (FM 6-02). SMO includes all activities in military operations to manage the electromagnetic spectrum. SMO is the management function of electromagnetic spectrum operations (EMSO). SMO aim to manage resources within the EMOE while resolving electromagnetic interference (EMI) by conducting EMI analysis and resolution activities. Figure 1-1 depicts the various responsibilities related to spectrum management operations as they pertain to the EMOE.

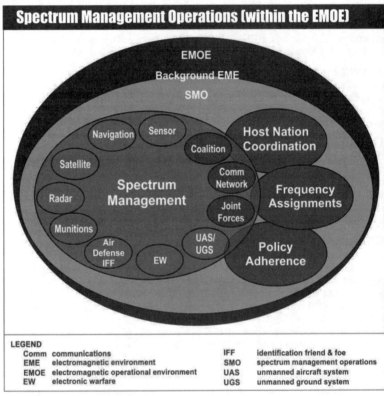

Ref: ATP 6-02.70, Techniques for Spectrum Management Operations (Oct '19), fig.1-1. Spectrum management operations within the EMOE.

Spectrum managers coordinate and collaborate with spectrum managers working in joint environments. Collaboration with joint personnel and coalition partners is common practice necessary for the Army spectrum manager while using the highly saturated and limited spectrum available. In the joint environment, joint electromagnetic spectrum operations encompass joint electromagnetic spectrum management operations and electronic warfare with the same intent as the Army's electromagnetic spectrum operations.

See pp. 5-9 to 5-14 for further discussion of spectrum management operations.

A. Objective of Spectrum Management Operations

SMO aims to ensure access to the electromagnetic spectrum in support of the Army's operational missions. SMO is a supporting function or enabler for unified land operations. SMO is an enabler for cyberspace electromagnetic activities (CEMA). Spectrum management is the operational, engineering, and administrative procedures to plan, coordinate, and manage use of the EMS and enables cyberspace, signal and electronic warfare (EW) operations.

SMO enables management of allotted and limited frequencies directly supporting operational forces throughout the world. The Army is dependent upon the use of the electromagnetic spectrum at all levels of unified land operations. An effective SMO program enables electronic systems to perform their functions in the intended environment without causing EMI.

Commanders must have the ability to see the use of their assigned spectrum resources so they can apply precise command and control (C2). The electromagnetic spectrum is a vital warfighting resource that requires the same planning and management as other critical resources such as fuel, water, and ammunition. Spectrum managers, with the appropriate expertise and tools, ensure that commanders have adequate knowledge of the utilization of the frequency spectrum to make decisions that positively influence the accomplishment of their missions.

B. Spectrum Management Operations Core Functions

SMO core functions determine the tasks and requirements of the spectrum manager. These four functions are—

- Spectrum management.
- Frequency assignment.
- Host nation coordination.
- Policy adherence.

See p. 2-13 for further discussion of the EMS and SMO from FM 3-12. For more information on Army SMO, refer to FM 6-02 and ATP 6-02.70.

III. Joint Electromagnetic Spectrum Operations (JEMSO)

JEMSO are military actions undertaken by a joint force to exploit, attack, protect, and manage the EMOE. These actions include/impact all joint force transmissions and receptions of electromagnetic (EM) energy. JEMSO are offensively and defensively employed to achieve unity of effort and the commander's objectives. JEMSO integrate and synchronize electromagnetic warfare (EW), EMS management, and intelligence, as well as other mission areas, to achieve EMS superiority.

JEMSO support military operations throughout the competition continuum to achieve desired objectives and attain end states. During peacetime, JEMSO are conducted to ensure adequate access to the EMS and may include deconflicting use of the EMS between joint users and coordinating with a host nation (HN). As a crisis escalates toward armed conflict, JEMSO shift from EMS access coordination to EMS superiority, with coordinated military actions executed to exploit, attack, protect, and manage the electromagnetic operational environment (EMOE).

See pp. 5-15 to 5-28 for discussion of planning joint electromagnetic spectrum operations (JEMSO).

A. JEMSO Actions

Ref: JP 3-85, Joint Electromagnetic Spectrum Management Operations (May '20); pp. I-2 to 1-3.

JEMSO actions to exploit, attack, protect, and manage the EMOE rely on personnel and systems from EW, EMS management, intelligence, space, and cyberspace mission areas. Instead of these mission areas being planned and executed in a minimally coordinated and stovepiped fashion, JEMSO guidance and processes prioritize, integrate, synchronize, and deconflict all joint force operations in the EMOE, enhancing unity of effort. The result is a fully integrated scheme of maneuver in the EMOE to achieve EMS superiority and joint force commander (JFC) objectives.

Exploitation

Exploitation takes full advantage of available information for tactical, operational, or strategic purposes. In a JEMSO context, exploitation refers to EMS systems capable of sensing the EMOE. Sensing systems support intelligence collection, SA, targeting, and warning. EMS sensors can be active (e.g., air-to-air radars, IFF interrogators) or passive (e.g., radar warning receivers, passive radars, IR weapons seekers). These sensing missions are typically executed through signals intelligence (SIGINT) and electromagnetic support (ES) operations.

Electronic Attack (EA)

JEMSO capabilities can directly produce effects in the EMOE. These capabilities can be used to deny (i.e., disrupt, degrade, destroy) and/or deceive an enemy's military EMS activities. EA is the division of EW involving the use of EM energy, including DE or antiradiation weapons, to attack personnel, facilities, or equipment with the intent of degrading, neutralizing, or destroying enemy combat capability. Typical EA capabilities include EM jamming and intrusion. EM jamming is the deliberate radiation, reradiation, or reflection of EM energy for the purpose of preventing or reducing an enemy's effective use of the EMS, to degrade or neutralize the enemy's combat capability. EM intrusion involves the intentional insertion of EM energy into transmission paths to deceive or confuse enemy forces. EA can be either active (i.e., radiating) or passive (i.e., non-radiating/reradiating). Examples of active EA systems (to include lethal and nonlethal DE) include lasers, electro-optical, IR, and RF weapons such as high-power microwave (HPM) or those employing an electromagnetic pulse (EMP). Examples of passive EA systems are chaff and corner reflectors. EA can also be used for offensive and defensive purposes.

- **Offensive EA.** Offensive EA describes the use of EA to project power in support of operations within the time and tempo of the scheme of maneuver. JEMSO planners use JFC guidance to integrate EA during joint planning through the joint planning group or operational planning group, coordinating effects and incorporating risk mitigation techniques to reduce collateral damage. In many cases, these activities suppress a threat for only a limited period of time. Examples include employing self-propelled decoys; jamming radar or C2 systems; using antiradiation missiles to suppress air defenses; using EM deception techniques to confuse intelligence, surveillance, and reconnaissance (ISR) systems; and using DE weapons to disable personnel, facilities, or equipment and disable or destroy materiel (e.g., satellites in orbit, airborne optical sensors, or massed land forces).

- **Defensive EA.** Defensive EA describes the use of EA to protect against threats by denying enemy use of the EMS to target, guide, and/or trigger weapons. EA used for defensive purposes in support of force protection or self-protection is often mistaken as EP. Although defensive EA actions and EP protect personnel, facilities, capabilities, and equipment, EP protects from the effects of EA or EMI, while defensive EA is primarily used to protect against lethal attacks by denying enemy use of the EMS to target, guide, and/or trigger weapons.

Protect

As joint forces are critically dependent on exploiting the EMOE, JEMSO facilitate the necessary EMS access by minimizing EMI from friendly, neutral, adversary, and enemy actions. JEMSO integrate EW and EMS management protection actions throughout planning and execution, enabling joint force EMS-dependent systems to operate in the EMOE as intended. EP refers to the division of EW involving actions taken to protect personnel, facilities, and equipment from any effects of friendly, neutral, adversary, or enemy use of the EMS, as well as naturally occurring phenomena that degrade, neutralize, or destroy friendly combat capability. EP focuses on system or process attributes or capabilities that eliminate or mitigate the impact of EMI. These inherent hardware features; processes; and dedicated tactics, techniques, and procedures (TTP) combine to enable friendly capabilities to continue to function as intended in contested and congested EMOEs.

Manage

All joint force operations in the EMS must be managed to facilitate unity of effort in executing the planned scheme of maneuver within the EMOE. EMS management's objective is to enable EMS-dependent capabilities and systems to perform their functions as designed, without causing or suffering unacceptable EMI. EMS management provides the framework to utilize the EMS in the most effective and efficient manner. EMS management is analogous to the airspace management function in air operations, coordinating and integrating joint EMS use in terms of time, space, and frequency.

- **Electromagnetic Battle Management (EMBM).** EMBM includes actions to monitor, assess, plan, and direct operations in the EMS in support of the commander's objectives. It is the coordinated direction of all joint functions in the EMS to enable the orderly conduct of friendly EMSO. When exercised, EMBM is a commander's mechanism for informing all actions that shape the OE. EMBM is accomplished through an EMBM system that consists of the facilities, equipment, software, communications, procedures, and personnel essential for a commander to plan, direct, and control operations in the EMS.

- **Frequency Management (FM).** FM encompasses interference analysis and requesting, nominating, coordinating, assigning, and promulgating frequencies for EMS-dependent capabilities and systems. FM assigns frequencies for non-EA EM transmissions, conducts frequency deconfliction, and mitigates EMI. FM is a key component for developing EMS operating instructions and coordination measures. FM includes spectrum analysis, engineering, and assessment of EMS-dependent systems and developing EMS products such as the JRFL, joint communications-electronics operating instructions (JCEOI), and others, as required.

- **Host-Nation Coordination (HNC).** HNC is the coordination with nation states for authorization to operate EMS-dependent systems within national borders (includes use of systems that emanate across the border from other AOIs). Coordination is required when operating within foreign nations as well as the United States. Granting approval to transmit EM energy within a nation is a sovereign right. HNC is normally accomplished through procedures established by CCMD agreements with HNs.

- **Joint Spectrum Interference Resolution (JSIR).** A contested and congested EMS, coupled with dynamic military operations, makes encountering EMI in the EMOE very likely. In fact, most system degradation can be attributed to EMI. As such, JSIR identifies, reports, analyzes, and mitigates or resolves incidents of EMI. JSIR uses a continuous systematic process to report and diagnose the cause or source of EMI. CCMDs should ensure incidents of EMI are reported immediately and are resolved or mitigated. EMI can be induced intentionally, as in EA, or unintentionally, as a result of harmonics, spurious emissions, intermodulation products, improper operation, or inadequate EMS management.

(SMO/JEMSO) I. Overview 5-7 *

B. Electromagnetic Environmental Effects (E3)

Ref: JP 3-85, Joint Electromagnetic Spectrum Management Operations (May '20); pp. I-11 to I-12.

The impact of the EMOE upon the operational capability of military forces, equipment, systems, and platforms is referred to as electromagnetic environmental effects (E3).

Examples of E3 include electromagnetic compatibility (EMC), EMI, EMP, and EM radiation hazards. EM radiation hazards include hazards of electromagnetic radiation to personnel (HERP); hazards of electromagnetic radiation to ordnance (HERO); hazards of electromagnetic radiation to fuels (HERF); and natural phenomena effects such as space weather, lightning, and precipitation static.

HERP
HERP is the potential hazard that exists when personnel are exposed to an EM field of sufficient intensity to heat the human body. Radar, communication systems, and EW systems which use high-power RF transmitters and high-gain antennas represent a biological hazard to personnel working on, or in the vicinity of, these systems. Therefore, stand-off areas around high-powered RF antennas should be clearly marked. Since it is not possible to visibly determine if an antenna is transmitting, personnel should avoid entering these stand-off areas at all times.

HERO
HERO is the danger of accidental actuation of electro-explosive devices or otherwise electrically activating ordnance because of RF EM fields. This unintended actuation could have safety (premature firing) or reliability (dudding) consequences. HERO may be induced through holes or cracks in the casing, wires, or fuses and is most susceptible during assembly, disassembly, loading, or unloading.

HERF
HERF is the potential hazard that is created when volatile combustibles, such as fuel, are exposed to EM fields of sufficient energy to cause ignition. HERF is most likely to occur when refueling operations are taking place. Care should be taken to separate fueling points and high-powered radar, radio, directed energy weapons, or jammers to reduce the possibility of RF induced arcs that could ignite fuel. Personnel must ensure proper grounding and static discharge procedures are adhered to and that RF transmissions be minimized or ceased during refueling operations.

Electromagnetic Pulse (EMP)
The interaction of gamma radiation with the atmosphere can cause a short pulse of electric and magnetic fields that may damage and interfere with the operation of electrical and electronic equipment and can cause widespread disruption. EMP is one of the primary ways that a nuclear detonation produces its damaging effects. The effects of EMP can extend to hundreds of kilometers depending on the height and yield of a nuclear burst.

High-Altitude Electromagnetic Pulse (HEMP)
A high-altitude electromagnetic pulse (HEMP) can generate significant disruptive field strengths over a continental-size area. The portion of the EMS most affected by EMP and HEMP is the radio spectrum. Planning for communication system protection is key when the potential for EMP is likely.

For more information on E3, refer to Department of Defense Instruction (DODI) 3222.03, DOD Electromagnetic Environmental Effects (E3) Program.

II. Spectrum Management

Ref: FM 3-12, Cyberspace and Electronic Warfare Operations (Apr '17), pp. 1-34 to 1-35 and ATP 6-02.70, Techniques for Spectrum Management Operations (Oct '19), chap. 2.

Spectrum management is the operational, engineering, and administrative procedures to plan, coordinate, and manage use of the electromagnetic spectrum and enables cyberspace, signal and EW operations. Spectrum management includes frequency management, host nation coordination, and joint spectrum interference resolution. Spectrum management enables spectrum-dependent capabilities and systems to function as designed without causing or suffering unacceptable electromagnetic interference. Spectrum management provides the framework to utilize the electromagnetic spectrum in the most effective and efficient manner through policy and procedure.

SMO are the interrelated functions of spectrum management, frequency assignment, host nation coordination, and policy that together enable the planning, management, and execution of operations within the electromagnetic operational environment during all phases of military operations. The SMO functional area is ultimately responsible for coordinating EMS access among civil, joint, and multinational partners throughout the operational environment. The conduct of SMO enables the commander's effective use of the EMS. The spectrum manager at the tactical level of command is the commander's principal advisor on all spectrum related matters.

The conduct of SMO enables and supports the execution of cyberspace operations and EW. SMO are critical to spectrum dependent devices such as air defense radars, navigation, sensors, EMS using munitions, manned and unmanned systems of all types (ground and air, radar, sensor), and all other systems that use the EMS. The overall objectives of SMO are to enable these systems to perform their functions in the intended environment without causing or suffering unacceptable electromagnetic interference. Understanding the SMO process in planning, managing, and employing EMS resources is a critical enabler for cyberspace and EW operations. SMO provides the resources necessary for the implementation of the wireless portion of net-centric warfare.

The spectrum manager should be an integral part of all electronic warfare (EW) planning. The SMO assists in the planning of EW operations by providing expertise on waveform propagation, signal, and radio frequency theory for the best employment of friendly communication systems to support the commander's objectives.

See chap. 3, Electronic Warfare.

Frequency Interference Resolution
Interference is the radiation, emission, or indication of electromagnetic energy (either intentionally or unintentionally) causing degradation, disruption, or complete obstruction of the designated function of the electronic equipment affected. The reporting end user is responsible for assisting the spectrum manager in tracking, evaluating, and resolving interference. Interference resolution is performed by the spectrum manager at the echelon receiving the interference. The spectrum manager is the final authority for interference resolution. For interference affecting satellite communications, the Commander, Joint Functional Component Command for Space is the supported commander and final authority of satellite communications interference.

I. Key SMO inputs to the MDMP

Ref: ATP 6-02.70, Techniques for Spectrum Management Operations (Oct '19), pp. 2-11 to 2-14.

Key inputs for the MDMP are actions, processes or information spectrum managers provide to the MDMP. SMO key outputs for MDMP are the completed CEOI, reports, frequency proposals or data call messages. Table 2-1 depicts the key SMO inputs and outputs for each step of the MDMP.

Key SMO inputs	Steps	Key SMO outputs
• Updated EMS database • Unit electronic order of battle • Library of EMS documents • HN allocation tables • Gather spectrum management tools	Step 1: Receive Mission	• Defined EMOE • Data call message • Identify EMS constraints • JFRL guidance
• Identified EMS capabilities pertaining to combat power • List of unit's SSDs • Frequency requests • JRFL requests	Step 2: Mission Analysis	• Prioritized EMS use • Completed JRFL • Frequency reuse plans • Initial EMS risk assessment
• Commander's intent • Frequency allotments • Initial frequency assignments • DD-1494 for unit's SDDs	Step 3: Develop COA	• M&S of EMS to develop multiple COAs • EMI/EW deconfliction • Initial Spectrum Plan • EMS COP
• Initial Spectrum Plan • Mitigating factors to decrease EMS risk	Step 4: COA Analysis (War Game)	• M&S shows EMS advantages/disadvantages for each COA • Continues analysis of EMS risk assessment • Recommend modifications
• Optional unit movement routes for planning COTM • Refines EMS COAs	Step 5: COA Comparison	• M&S depicts EMS use to compare COAs • Recommended EMS COAs
• Recommended EMS COA • Coordinated frequency conflicts • Frequency proposals	Step 6: COA Approval	• Commander selected EMS COA and any modifications • Frequency assignments
• Frequency assignments/allotments from higher echelon ESM • HN frequency clearance • CREW loadsets	Step 7: Orders Production, Dissemination and Transition	• The Spectrum Plan • CEOI/JCEOI • Annex H of OPORD • Distribute frequency assignments to requestors • CNR loadsets

SMO supports the commander's SMO objectives during each step of the MDMP. The following are some responsibilities expected of the spectrum manager for each step—

Step 1: Receipt of Mission

- The spectrum manager conducts data calls to attain a list of SDDs and their spectrum requirements.
- Using spectrum management tools, the spectrum manager models the operational area with digital topography and electromagnetic environmental effects information to analyze spectrum supportability.
- Using governmental and host nation spectrum allocation tables, the spectrum manager determines frequencies used in an AO.
- The spectrum manager compiles restrictions or constraints of spectrum use that may prevent planning and use of protected, taboo, and guarded frequencies in the AO. For a listing of the worldwide-restricted frequency list, see CJCSM 3320-01C.
- The spectrum manager should understand the EMOE for awareness of the spectrum occupancy in the AO. Colors representing users of the spectrum are—blue (friendly), red (enemy), and gray (neutral and civil).

Step 2: Mission Analysis
- The spectrum manager analyzes the EMOE, highlighting unified action partners' spectrum users, and aid the commander in determining spectrum priorities.
- The spectrum manager conducts an initial spectrum risk assessment identifying the spectrum impact mission on unified action partners in the operational area. This process also identifies frequency usage conflicts such as EMI and frequency fratricide.
- The spectrum manager generates a frequency reuse plan for spectrum optimization and increased spectrum capabilities.
- The spectrum manager identifies spectrum constraints where certain frequencies are either taboo, protected, or guarded. Constraints include those frequencies not allocated for use by the host nation.
- The spectrum manager, with guidance from the CEWO, determines spectrum capabilities of combat power, such as EW and counter radio-controlled improvised explosive device electronic warfare (CREW) systems.

Step 3: Course of Action Development
- Using spectrum management tools, the spectrum manager models the unit's boundaries and movement formations. The use of these models is for developing COA recommendations.
- Using spectrum management tools, the spectrum manager performs EMI and EW frequency deconfliction for both COA development and spectrum supportability.
- The spectrum manager generates frequency allotment and allocation tables for subordinate units.
- The spectrum manager identifies spectrum impact on civilian spectrum users in the AO.
- The spectrum manager evaluates primary, alternate, contingency, and emergency communications for each COA based on unit capabilities, software simulation, and spectrum supportability.

Step 4: Course of Action Analysis (War Game)
- The spectrum manager identifies the spectrum advantages and disadvantages throughout the AO for each COA.
- The spectrum manager identifies mitigating factors for the spectrum risk assessment to reduce or eliminate risks.
- The spectrum manager recommends modifications to the COA based on newly identified spectrum requirements and supportability during the wargame.

Step 5: Course of Action Comparison
- Using spectrum management tools, the spectrum manager develops multiple COAs. The commander determines the COA best suited for the mission.
- The spectrum manager analyzes routes used for movement of forces and advises the commander on routes with the least likelihood of spectrum interference or loss of spectrum coverage.

Step 6: Course of Action Approval
The spectrum manager consolidates units' submission of frequency proposals and provides the units with frequency assignments.
- The spectrum manager modifies the spectrum management portion of COAs according to the commander's guidance

Step 7: Orders Production Dissemination, and Transition
- The spectrum manager produces the CEOI and disseminate to units.
- The spectrum manager provides input to Annex H (Signal) of the operations order (OPORD) that addresses all signal concerns, to include spectrum use information.

II. SMO Support to the Warfighting Functions

Ref: ATP 6-02.70, Techniques for Spectrum Management Operations (Oct '19), chap. 3.

SMO enables and supports the Army's warfighting functions described in ADP 3-0, Unified Land Operations. A warfighting function is a group of tasks and systems (people, organizations, information, and processes) united by a common purpose that commanders use to accomplish missions and training objectives. The Army's warfighting functions are—movement and maneuver, intelligence, fires, sustainment, command and control, and protection. This chapter links Army SMO to the warfighting functions, also describes how SMO supports and enables the commander's efforts as they exercise command and control.

Movement and Maneuver

SMO enables movement and maneuver by maintaining freedom of action within the electromagnetic spectrum. Commanders can leverage information derived from SMO to provide lethal and non-lethal effects against enemy combat capabilities while ensuring protection from adversary's use of the spectrum. SMO supports movement and maneuver by—

- Spectrum resource planning, analysis, and simulation to determine spectrum supportability over a projected movement of forces.
- Analysis, location, and direction finding of unknown and unplanned signals.
- Planning and simulating spectrum within the AO.
- Frequency deconfliction planning during movement of forces.

Intelligence

SMO supports intelligence through the provision of spectrum situational understanding and the ability to gain a greater understanding of the EMOE. Understanding the EMOE results in successful frequency deconfliction of SDD, greater fidelity in threat recognition, and provision in support to the denial and destruction of enemies' counter-intelligence, counter-surveillance, and counter-reconnaissance systems. SMO supports intelligence by—

- Spectrum situational awareness using measurement, analysis, and assessment of signals in the AO.
- Providing a detailed caption of the EMOE for situational awareness.
- Production and promulgation of JRFL identifying protected frequencies used by friendly forces that are of critical importance, to include intelligence operations, including guarded frequencies on the JRFL to exploit an adversary's intelligence.
- Centralized databases facilitate collection management through subordinate and adjacent units.
- Deconflicting frequencies that create EMI with unmanned aircraft systems that may be conducting intelligence operations in the AO.

Fires

SMO provides crucial support to the fires warfighting function through spectrum awareness and direct support to EW. Electromagnetic environmental effects influence the operational capability of military forces, equipment, systems, and platforms. Spectrum management operations support the fires warfighting function through mitigation of EMI amongst fires systems. SMO supports fires by—

- Coordination throughout the EMOE to prevent EMI to and from firing devices, sensors, and data links that use the spectrum.
- Coordination with the CEMA element that allows effective use of spectrum resources for EW operations.
- Integration and synchronization of CEMA by assignment and allocation of spectrum use in joint environments.

Sustainment

The sustainment warfighting function is the related tasks and systems that provide support and services to ensure freedom of action, extend operational reach, and prolong endurance. SMO ensures that all SDDs used for sustainment have necessary frequencies and minimal EMI.

Through coordination with EW, SMO contributes to overall sustainment in a hostile EMOE. SMO supports sustainment by—

- Providing the necessary frequencies for logistics SDDs within the EMOE conducting sustainment operations.
- Obtaining frequency clearance for logistics SDDs to conduct sustainment operations for the duration of the mission.
- Frequency deconfliction and emissions control procedures in support of sustainment operations.

Command and Control

The command and control (C2) warfighting function develops and integrates those activities, enabling a commander to balance the art of command and the science of control. C2 emphasizes the centrality of the commander. Commanders exercise C2 by driving the operations process, knowledge management and information management, synchronization of information-related capabilities, and conducting CEMA. SMO enables C2 through the mitigation of EMI resulting from both frequency fratricide and enemy attack actions. In a contested, congested, and competitive EMOE, the C2 function must remain effective. SMO plays a vital part in the planning and management process that results in situational awareness of the EMOE.

SMO supports C2 by—

- Planning and preparing the spectrum in response to a mission.
- Assessment of the EMOE in response to the commander's intent.
- Preparation and maintenance of the EMOE database.
- Understanding the impact of a mission on friendly, neutral, adversary, enemy, joint, interagency, intergovernmental, and multinational entities.
- Collecting spectrum information and visualizing this information in quick and easy to understand formats for completion of the COP.
- Control of the spectrum through force tracking and visualization, frequency deconfliction, reprogramming, registration of SDDs.
- Development of SMO planning and management tools that support the network-centric environment (NCE) and become interoperable with Army and joint task force spectrum users.

Protection

The protection warfighting function is the related tasks and systems that preserve the force so the commander can apply maximum combat power. SMO supports the protection warfighting function through the conduct of frequency deconfliction, interference mitigation, and support to EW defensive actions. SMO supports protection by—

- Network and frequency fratricide avoidance, detection, and mitigation.
- Developing of the JRFL to prevent frequency fratricide and mission degradation.
- Coordinating with CEMA Element to protect against blue force EMI during EW operations, such as counter radio-controlled improvised explosive device EW use.
- The spectrum manager also protects the force by recognizing the potential of electromagnetic environmental effects.

(SMO/JEMSO) II. Spectrum Management 5-13 *

II. The Common Operational Picture (COP)

Ref: ATP 6-02.70, Techniques for Spectrum Management Operations (Oct '19), pp. 2-14.

The COP is a single display of relevant information within a commander's AO tailored to user requirements and based on shared data and information shared by more than one command. The spectrum manager assists with the information collection efforts by providing detailed data of the EMOE for the commander's COP.

SMO planning tools, used in conjunction with Intelligence and EW information, allow the spectrum manager to collect spectrum-related details tailored to the commander's AO. These tools provide a visual depiction of force structure and geographical locations in a three-dimensional picture that personnel can understand quickly and easily. The following are some examples of SMO supports to the COP—

Live Spectrum Analysis

The spectrum manager uses SMO planning tools to analyze spectrum emissions within the commander's AO. Use of information attained from the spectrum analysis is to perform EMI mitigation. SMO planning tools include—spectrum analyzers or monitoring receivers, direction-finding antennas, and analysis software. SMO planning tools can be used to show or model persistent unplanned signals that interfere with assigned frequencies during detection of EMI. SMO planning tools provide a three-dimensional picture of the EME to the commander and includes a graphical depiction of the spectrum footprint, along with recommendations for frequency reassignment to maintain communications in the AO. Using information provided by SMO planning tools and mission priorities, the commander may deem it necessary to obtain new frequencies for mission accomplishment.

Movement of Forces to a New Location

When the commander orders movement of forces to a new area, the spectrum manager creates the proposed movement route with the SMO planning tools. The spectrum manager collaborates with adjacent units to minimize EMI with friendly forces' communications systems, sensors, and receivers throughout the movement. The SMO planning tools perform a simulation and provide COAs to determine if communication systems remain operational during movement. The SMO planning tools determine if a specific movement route with active EW systems can cause interference of friendly communications along that route. The SMO planning tools produce a report with actionable information such as sources, victims, levels, and duration of interference. This information provides the commander with supplementary information to make knowledgeable decisions.

III. Planning Joint EMS Operations (JEMSO)

Ref: JP 3-85, Joint Electromagnetic Spectrum Management Operations (May '20); chap. III.

JFCs centralize JEMSO planning under the designated EMSCA and decentralize execution to ensure JEMSO unity of effort while maintaining tactical flexibility. Operations in the EMS cross all joint functions, span the OE, and are often complex and interwoven. This requires detailed prioritization, integration, and synchronization to attain EMS superiority, achieve the commander's objectives, mitigate EMI, and avoid friendly fire EA incidents (involving personnel or equipment). JEMSO planning provides the basis for the prioritization, integration, and synchronization of joint force EMS operations between the staff functions (primarily J-2, J-3, and J-6), components, and multinational partners across all phases of military operations. The CCMD JEMSOC is the lead staff element for JEMSO planning. JEMSO planning uses the joint planning process (JPP) to frame the problem; examine mission objectives; develop, analyze, and compare alternative courses of action (COAs); select the best COA; and produce the JEMSO plan or order. The JPP normally results in the development of CONOPS, OPLANs, and OPORDs. The JEMSOC ensures JEMSO are integrated throughout the command's planning process.

Planning Process

The commander's guidance and estimate form the basis for determining components' objectives. During mission analysis, JEMSO planners develop a JEMSO staff estimate, which forms the basis for an EMS superiority approach. The staff estimate is used during COA development and analysis to determine the EMS activities and capabilities required to accomplish the mission, the JEMSO capabilities required to support operations, and the risk to the operation if EMS superiority is not achieved. When a COA is chosen, it becomes the basis for developing the JEMSO appendix, which outlines JEMSO missions, priorities, policies, processes, and procedures across all phases of the operation. The joint force components will develop component EMSO plans and submit them to the JEMSOC for integration into the JEMSO appendix under annex C (Operations). The JEMSO planning process is a formal, top-down, centralized process that integrates EMSO into the JFC's plan.

Figure III-1 (following page) shows the types of tasks and products the JEMSOC should develop during each JPP step.

I. Electromagnetic Order of Battle (EOB)

The EOB is a key product the JEMSOC updates to support planning. The EOB details the strength, command structure, disposition, and operating parameters of friendly force, threat, and neutral EMS-dependent systems identified in the order of battle. This includes the identification of transmitters and receivers in an AOI, a link to systems and platforms supported, determination of their geographic location or range of mobility, characterization of their signals, EMS parameters, and, where possible, a determination of their role in the broader organizational order of battle. While the J-2 provides the information required to build the threat and neutral EOBs, the J-3, J-6, components, and supporting units provide the information necessary to build the friendly force EOB. The J-2 will also contribute to the friendly force EOB by providing information regarding ISR within the EMOE.

II. Electromagnetic Operational Environment (EMOE) Estimate

The JEMSOC defines and characterizes the EMOE within the AOI associated with a given OA. The EMOE estimate includes sections that describe the background EME; identify factors that affect signal propagation (e.g., environmental characteristics and terrain); create a database of the known spectrum-use information; review historic EMI events within the area; and integrate the friendly, neutral, and threat EOBs.

Define and Characterize the Electromagnetic Operational Environment (EMOE)

The situation analysis portion of the JEMSO staff estimate is where the EMOE is initially defined and characterized, forming the foundation for the JEMSO aspects of COA development, analysis, and selection.

Characterizing the EMOE is an iterative process that employs many of the tasks and methodologies associated with JIPOE. An EMOE tends to be dynamic, requiring the associated databases and analyses be updated periodically, often on a very short timeline. The physics of the EMS dictate that the military usefulness and properties of a given set of frequencies may vary periodically, based on environmental factors outside of JFC control. JEMSO planners not only must anticipate changes in both neutral and threat operations in the EMS but also need to consider potential naturally occurring EMOE changes as well. Sources and areas subject to EMI (e.g., local civilian infrastructure such as airports) should be identified as part of the EMOE.

EMOE information should be current, accurate, and accessible to authorized users. JEMSO planners should designate primary EMOE data sources to facilitate this. This source designation should be accompanied by information on the organization(s) responsible for maintaining the data sources, the associated processes and timelines for source population, requirements for access (user clearances and timelines), and the processes for dealing with data source conflicts.

Meteorological, oceanographic, and space conditions should be considered. JEMSO planners should include the effects of atmospherics and space weather on both the EMOE and all EMS-dependent systems. The various types of atmospheric conditions and phenomena can positively or negatively affect these systems. For example, atmospheric temperature inversions can increase the propagation of radio signals with frequencies in excess of 30 megahertz; high humidity and rainy climates are detrimental to IR systems; and ionospheric scintillation can adversely affect GPS, high frequency, and ultrahigh frequency transmissions. Some atmospheric effects are well known and are categorized by season and location. Planners should consult with the CCMD meteorological, oceanographic, and space staffs to determine the type of support available for their operation.

The JEMSOC will use this information to create EMOE estimates that support each step of the JPP. These EMOE estimates describe the predicted state of the EMOE at a future time and location. Components of an EMOE estimate include:

   (a) Expected state of the physical environment (e.g., METOC predictions).

   (b) Threat, neutral, and friendly force EMS-dependent systems expected to be active during that time.

   (c) Level of readiness and predicted role of the EMS-dependent systems in support of operations.

   (d) Most likely locations and range of operation of the EMS-dependent systems.

   (e) Predicted set of EMS parameters to be used.

JEMSMO Cell Actions and Outputs as Part of Joint Planning

Ref: JP 3-85, Joint Electromagnetic Spectrum Management Operations (May '20), fig. III-1.

Planning Process Steps	Joint Electromagnetic Spectrum Operations Cell (JEMSOC) Planning Actions	JEMSOC Planning Outputs
Planning Initiation	• Review appropriate documents such as warning order and strategic assessment. • Review joint intelligence preparation of the operational environment, desired end state, strategic effects and objectives • Obtain order of battle and begin building electromagnetic order of battle (EOB) • Review rules of engagement (ROE), guidance, and operational estimates • Review operational factors within theater to identify risk to mission • Identify organizational construct for the JEMSOC • Identify US/multinational electromagnetic spectrum (EMS) considerations • Identify EMS-use restrictions • Disseminate electromagnetic interference reporting procedures • Disseminate joint restricted frequency list requirements • Disseminate EMS management tools and procedures	• Initial EOB • Requests for information (RFIs) on threats • Friendly force information requirements (FFIRs) • Data call message • EMS management concept • Multinational frequency assignment agreement(s) initiated • Initiate host nation (HN) frequency coordination
Mission Analysis	• Support development of intelligence estimate • Describe how threat uses the electromagnetic operational environment (EMOE) to support operations • Describe threat capability to deny friendly force EMS use • Identify specified, implied, and essential electromagnetic spectrum operations (EMSO) tasks • Identify assumptions, constraints, and restraints relevant to EMSO • Identify planning support requirements, issue support requests • Review available EMSO assets, identify employment authorities • Define the EMOE area of interest • Describe EMOE physical and environmental characteristics • Provide EMS perspective in support of mission requirements • Identify EMSO opportunities for EMSO and risk to mission • Support center of gravity (COG) decomposition and analysis • Determine EMSO role in defeating COG	• Updated EOB • Draft initial EMS staff estimate • List of EMSO tasks • Assumptions, limits, constraints, and restraints • EMSO planning guidance • JEMSOC augmentation request • List of EMSO capabilities potentially required
Course of Action (COA) Development	• Build EOB and EMOE estimate for each COA • Review intelligence estimate of threat and friendly force COAs • Identify electromagnetic warfare (EW) requirements and opportunities for each COA • Determine how EMOE must be shaped to support the COA • Identify EMSO capabilities required to meet EW tasks • Revise EMSO portion of COA to develop staff estimate • Analyze COA from EMSO perspective, build mitigation methods	• List of objectives, tasks, capabilities • EOB and EMOE for each COA • Threat and friendly force targets vulnerable to electromagnetic attack (EA) • Initial joint task force (JTF) EMS requirements summary developed
COA Analysis and Wargaming	• Analyze each COA from EMSO perspective • Identify operations in the EMS supporting all component missions • Identify threat capabilities that impact friendly force operations • Identify opportunities to exploit or attack threat electromagnetic operations • Identify possible targets for EA • Recommend EMSO critical information requirements • Identify the activities required to shape the electromagnetic environment to support operations and the risk to COA if EMOE is not shaped accordingly	• List of EMSO assets • Assessment of COA risk from EMSO view • List of EA vulnerable targets • List of targets to enable friendly force EMSO • JTF EMS requirements summary developed
COA Approval	• Compare each COA based on mission and EMSO tasks • Compare EMSO requirements from each COA • Review EMSO assets and capabilities needed to execute COAs • Identify risk to COA execution from EMSO perspective • Prepare EMSO risk assessment portion of decision brief • Obtain EMS resources and HN approval	• COA EMSO strengths and weaknesses • EMSO risk assessment for each COA • Risk mitigation methods • JTF allotment plan
Plan or Order Development	• Update EOB and EMS staff estimate based on COA decision • Develop EMSO guidance • Review joint and component concept of operations and schemes of maneuver • Develop EMSO portion of a synchronization matrix • Submit EMSO-related information requests and ROE • Refine EMSO tasks from the approved COA • Identify EMSO capability shortfalls and recommend solutions • Update EMSO portions of operations plan • Advise commander on EMSO issues and concerns	• Initial EMOE estimate • EMS staff estimate • Joint electromagnetic spectrum operations (JEMSO) appendix • Request for EMS forces • EMSO ROEs, RFIs, and FFIRs • Initial master net list • JEMSO plan (includes EMS plan)

See also pp. 4-41 to 4-44, cyberspace integration in the joint planning process.

Information (Planning Considerations)

Ref: JP 6-01, Joint Electromagnetic Spectrum Management Operations (Mar '12), pp. III-13 to III-16.

Information Function

The information function encompasses the management and application of information and its deliberate integration with other joint functions to change or maintain perceptions, attitudes, and other elements that drive desired behaviors and to support human and automated decision making. The information function helps commanders and staffs understand and leverage the pervasive nature of information, its military uses, and its application during all military operations. This function provides JFCs the ability to integrate the generation and preservation of friendly information while leveraging the inherent informational aspects of military activities to achieve the commander's objectives and attain the end state. JEMSO enable information activities by coordinating and integrating EMS-use requirements to eliminate or mitigate EMI caused by friendly or threat forces. JEMSO also provide information activities with the means of transmitting information through the EMS.

The JFC or designated staff element may establish an information cell to coordinate the inherent informational aspects of activities that support the CONOPS. Nearly all information activities depend on, use, or exploit the EMS for at least some of their functions. JEMSO prioritization, integration, and synchronization are continuous processes and a constant consideration in information planning efforts.

EA can create decisive and enhanced effects in the information environment that provide the JFC with an operational advantage by contributing to the gaining and maintaining of information superiority. Information superiority is the operational advantage derived from the ability to collect, process, and disseminate an uninterrupted flow of information while exploiting or denying a threat's ability to do the same.

When EA is employed as nonlethal fires, it can often be employed with little or no associated physical destruction. EA in support of information activities is integrated at the JFC level, through the joint targeting coordination board (JTCB) or like body, to predict collateral damage and/or effects and incorporate risk mitigation techniques.

Military Information Support Operations (MISO)

JEMSO support and enable the joint MISO communications plan by ensuring frequencies are available for broadcast services when these are controlled by the CCDR. MISO units depend on information gathered through JEMSO (e.g., ES) and intelligence (e.g., SIGINT) sensors to warn them of threats and provide feedback about reaction to MISO broadcasts and other activities. MISO uses EP and JSIR processes to eliminate or mitigate threat EA activities or inadvertent EMI from disrupting their efforts. MISO and JEMSO coordination, especially with regards to EA, depends on timely updates to EMS operating instructions.

Operations Security (OPSEC)

JEMSO support OPSEC by degrading threat intelligence collection against friendly units and activities. ES supports OPSEC by providing information about threat capabilities and intent to collect intelligence on friendly forces through the EMS. ES can also be used to evaluate the effectiveness of friendly force EMCON measures and recommend modifications or improvements. An effective and disciplined EMCON plan and other appropriate EP measures are important aspects of good OPSEC. OPSEC supports EMSO by concealing units and systems to deny information on the extent of EMSO capabilities. During operations, OPSEC and JEMSO staff personnel should frequently review the CCIRs in light of the dynamics of the operation. Adjustments should be recommended to the EMCON posture and other EP measures as necessary to maintain effective OPSEC.

Military Deception (MILDEC)
JEMSO support MILDEC by using EA as deception measures; degrading threat capabilities to see, report, and process competing observables; providing threats with information received by EM means that is prone to misinterpretation; and using EP and EMCON to control EM activity observable by a threat. MILDEC frequently relies on the EMS to convey the deception to threat intelligence or tactical sensors. JEMSO planners should ensure EMS frequencies necessary to support deception plans are accounted for in EMS management databases and in the EMS operating instructions without disclosing that specific frequencies are related to deception.

Designated JEMSO planners work through the J-3 staff to coordinate and integrate JEMSO support to MILDEC operations.

Suppression of Enemy Air Defenses (SEAD)
SEAD is a specific type of mission intended to neutralize, destroy, or temporarily degrade surface-based enemy air defenses with destructive and/or disruptive means. Joint SEAD is a broad term that includes all SEAD activities provided by one component of the joint force in support of another. SEAD missions are of critical importance to the success of any joint operation when control of the air is contested. SEAD relies on a variety of EW platforms to conduct ES and EA in its support, and JEMSO planners should coordinate closely with joint and component air planners to ensure support to SEAD missions is integrated into the overall JEMSO plan.

EW Reprogramming
EW reprogramming is the deliberate alteration or modification of EW or target sensing systems (TSSs), or the tactics and procedures that employ them, in response to validated changes in equipment, tactics, or the EME. The purpose of EW reprogramming is to maintain or enhance the effectiveness of EW and TSS equipment. EW reprogramming includes changes to self-defense systems, offensive weapons systems, and ES systems. The reprogramming of EW and TSS equipment is the responsibility of each Service or organization through its respective EW reprogramming support programs. The swift identification and resolution of reprogramming efforts is vital in gaining EMS superiority in a rapidly evolving, congested, and contested EMOE. Service reprogramming efforts include coordination with the JEMSOC to ensure those reprogramming requirements are identified, processed, deconflicted, and implemented in a timely manner by all affected friendly forces. The JEMSOC includes the status of EW reprogramming efforts during planning to account for potential platform vulnerabilities.

Cybersecurity
The DOD cybersecurity program is concerned with preventative, protective, and restorative measures for information systems and the information contained therein. Many of these measures involve the use of the EMS. EP equipment, attributes, and processes assist in assuring the availability and integrity of modulated data traversing the EMOE. EA TTP assist in compromising those same qualities which threat cybersecurity seeks to protect. EMS management procedures, particularly EMI resolution, assist the application of cybersecurity policy in overcoming the problem of EM friendly fire incidents.

Refer to INFO1: The Information Operations & Capabilities SMARTbook (Guide to Information Operations & the IRCs). INFO1 chapters and topics include information operations (IO defined and described), information in joint operations (joint IO), information-related capabilities (PA, CA, MILDEC, MISO, OPSEC, CO, EW, Space, STO), information planning (information environment analysis, IPB, MDMP, JPP), information preparation, information execution (IO working group, IO weighted efforts and enabling activities, intel support), fires & targeting, and information assessment.

III. JEMSO Staff Estimate

The purpose of the JEMSO staff estimate is to inform the commander, staff, and subordinate commands how EMSO support mission accomplishment. The commander and staff use this information to support COA development and selection. JEMSO planners use the staff estimate (a primary product of mission analysis) to prepare evaluation request messages to solicit COA input from subordinate components and units to subsequently develop preliminary COAs. The JFC's JEMSOC uses the CCMD's mission, commander's estimate, objectives, intent, and CONOPS to develop COAs. During COA development and selection, JEMSO planners fully develop their estimate, providing an EMS analysis of the COAs, as well as recommendations on which COAs can be adequately supported by JEMSO. Planners should identify critical shortfalls or obstacles that impact mission accomplishment. The JEMSO staff estimate is continually updated, based on changes in the situation.

For information on JEMSO staff estimates, refer to JP 6-01, appendix G, "Joint Electromagnetic Spectrum Operations Staff Estimate Template."

EMS Superiority Approach

The EMS superiority approach ensures joint forces achieve the advantage in the EMS that permits the conduct of operations at a given time and place without prohibitive interference, while affecting an enemy's ability to do the same. The approach is comprised of the mission analysis and mission statement portions of the JEMSO staff estimate and should be documented in the EMSO section of the CONPLAN OPLAN/OPORD. This approach outlines the key missions and tasks the joint force components will carry out to achieve EMS superiority and establishes the basic relationships between the exploit, attack, protect, and manage activities the joint force will accomplish. The approach identifies key EMS users throughout the OE. It provides the framework for detailed JEMSO planning.

Determine Friendly EMS-Use Requirements

A joint force employs EMS-dependent systems across all functions and activities. The JEMSOC establishes the process to solicit, compile, and process joint EMS-use requirements. Components identify the EMS-dependent systems they will employ in the OE, describe the capabilities and associated EMS-use requirements, and request EMS support. The resultant data is used to build the friendly force EOB, develop the EMS superiority approach, define and characterize the EMOE, determine the supportability of each COA, build the joint EMSO plan (i.e., identifies all component and supporting unit military activities in the EMS), and provide EMSO input to OPLAN or OPORD (i.e., authorizes component military EMS activities).

IV. JEMSO Appendix to Annex C

Once a COA is chosen, the JEMSOC develops the JEMSO appendix within annex C (Operations) for the JFC's approval. This appendix establishes procedures for C2 of forces conducting JEMSO in the JOA and includes EMS coordination measures, specifying procedures, and ROE for joint force EMS use. To provide effective operational procedures, the JEMSO appendix is integrated across all portions of the JFC's COPLANs, OPLANs, and orders. The appendix considers procedures and interfaces with the international or national frequency control authorities/systems necessary to effectively support JEMSO, augmenting forces, and JFC objectives.

For more information, refer to JP 6-01, appendix A, "Electromagnetic Spectrum Management." Consequently, the JEMSO appendix should be planned in advance to the highest degree possible and maintained in a basic, understandable format.

Chap 6

I. Dept of Defense Info Network (DODIN) Ops

Ref: ATP 6-02.71, Techniques for Department of Defense Information Network Operations (Apr '19), chap. 1.

Department of Defense information network (DODIN) operations are operations to secure, configure, operate, extend, maintain, and sustain Department of Defense cyberspace to create and preserve the confidentiality, availability, and integrity of the Department of Defense information network (JP 3-12).

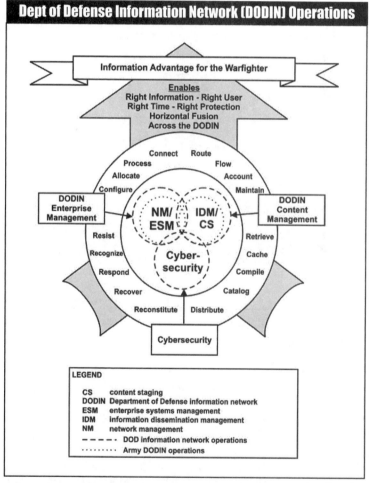

Ref: ATP 6-02.71, Techniques for Department of Defense Information Network Operations (Apr '19), fig. A-1. DODIN Operational Construct.

DODIN operations are one of the three cyberspace missions. The other cyberspace missions are defensive cyberspace operations and offensive cyberspace operations.

(DODIN) I. Overview 6-1

II. Department of Defense Information Network Operations in Army Networks (DODIN-A)

Ref: ATP 6-02.71, Techniques for Department of Defense Information Network Operations (Apr '19), pp. 1-8 to 1-9.

The Army conducts distributed DODIN operations within the DODIN-A, from the global level to the tactical edge. DODIN operations personnel install, operate, maintain, and secure from post, camp, or station to deployed tactical networks. DODIN operations provide assured and timely network-enabled services to support DOD warfighting, intelligence, and business missions across strategic, operational, and tactical boundaries. DODIN operations enable system and network availability, information protection, and information delivery.

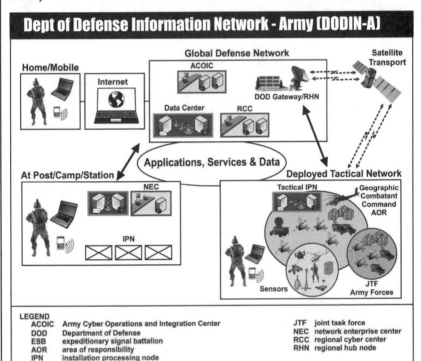

Ref: ATP 6-02.71, Techniques for Department of Defense Information Network Operations (Apr '19), fig. 1-2. Department of Defense information network-Army.

Army personnel implement enterprise DODIN operations through an established hierarchy. The DODIN-A enables access to the right information at the right place and time, so commanders, staffs, Soldiers, civilians, and joint, inter-organizational, and multinational mission partners can meet mission requirements. The DODIN-A segments are home or mobile; post, camp, or station; and deployed tactical network. These segments allow operating and generating forces to access centralized resources from any location during all operational phases. Network support is available at the home post, camp, or station and throughout the deployment cycle.

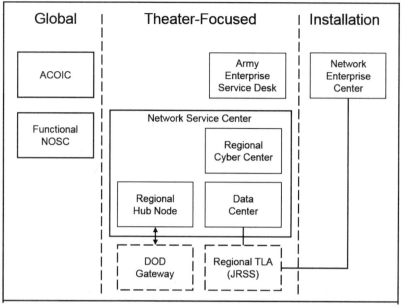

The network service center includes data centers, DOD gateway facilities, and regional hub nodes. The network service center is the DODIN-A interface that connects Army users with joint services and applications. The network service center includes both joint and Army-unique capabilities. The DOD gateway and long-haul satellite transport are joint capabilities that provide backbone connectivity and connection to DISN services. The data center and regional hub node are Army capabilities. The regional hub node provides the connection between deployed Army enclaves and the DODIN.

For more information about the regional hub node, refer to ATP 6-02.60. For more information on DOD gateway, refer to JP 6-0 and ATP 6-02.54.

The data center provides a data repository for content staging, continuity of operations, and redeployment support. It also provides access for those users who access the network from home or a temporary duty location.

The post, camp, and station segment is the primary network access point while in garrison. The post, camp, and station segment connects through the data center and provides access to the other network segments in both secure and nonsecure modes. The post, camp, and station segment allows users to train, collaborate, and conduct mission rehearsals. The installation processing node hosts enterprise services and applications associated with garrison operations. The installation processing node also connects users to installation-level services. The local NEC centrally manages these services. Applications and services within installation processing nodes provide either a temporary processing center presence until data center service is available or a permanent computing presence where technical or operational considerations dictate.

The deployed tactical network enables real-time employment of battle command common services, automated information systems, and information collection assets by deployed forces. The deployed tactical network enables the GCC and commander, joint task force (CJTF) to conduct joint, distributed operations with units in dispersed geographic locations. It allows commanders to conduct collective training with their units. The deployed tactical network connects to the DOD gateway to access DISN services, and to facilitate data replication at the data center for continuity of operations. This allows the unit to maintain its operational tempo with minimal mission impact.

(DODIN) I. Overview 6-3

III. DODIN Critical Tasks

Ref: ATP 6-02.71, Techniques for Department of Defense Information Network Operations (Apr '19), pp. 1-7 to 1-8.

DODIN operations enable staffs at each echelon to execute commanders' priorities throughout the enterprise. DODIN operations allow commanders to communicate and collaborate effectively, and to share, manage and disseminate information using automated information systems. DODIN operations consist of three critical tasks:

DODIN Enterprise Management
DODIN enterprise management is the technology, processes, and policies necessary to execute the DODIN operations functions to install, operate, maintain, secure, and restore communications networks, information systems, and applications. Enterprise management merges baseline IT services with DODIN operations capabilities. For more information on baseline services, see FM 6-02. Within Army networks, enterprise management is composed of network management and enterprise systems management.

Cybersecurity
Cybersecurity is prevention of damage to, protection of, and restoration of computers, electronic communications systems, electronic communications services, wire communication, and electronic communication, including information contained therein, to ensure its availability, integrity, authentication, confidentiality, and nonrepudiation (DODI 8500.01). Cybersecurity satisfies the DODIN operations function of securing DOD networks. Through cybersecurity, DODIN operations providers protect, monitor, analyze, detect, respond to, and report unauthorized activity within DOD information systems and computer networks.

See chap. 7, Cybersecurity.

DODIN Content Management
DODIN content management allows DODIN operations personnel to optimize the flow and location of information over the DODIN by positioning and repositioning data and services to optimum locations on the DODIN relative to information producers, information consumers, and mission requirements. Content management enables knowledge management. In Army networks, content management is information dissemination management and content staging.

For detailed information on information dissemination management and content staging refer to ATP 6-02.71, appendix A.

Shared visibility of DODIN operations status across the DODIN is critical to situational understanding and decision making. Shared network situational understanding, along with coordination between stakeholders on potential events, helps commanders and non-IT staff understand the impact of DODIN operations on their operational mission. Comprehensive network situational awareness identifies policy violations and facilitates network troubleshooting and restoral. Anomalous network activity may give the first indication of a cyberspace attack.

In the past, the Army and DOD have treated network operations as a task performed to manage the network. DODIN operations are not an individual or crew task, but multifaceted military operations that take place at all echelons. DODIN operations are arguably the most important and most complex type of operation the Army performs on a daily basis. The network is the foundational capability for all other Army warfighting functions and capabilities, including mission command; intelligence, surveillance, and reconnaissance; precision fires; logistics; and telemedicine. Commanders and their staffs conduct DODIN operations to leverage the network as a warfighting platform.

Besides providing the network to support mission command and all other warfighting functions, DODIN operations support and enable defensive and offensive cyberspace operations. Successful cyberspace operations require integrating and synchronizing all three types of cyberspace missions.

IV. DODIN across the Operational Phases

The joint force plans and conducts operations according to the six operational phases defined in the phasing model. During execution, a transition marks a change between phases or between the ongoing operations and execution of a branch or sequel. This shift in focus by the joint force is often accompanied by changes in command or support relationships and priorities of effort.

Joint Phasing Model

Despite the numbering of the phases, the phasing model depicts the level of effort applied to each military activity, not necessarily a chronological sequence. While a large-scale combat operation may progress through all of the phases, most phase 0 shaping operations never transition to another phase. Phase IV stability operations or phase V operations to enable civil authority may also be standalone operations.

Theater DODIN operations roles, responsibilities, and relationships at the operational level shift as forces transition between phases. The DODIN operations framework adapts to the operational commander's requirements by phase. The framework indicates the DODIN operations priorities and support relationships across the joint operational phases. DODIN operations control and responsibilities shift as an operation matures and units arrive in the affected theater.

See following pages (pp. 6-6 to 6-7) for an overview and further discussion.

Refer to JFODS5-1: The Joint Forces Operations & Doctrine SMARTbook (Guide to Joint, Multinational & Interorganizational Operations). Updated for 2019, topics include joint doctrine fundamentals (JP 1), joint operations (JP 3-0 w/Chg 1), an expanded discussion of joint functions, joint planning (JP 5-0), joint logistics (JP 4-0), joint task forces (JP 3-33), joint force operations (JPs 3-30, 3-31, 3-32 & 3-05), multinational operations (JP 3-16), interorganizational cooperation (JP 3-08), & more!

The mission and the level of development of the gaining command and theater dictate the actual transition of DODIN operations responsibilities. Units deploying into austere environments on contingency missions transition gradually until the gaining command is ready to accept them.

Theater DODIN Operations Construct

Phases of Joint Operations	Phase 0 Shape	Phase I Deter	Phase II Seize Initiative	Phase III Dominate	Phase IV Stabilize	Phase V Enable CivAuth
Mission Sets	CCDOR	Expeditionary		Campaign		
Supported Commander	Multiple	GCC	CJTF			
DODIN operations Framework (Tenets)	Theater Based Global Enterprise	Increased Decentralization Theater Network to GCC	Begin Decentralization of JOA Network to CJTF	Increased Decentralization of JOA Network to CJTF	Complete Decentralization of JOA Network to CJTF	Begin Transition Back to Theater Based Global Enterprise
Network Main and Supporting Effort	ME:IT Services - SE: Network Enabled Capabilities			ME: Network Enabled Capabilities - SE: IT Services		
Allocation of Network Resources	Functional Supporting Commander			CJTF		
Network Focus	Maintain Infrastructure and Extend Services			Integrated Joint Capabilities		
Sustaining a Campaign	Phase 0 - TOA Rehearsal Do not Execute Phases I - II			Admin Deployment - RSOI - Employment - TOA Directly into Phases III/IV or V - METT-TC Dependent		

Legend
CCDOR — combatant commander's daily operational requirements
CJTF — commander, joint task force
DODIN — Department of Defense information network
JOA — joint operations area
GCC — geographic combatant commander
IT — information technology
ME — main effort
METT-TC — mission, enemy, terrain and weather, troops and support available, time available, and civil considerations
RSOI — reception, staging, onward movement, and integration
SE — supporting effort
TOA — transfer of authority

Ref: ATP 6-02.71, fig. 1-4. Theater Department of Defense Information Network (DODIN) operations framework.

Phase 0—Shape

The purpose of shape phase missions, tasks, and actions is to dissuade or deter adversaries and assure friends, as well as set conditions for the contingency plan. Shaping missions are generally conducted through security cooperation activities. The shape phase generally aligns with the Army strategic role of shape. During the shape phase, the network is theater-focused. DODIN operations focus on meeting local commanders' requirements. The priorities for network resource allocation support functional commands. The network's main effort is access to IT services; the supporting effort is network-enabled capabilities. Units have daily access to automated information systems and maintain a SIPRNET presence for training and situational understanding. DODIN operations support the theater-focused global enterprise with priorities derived from the unit's interaction with the strategic network. The mission sets support the generating force and the combatant commander's daily operational requirements.

The regional hub node and data center connect units to the DODIN through both NIPRNET and SIPRNET and provides battle command common services to support mission command information systems. The RCC provides theater-wide DODIN operations oversight and situational awareness. Units have a single identity that follows them through all phases. The NEC administers this in the post, camp, and station network through a technique called installation as a docking station. Installation as a docking station connects tactical information systems to the DODIN-A through the installation campus area network. When operating on installation as a docking station, the unit performs all DODIN operations for its automated information systems as they would in a deployed tactical network, since the NEC has no visibility of these systems. Units can also disconnect from the installation campus area network and use their WIN-T equipment for training, employing the regional hub node to connect with DISN services.

Phase I—Deter

The intent of this phase is to deter an adversary from undesirable actions because of friendly capabilities and the determination to use them. Deterrence leans toward security activities characterized by preparatory actions to protect friendly forces and indicate the intent to execute subsequent phases of the planned operation. In the deter phase, Army forces conduct tactical tasks to prevent conflict. In this phase, decentralization from a global enterprise to a theater network begins. Designated units receive warning orders, shifting their activities from the shape phase to the deter phase. The warning order triggers a series of actions that affect the DODIN and post, camp, and station network environments as the deployed tactical network takes shape. DODIN operations control begins to shift from the GCC to the CJTF as units rotate into the operational theater.

The DODIN-A and the other Services' network capabilities allow joint force commanders to package and examine available intelligence estimates in a crisis or on other indication of potential military action. The DODIN-A enables the tools and processes to focus intelligence efforts and refine estimates of enemy capabilities, dispositions, and intentions. These operational estimates allow staffs to develop courses of action within the context of the current situation and identify additional information requirements. Enterprise services facilitate planning and coordinating flexible deterrent options, and enable situational understanding and the command and control capabilities the GCC requires to resolve a crisis without armed conflict, or to deter further aggression.

During predeployment, the signal command (theater) [SC(T)] with DODIN operations responsibility for the deploying unit receives the warning order concurrently with the unit. The SC(T) prepares to transfer DODIN operations responsibility to the gaining theater as the unit progresses through the deployment process. The SC(T) alerts the signal brigade with regional responsibility for the unit. The signal brigade provides technical support to facilitate the unit's transition. The network service center supports the deploying unit's transition to the gaining theater network during this phase. The network service center servicing the unit's home station coordinates to replicate necessary data and services to the gaining network service center and integrates the unit into the theater network infrastructure. The gaining theater's RCC assumes DODIN operations responsibility for the unit's automated information systems and aligns DODIN operations with the gaining commander's priorities. The unit continues predeployment training and takes part in a mission readiness exercise at a combat training center with their automated information systems integrated into the gaining theater network. At this point, based on the situation, the GCC may have established a joint task force to conduct operations. The joint task force's joint network operations control center assumes DODIN operations responsibility when designated by the GCC.

Units deploying into the gaining theater provide the GCC with expeditionary capabilities—the ability to deploy combined arms forces quickly into an operational area and conduct operations upon arrival. The DODIN operations framework increasingly decentralizes to support the GCC. The network's main effort is providing IT services; the supporting effort is network-enabled capabilities. Network resource allocation supports the functional supporting commander's requirements. Units in this phase normally access DISN services using their organic network capabilities or a tailored, limited package from their organic capabilities. Commanders establish priorities of service to align the limited capabilities with their most critical mission requirements.

Phase II—Seize Initiative

Joint force commanders seek to seize the initiative in all situations through the decisive use of joint capabilities. In combat, this involves both defensive and offensive operations at the earliest possible time, forcing the enemy to culminate offensively and setting the conditions for decisive operations. During this phase, Army forces begin large-scale combat operations. By the time an operation reaches this phase, the unit conducts DODIN operations as part of the theater network. The unit has replicated all services and data to the gaining theater, disconnected from their post, camp, or station NEC and network service center, and deployed to the theater. The gaining network service center and supporting SC(T) verify the transfer of DODIN operations responsibility.

Upon arrival in the gaining theater, the unit aligns with its gaining command, reclaims its equipment, and coordinates with its DODIN operations authority. The gaining DODIN operations authority ensures the unit can access its replicated data and services and resolves issues integrating the unit into the theater network. Beginning with the seize initiative phase, the CJTF is the supported commander. The CJTF sets mission priorities and aligns DODIN operations to support these priorities. DODIN operations responsibility and authority transition from the GCC to the CJTF as they posture the joint task force. In this phase, expeditionary units receive their network support using their organic signal assets connecting to the DODIN through the regional hub node.

Phase III—Dominate

This phase focuses on breaking the enemy's will to resist or, in noncombat situations, controlling the operational environment. Success in the dominate phase depends on overmatching enemy capabilities at the critical time and place. The network's primary focus is supporting the CJTF, who establishes mission priorities that may change over the course of operations. DODIN operations align to meet and support these priorities and adapt to mission changes. The deployed tactical network is a joint network over which the CJTF exercises DODIN operations control.

The CJTF is the supported commander beginning with the dominate phase. The mission sets are campaign-oriented. The network main effort is network-enabled capabilities to support these mission sets. The supporting effort is IT services. The GCC allocates network resources to the CJTF. The CJTF uses the network to integrate joint capabilities. The CJTF continues to control the network for the rest of the phases. As larger force packages enter the theater, signal support units deploy to augment the theater portion of the network and establish more robust network infrastructure on forward operating bases.

Phase IV—Stabilize

Forces enter the stabilize phase as they shift from sustained combat operations to stability activities. The CJTF still exercises DODIN operations control of the network. DODIN operations remain aligned to joint task force mission priorities, with responsibilities delegated according to the DODIN operations framework. In this phase, Army forces focus their efforts on consolidation of gains.

Phase V—Enable Civil Authority

This phase primarily consists of joint force support to legitimate civil governance. The desired end state is terminating operations and redeploying the joint forces. In this phase, the network reverts to theater-focused to meet local commanders' information requirements. Redeploying units reverse the transition from the earlier phases, return to their home stations, and reintegrate into their post, camp, and station networks. In this phase, Army forces continue to consolidate gains until the transition to legitimate authorities.

Chap 6
II. DODIN Roles & Responsibilities

Ref: ATP 6-02.71, Techniques for Department of Defense Information Network Operations (Apr '19), chap. 2.

DODIN operations ensure users' network and information systems connectivity and security throughout the DODIN. This section explains the distributed roles and responsibilities of DODIN operations entities and network managers from the global, strategic level to the theater army, corps, division, and brigade. It further identifies and describes the control centers that perform DODIN operations functions to manage, control, and secure tactical networks, and their interfaces into the DODIN.

I. Global Level

A. United States Cyber Command (USCYBERCOM)

United States Cyber Command (USCYBERCOM) has the sole authority and responsibility to secure, operate, and defend the DODIN. In this capacity, USCYBERCOM carries out operational and tactical level planning and day-to-day management responsibilities for DODIN operations and defense. Combatant commanders coordinate through USCYBERCOM to consider global impacts to the DODIN. USCYBERCOM plans, coordinates, integrates, and synchronizes activities to direct DODIN operations and defense, and conducts offensive cyberspace operations when directed.

The USCYBERCOM Joint Operations Center performs crisis planning, synchronization, direction, and execution of current cyberspace operations. The USCYBERCOM Joint Operations Center—

- Directs DODIN operations and defense.
- Develops processes and policies to enable comprehensive cyberspace situational understanding.
- Establishes partnerships to develop network defense tools.
- Centralizes operation and maintenance of cross domain solutions.
- Analyzes cyberspace risk.
- Develops and recommends countermeasures to events.
- Executes continuity of operations.

B. Defense Information Systems Agency (DISA)

The Defense Information Systems Agency is a combat support agency of the DOD. The agency provides, operates, and ensures information sharing capabilities and a globally accessible enterprise information infrastructure to support joint warfighters, national-level leaders, and joint, inter-organizational, and multinational elements across the range of military operations. The Director, Defense Information Systems Agency reports to the DOD Chief Information Officer.

The Director, Defense Information Systems Agency has dual responsibilities as Commander, Joint Force Headquarters-DODIN under the operational control (OPCON) of USCYBERCOM. Joint Force Headquarters-DODIN provides operational-level network command and control to direct and verify the DODIN's defensive posture. Joint Force Headquarters-DODIN exercises tactical control of the Service cyber components and supports the geographic combatant commands to synchronize global and theater DODIN operations. Joint Force Headquarters-DODIN exercises directive authority to ensure all Services, combatant commands, agencies, and field activities actively implement the security measures necessary to secure their portions of the DODIN and minimize shared risk.

Joint Force Headquarters-DODIN does not duplicate the DODIN operations activities performed by the Services and defense agencies. It engages to perform—

- Activities specifically directed by USCYBERCOM.
- DODIN operations activities the Service components and agencies cannot perform.
- DODIN operations activities more effectively executed at the joint level.

C. Chief Information Officer/G-6

The Chief Information Officer (CIO)/Assistant Chief of Staff, Signal (G-6) establishes policy for Army use of information technology systems and networks. This responsibility includes evaluating existing Army information management and information technology policies and overseeing their implementation. The CIO/G-6 sets the strategic direction for, and supervises the implementation of, Army information management programs and policy. These programs and policies include network architecture, information sharing policy, cybersecurity policy, the Army cybersecurity program, resource management, process modernization, and synchronization of the Army's network activities.

D. United States Army Cyber Command (ARCYBER)

United States Army Cyber Command (ARCYBER) is an Army Service component command to USCYBERCOM. ARCYBER is the primary Army headquarters responsible for cyberspace operations to support joint requirements. ARCYBER is the single point of contact for reporting and assessing cyber incidents, events, and operations in Army networks, and for synchronizing and integrating Army responses. When directed, ARCYBER conducts offensive and defensive cyberspace operations to ensure U.S. and allied freedom of action in cyberspace, and to deny the same to adversaries. ARCYBER provides appropriate-level interactions both as a supported and as a supporting commander to other Army Service component commands (including theater armies), Army commands, direct reporting units, and joint, inter-organizational, and multinational elements.

Commander, USCYBERCOM has designated the ARCYBER commander as the Commander, Joint Force Headquarters-Cyber to provide command and control of joint and coalition cyberspace forces.

Figure 2-1 (facing page) illustrates ARCYBER's structure and DODIN operations relationships.

1st Information Operations Command

The 1st Information Operations Command is under OPCON of ARCYBER and administrative control of United States Army Intelligence and Security Command. It provides information operations planning, intelligence, and training support to Army forces and other Services.

United States Army Network Enterprise Technology Command (NETCOM)

NETCOM engineers, installs, operates, maintains, and secures the DODIN-A to enable command and control and support the other warfighting functions through all Army operations and missions. NETCOM is the Army's global enterprise network service provider and performs DODIN operations on behalf of ARCYBER to gain information advantage. NETCOM's standardized DODIN operations ensure interoperability in a joint, inter-organizational, and multinational enterprise network.

United States Army Cyber Command (ARCYBER)

Ref: ATP 6-02.71, Techniques for Department of Defense Information Network Operations (Apr '19), pp. 2-2 to 2-4.

United States Army Cyber Command (ARCYBER) is an Army Service component command to USCYBERCOM. ARCYBER is the primary Army headquarters responsible for cyberspace operations to support joint requirements. ARCYBER is the single point of contact for reporting and assessing cyber incidents, events, and operations in Army networks, and for synchronizing and integrating Army responses.

Commander, USCYBERCOM has designated the ARCYBER commander as the Commander, Joint Force Headquarters-Cyber to provide command and control of joint and coalition cyberspace forces.

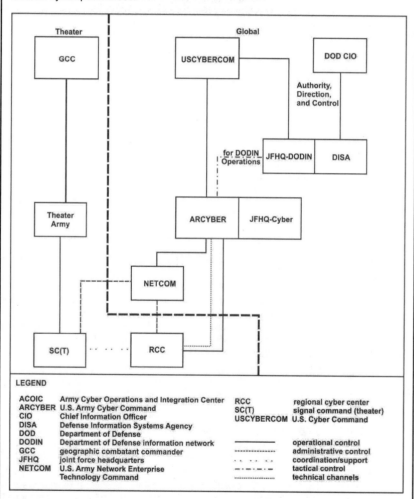

Ref: ATP 6-02.71 (Apr '19), fig. 2-1. United States Army Cyber Command Department of Defense information network operations relationships.

(DODIN) II. Roles and Responsibilities 6-11

NETCOM is the authorizing official for the Army enterprise, as directed by the Department of the Army CIO/G-6. NETCOM integrates Army IT to achieve a single, virtual enterprise network by overseeing end-to-end management of the Army enterprise service area, including service delivery, service operations, and infrastructure management. NETCOM prescribes all service delivery activities, policies, processes, procedures, and protocols for configuration management, availability management, capacity management, change management, and release management for Army networks and functional processing centers. To ensure unity of effort in DODIN operations, NETCOM has direct liaison authority to the Chief Information Officer/G-6, with notification to ARCYBER.

NETCOM manages the DODIN-A, including enforcing cybersecurity, technical, and configuration management programs and policies according to AR 25-1 and AR 25-2. NETCOM is the single entry point to submit validated telecommunications requirements to the Defense Information Systems Agency for coordination and implementation.

E. Global Department of Defense Information Network Operations Organizations

The DODIN operations architecture focuses on central management from higher-level echelons and decentralized execution. Within the DODIN, many organizations perform network and information systems management, cybersecurity, and physical and operational management functions. These organizations manage the DODIN through an established hierarchy of DODIN operations control centers. To ensure worldwide interoperability of the network and reduce redundant efforts, some DODIN operations activities take place at the global level. Global oversight also allows holistic network situational understanding. These control centers integrate DODIN operations to support communications and information systems. USCYBERCOM, joint and combatant commands, and Service components operate and manage these centers to control their respective portions of the DODIN. DODIN operations facilities at every level should be prepared to assume the responsibilities and functions of the next higher, lower, and adjacent DODIN operations elements in case of catastrophic failure, for example, battle damage, terrorist attack, or natural disaster.

Army Cyber Operations and Integration Center (ACOIC)

The ACOIC is an operational element of the ARCYBER headquarters. The ACOIC is the top-level control center for all Army cyberspace activities. The ACOIC provides DODIN operations reporting for Army networks and the wider DODIN, and situational understanding for Army networks. The ACOIC also provides worldwide operational and technical support for the DODIN-A across the strategic, operational, and tactical levels, in coordination with the theater armies. The ACOIC interfaces with the RCC, functional NOSCs, and other Service and agency DODIN operations centers.

The ACOIC analyzes threat information and directs network security actions to the RCCs in coordination with the theater armies. The ACOIC develops technical solutions to secure Army networks, and helps subordinate units implement network security measures.

Army Enterprise Service Desk

The Army Enterprise Service Desk provides user support to Army IT customers. The Army Enterprise Service Desk is the central agent for tier 0 and tier 1 service and application support.

The Army Enterprise Service Desk delivers a consistent set of centralized service desk processes to reduce manpower requirements. It uses cost effective assets and aligns with the Army's strategic vision. Army Enterprise Service Desk responsibilities include—

- Detecting incidents or reports of incidents.
- Accepting incident assignments.
- Performing initial diagnosis of an incident.
- Validating assignment of an incident.
- Troubleshooting an incident.
- Determining if incident functional escalation is required.
- Determining if hierarchical escalation is required.
- Escalating incidents as required.
- Reassigning an incident internally to another cyber center support group.
- Taking corrective actions to restore service.
- Performing incident resolution.
- Participating in post incident reviews.
- Performing process overview and review.
- Recommending process changes to the process lead.

Functional Network Operations and Security Centers (NOSC)

A functional NOSC performs the same enterprise manager functions as an RCC. While the RCCs' operational areas align with a geographic location, the functional NOSCs responsibilities align with a function, such as Army National Guard or special operations.

The Army National Guard operates a functional NOSC that provides the RCC functions for the states and territories, and administers GuardNet. The Army National Guard Information Networks Division manages GuardNet as an enterprise network that connects all the state and territory intranets.

The special operations signal battalion operates a functional NOSC to support the Army special operations forces communications system in a theater.

Refer to ATP 3-05.60 for more information about the Army special operations communications system.

II. Theater Level

From an operational perspective, the theater network consists of that portion operated by a GCC, its sub-unified and component commands, its joint and single-Service task forces, and installations and activities within the AOR. From a technical perspective, it is a subset of DODIN assets, resources, and services supporting that theater.

The GCCs direct joint DODIN operations in their respective theaters with support from Commander, USCYBERCOM, the USCYBERCOM Joint Operations Center, and the SC(T) associated with the theater army. The Defense Information Systems Agency provides an enterprise operations center for each theater under the tactical control of the GCC. Each GCC also has a theater network operations control center (TNCC) to maintain network situational understanding and provide operational and tactical control of their respective systems and network. Comprehensive network situational awareness allows commanders to make informed decisions to align network assets and capabilities to mission priorities and secure the network.

The joint cyberspace center (JCC) and TNCC collaborate with the USCYBERCOM cyberspace support element, the Defense Information Systems Agency enterprise operations center, and Service component DODIN operations centers, as appropriate, to ensure effective operation and defense of the DODIN in the theater. The enterprise operations center also offers onsite support teams for the theater. The enterprise operations center develops, monitors, and maintains the theater network situational awareness view. The geographic combatant command communications system directorate of a joint staff (J-6) provide requirements to aggregate and segment the theater network situational awareness view, as derived from the DODIN common network management data exchange standards (see DODI 8410.03). The network situational awareness view includes operational and tactical enterprise management, cybersecurity, and content management status. The enterprise operations center coordinates reporting requirements and the network situational awareness view with the JCC and TNCC.

The GCC exercises OPCON over all assigned DODIN operations elements and the theater network. The enterprise operations center is under the tactical control of the GCC for DODIN operations in the theater. The TNCC operates the theater network. The TNCC coordinates with the enterprise operations center and directs the Service component DODIN operations organizations to ensure the theater network supports the mission. The USCYBERCOM Joint Operations Center may provide support to ensure the theater network supports the GCC's requirements.

Figure 2-2 (facing page) depicts a notional theater DODIN operations structure.

USCYBERCOM adjudicates conflicts or resource contentions that arise due to a GCC's requirements. USCYBERCOM forwards conflicts they cannot resolve to the Chairman of the Joint Chiefs of Staff for adjudication. Services and agencies may establish theater-level operations centers or provide an uninterrupted theater-level network situational awareness view to support the GCCs' and their Service components' requirements. The global or theater operations center provides theater network visibility to the enterprise operations center and DOD component operations centers, as required. This Service or agency DODIN operations center also serves as the central point of contact for operational matters and emergency provisioning for a supported GCC to improve network situational understanding at all levels of command and enable end-to-end management of the DODIN.

A. Geographic Combatant Commander (GCC)

The GCC exercises combatant command authority over the Service and functional component commands in theater. The GCC exercises OPCON over all signal forces and associated DODIN operations elements within their AOR. This responsibility includes organizations and systems the DOD Services and agencies provide to extend the DODIN into the theater. The GCC's supporting DODIN operations organizations manage the theater network situational understanding data stores, databases, and graphical views. The GCC establishes information collection, filtering, display, and dissemination priorities. The GCC controls release of theater network status information to supporting and multinational forces, consistent with these priorities. The GCC aggregates theater network event, performance, and fault reporting data from subordinate and supporting Service and functional component commands and joint task forces relating to all systems and networks within their AOR to develop the network situational awareness view.

Combatant Command J-6

The J-6 establishes policy and guidance for all communications assets supporting the joint force commander, and develops communications system architectures and plans to support the GCC's mission. The J-6 also advises the GCC of the network's ability to support operations.

The J-6 develops policy and guidance for integrating and configuring operational networks and controls the joint information systems infrastructure. The J-6 exercises staff supervision and controls theater assets that the Defense Information Systems Agency, other Services, and other DOD agencies provide. The J-6 oversees theater DODIN operations through technical channels. Technical channels are the chain of authority for ensuring the execution of clearly delineated technical tasks, functions, and capabilities to meet the dynamic requirements of Department of Defense information network operations.

Theater DODIN Operations Structure

Ref: ATP 6-02.71, Techniques for Department of Defense Information Network Operations (Apr '19), pp. 2-2 to 2-3.

The GCC exercises OPCON over all assigned DODIN operations elements and the theater network. The enterprise operations center is under the tactical control of the GCC for DODIN operations in the theater. The TNCC operates the theater network. The TNCC coordinates with the enterprise operations center and directs the Service component DODIN operations organizations to ensure the theater network supports the mission. The USCYBERCOM Joint Operations Center may provide support to ensure the theater network supports the GCC's requirements.

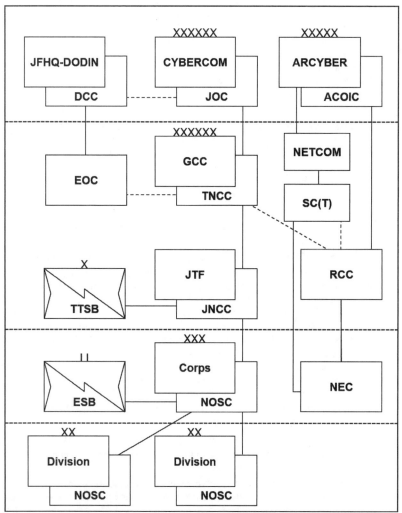

Ref: ATP 6-02.71 (Apr '19), fig. 2-2. Theater Department of Defense information network operations structure.

(DODIN) II. Roles and Responsibilities 6-15

Awareness of all current, future, or contemplated DODIN operations allows the JCC to advise the GCC of the network's ability to support assigned missions and operations. Maintaining situational awareness requires continual coordination with the USCYBERCOM cyberspace support element, the Defense Information Systems Agency, and the Defense Intelligence Agency.

Note. Signal personnel commonly incorrectly refer to a 'network operations chain of command.' The correct term to describe the chain of authority for the conduct of DODIN operations is technical channels.

Joint Cyberspace Center (JCC)

The JCC is the operational element of the combatant command that integrates DODIN operations, defensive cyberspace operations, and offensive cyberspace operations in the theater. As an operational extension of the GCC's command center, the JCC provides the commander and the enterprise operations center with the theater network situational awareness view and operational impact assessments.

In coordination with the TNCC, the JCC establishes the network situational awareness view based on commander and J-6 guidance and subordinate commands' requirements. The commander decides the minimum status information to ensure consistent, common situational understanding. Standardizing the status information simplifies integrated and roll-up views generated by different theaters or organizations.

Theater Network Operations Control Center (TNCC)

The TNCC controls all theater systems and networks operated by forces assigned to or supporting the GCC through technical channels. The TNCC, in coordination with the JCC, responds to USCYBERCOM direction for global DODIN operations issues. The TNCC's roles include—

- Monitoring the theater network.
- Determining operational impact of degradations and outages.
- Coordinating responses to degradations and outages that affect joint operations.
- Coordinating network actions to support changing operational priorities.

The TNCC aggregates the network situational awareness view from the enterprise operations center, Service component DODIN operations organizations, and joint network operations control center. Shared network situational understanding enables the success of the GCC's missions. The network situational awareness view application is part of an enterprise-wide software toolset, but the input data requirements and output reports are user-definable to meet each commander's needs.

The TNCC prioritizes and directs operational actions through the supporting enterprise operations center and DODIN operations personnel. The TNCC directs system and network management activities throughout the theater to support the GCC's DODIN operations decisions. To carry out its mission, the TNCC—

- Collaborates with the DODIN operations community of interest to ensure effective DODIN operation and defense.
- Tracks system and network outages and customer service shortfalls.
- Consolidates and analyzes reports from Service components, agencies, joint task forces, and deployed units.
- Directs DODIN operations event reporting, analyzes the impact of events on the operational mission, develops alternate courses of action, and advises the commander and other senior decision makers on the status of network degradations, outages, events, and areas requiring improvement.
- Establishes priorities for installing, configuring, and restoring systems and network services for the enterprise operations center and subordinate organizations.

- Directs, coordinates, and integrates response to network attacks and intrusions affecting the theater network.
- Directs the theater response to USCYBERCOM directives for global DODIN operations issues.
- Coordinates with USCYBERCOM to reconcile the GCC's DODIN operations priorities with the global priorities.

B. Enterprise Operations Center

The Defense Information Systems Agency operates the enterprise operations center to provide near real-time monitoring, coordination, control, and management of the theater network. The enterprise operations center aggregates and disseminates the consolidated network situational awareness view in a combatant command AOR. This capability includes shareable, look-up and lookdown views of Service component and joint task force elements in the theater.

The enterprise operations center develops, monitors, and maintains the network situational awareness view based on GCC or global enterprise operations center requirements. The situational awareness view includes pertinent theater, operational, and tactical system, network, and content management status information. To carry out its mission, the enterprise operations center—

- Operates and maintains the DISN backbone services in their theater.
- Coordinates theater network support, in collaboration with the TNCC.
- Collaborates with the DODIN operations community of interest to ensure effective operation and defense of the DODIN.
- Issues technical directives to Service DODIN operations centers to ensure compliance with GCC and USCYBERCOM direction.
- Supports the GCC, Services, and agencies by creating and disseminating the theater network situational awareness view. The network situational awareness view includes wireless and terrestrial links, satellite communications systems, and enterprise services.
- Maintains situational understanding to support current and near-term operations and deliberate plans.
- Coordinates reporting requirements and view specifications for network situational understanding with the JCC and TNCC.
- Continuously monitors and collects performance data for information resources based on GCC and global enterprise operations center priorities.
- Provides information security services to the TNCC or global enterprise operations center, including—
  - Monitoring, reporting, and analysis of intrusions and physical threats.
  - Correlating intrusion incidents with Service components, sub-combatant commands, and joint task forces.
- Helps identify the mission effects of degradations, outages, and DODIN events.
- Identifies and resolves security anomalies affecting theater network assets.
- Performs incident and intrusion monitoring and detection, strategic vulnerability analysis, computer forensics, and theater network-related activity response.
- Identifies courses of action and directs restoration of capabilities and services, when required.
- Directs courses of action and coordinates incident response to secure networks under attack.
- Coordinates with, and receives support from, the law enforcement and counterintelligence center.

- Manages theater radio frequency interference resolution.
- Supports satellite anomaly resolution.
- Supports satellite communications interference resolution.

C. Joint Task Force

If a theater army, corps, or division is the designated joint task force headquarters, its G-6 usually becomes the joint task force J-6. If another Service is the designated joint task force headquarters, Army G-6s function as subordinate organizations to the joint task force J-6. The CJTF controls joint force systems and networks through a joint network operations control center. Refer to JP 3-33 for more information on joint task force operations.

D. Theater Army G-6

Theater army signal support consists of the theater army G-6 staff and the SC(T). The SC(T) commander may also act as the theater army G-6. The G-6 is the principal advisor to the theater army commander on information management and information system matters across the AOR. The G-6 staff focuses on Army requirements in the theater.

The G-6 staff plans, manages, and controls communications systems input, information systems architecture, and long-range modernization plans for the theater army. It manages network enterprise initiatives and ensures the theater network architecture complies with DOD and Army standards.

E. Signal Command (Theater) SC(T)

The SC(T) provides DODIN operations capabilities to support Army, joint, and multinational forces in theater through its associated RCC. These capabilities use the DODIN-A for network extension and reachback to support the GCC. In coordination with the RCC, the SC(T) operates Army networks in the theater and delivers common user services to support the GCC and the theater army. With joint augmentation, the SC(T) may also assume joint or multinational DODIN operations responsibility for a joint task force. As the theater's senior Army signal commander, the SC(T) commander may be designated to serve as the J-6 of an Army-led joint task force, or the Army forces (ARFOR) G-6.

The theater army may task the SC(T) to provide overall control of theater signal assets. All or a portion of the SC(T) may be tasked to establish or augment the joint network operations control center, or provide land forces network control when tasked as part of an ARFOR. The SC(T) is comprised of a headquarters and one or more theater strategic signal brigades. The SC(T) DODIN operations responsibilities include—

- Providing centralized management and control for the theater army's data, voice, and video networks, including interfaces with joint, combined, and multinational systems via the RCC.
- Facilitating or establishing the joint network operations control center to support the CJTF.
- Enforcing cybersecurity policies and directions to support the GCC and theater army commander.
- Tailoring Army signal support to meet operational requirements.
- Providing direct support to Army signal formations supporting the ARFOR and joint task force.
- Providing guidance and governance through technical channels to Army NOSCs within the theater.
- Supervising DODIN operations tool employment.
- Providing DODIN operations and operational management for network and

automation assets provided by external organizations and agencies according to applicable service level agreements.
- Ensuring the cybersecurity tools are in place to maintain the integrity of the network, and to support secure access controls and connectivity.
- Implementing plans, policies, and procedures to install, operate, maintain, and secure assigned portions of the DODIN.
- • Establishing, or augmenting and staffing the Army's portion of, the joint network operations control center, as required.
- Conducting spectrum management operations for Army, joint, and multinational elements throughout the theater. Spectrum management operations are the interrelated functions of spectrum management, frequency assignment, host nation coordination, and policy that together enable the planning, management, and execution of operations within the electromagnetic operational environment during all phases of military operations (FM 6-02).
- Validating satellite communications requirements and managing ground mobile force tactical satellite communications equipment in the theater.

Note. In a theater with no assigned SC(T), the strategic signal brigade commander and staff carry out these functions as the senior signal organization in theater.

F. Theater Tactical Signal Brigade

The theater tactical signal brigade provides command and control and staff supervision for its subordinate expeditionary signal battalions, expeditionary signal companies, and joint/area signal companies. The brigade staff develops local area network and wide-area network (WAN) architectures necessary to accomplish the brigade's current and future missions. The brigade staff oversees the installation, operation, and maintenance of tactical communications systems by their subordinate battalions and companies.

The theater tactical signal brigade staff may also reinforce the supported G-6's DODIN operations. The supported G-6 makes network and augmentation requirement recommendations to the commander for effective DODIN operations. Augmenting the supported G-6 is situation-dependent, and requires close coordination between maneuver commanders, G-6s, and signal unit commanders to ensure the theater tactical signal brigade provides the necessary personnel and equipment to support the mission. When elements of the theater tactical signal brigade support a unit with an organic NOSC capability, they configure their systems to report status to the supported unit NOSC. When theater tactical signal brigade elements support a unit with no DODIN operations capability, they configure their systems to report status to the RCC through the regional hub node or DOD gateway.

Refer to FM 6-02 for more information on the theater tactical signal brigade and supporting units.

G. Strategic Signal Brigade

The strategic signal brigade provides the fixed theater communications infrastructure and services. Each strategic signal brigade is unique and tailored to support theater-specific communications infrastructure requirements. The RCC supports the strategic signal brigade in performing DODIN operations for theater networks and information systems. These functions include backbone networks, e-mail, spectrum management, communications circuitry, gateway routing to multinational networks, commercial telephone access in theater, and Defense Switched Network access to outside of the theater. The strategic signal brigade provides NEC services and support throughout its area of operations using its organic signal battalions, companies, and detachments.

H. Theater Department of Defense Information Network Operations Organizations

An interoperable theater network enables unity of command and unity of effort within a geographic AOR, as well as between AORs. Theater-level DODIN operations help joint force commanders exercise their DODIN authorities in their AOR.

Network Service Center

The overarching technique for implementing the Army enterprise network is the employment of the network service center. The network service center integrates three critical components—

- Enterprise manager (RCC)—provides DODIN operations oversight for Army forces in theater.
- Regional hub node—extends DODIN access and reachback to deployed units.
- Data center—provides enterprise IT services, application hosting, and backup for mission command, intelligence, and business systems.

Network service centers implement standardized policies to integrate voice, data, imagery, and DODIN operations capabilities, down to Soldier level, across all echelons. The network service center is not a facility, but a logical collection of capabilities. The network service center enhances the

Army's ability to—

- Maintain a continental United States-based Army that deploys to, and operates successfully in, remote operational areas.
- Rapidly and dynamically task-organize to enable operational flexibility.
- Train as it fights.
- Fight on arrival.

Enterprise Manager

The enterprise manager is the Army's lead DODIN operations authority throughout their service area. Enterprise managers' service areas are defined either geographically (the RCC) or functionally (Army National Guard NOSC or special operations forces NOSC).

Regional Hub Node

The regional hub node is a component of the network service center, which provides a transport connection between the Warfighter Information Network-Tactical and the wider Department of Defense information network (ATP 6-02.60). The regional hub node's network transport extends DISN services to deployed WIN-T enabled units. Network transport is the processes, equipment, and transmission media that provide connectivity and move data between networking devices and facilities. NETCOM maintains a limited amount of leased commercial satellite bandwidth for the regional hub nodes to support contingencies. Regional hub nodes provide the primary communications hub capability for operational forces.

A typical regional hub node can support up to 3 Army divisions and 12 separate enclaves, such as a brigade combat team (BCT), support brigade, or joint user, or up to 56 discrete missions simultaneously. Regional hub nodes' geographic distribution provides global coverage. Regional hub nodes are located at DOD gateway sites to provide a connection to DISN services. See ATP 6-02.60 for more information on the regional hub node.

Data Center

The data center is the Army's DODIN operations element that provides the primary information services capability within the enterprise, relaying SIPRNET and NIPRNET services within all theaters. Data centers deliver standardized local, regional, and global network services from centrally managed locations. The data center concentrates interconnectivity, hosts common servers and services, and interfaces users with the DODIN-A through an enhanced security gateway. Data centers improve the Army's DODIN operations posture by reducing the number of access points into the DODIN-A and by employing standardized DODIN operations tools and processes.

An installation processing node is a fixed data center serving a single DOD installation and its local area. It provides local services that a DOD core data center cannot technically or economically provide. There is no more than one installation processing node per installation, but an installation processing node may have multiple enclaves to accommodate unique installation needs, such as joint bases.

The installation security router forms the boundary of the installation processing node. The installation security router is the security perimeter to the local processing center that houses applications and servers, and the interface to the installation's local area networks.

Data Center Consolidation

Army systems and networks face the constant risk of compromise and disruption. Each IT system and application presents a potential attack surface for enemies and adversaries. For this reason, the Army is reducing the number of Army IT systems and applications and optimizing those remaining to operate in modern, cloud-enabled computing environments. The Army is consolidating its data centers to one installation processing node for each post, camp, or station as part of the larger DOD goal to reduce data center infrastructure by at least 60 percent. Reducing the number of data centers will enable transition to the long-term end state of four Army enterprise data centers in the continental United States and six outside the continental United States.

Regional Top Level Architecture (Joint Regional Security Stack)

A top level architecture provides perimeter protection in the distributed DOD network. Joint information environment is replacing over 1,000 installation-based top level architectures across the DOD with fewer than 50 regional top level architectures (joint regional security stacks). Reducing the number of top level architectures minimizes the DODIN's threat surface and cybersecurity workload. In the Army's consolidated security architecture, the post, camp, or station network perimeter logically extends to the joint regional security stack hosted at a DOD core data center, defense enterprise computing center, or at a Service base, post, camp, or station. The installation processing node and the joint regional security stack support defense-in-depth. Defense-in-depth is an information security strategy integrating people, technology, and operations capabilities to establish variable barriers across multiple layers and missions of the organization (CNSSI 4009). The installation processing node provides perimeter protection for the local processing center at selected posts, camps, and stations according to security technical implementation guides and isolates the Army intranet from external traffic on a post, camp, or station.

Note. On full implementation of the joint information environment, the Army will eliminate all installation-based top level architectures.

Regional Cyber Center (RCC)

The RCC is the single DODIN operations point of contact in the theater for Army network services, operational status, service provisioning, and service interruption resolution and restoral. The RCC provides network visibility and status information to the Defense Information Systems Agency enterprise operations center in theater. In some theaters, the RCC provides network visibility to other Service component DODIN operations centers. RCCs perform the same functions and use the same tactics, techniques, and procedures across all theaters.

While all units are responsible for securing their respective portions of the network, the RCC exercises overall responsibility for securing the Army's portion of the theater network. The RCC develops technical solutions to ensure network security, and helps subordinate units implement network security. The RCC helps develop theater cybersecurity policy and implements that policy.

The RCC performs or coordinates DODIN operations tasks that span the theater or multiple regions to provide consistent service among regions. This places the operational function at the only location in the theater with visibility or awareness of multiple regions. The RCC's DODIN operations responsibilities include—

- Providing event and incident management capabilities, such as analysis and correlation of event data, to all units in the theater, as required.
- Disseminating NETCOM-developed software distribution packages for all units in the theater.
- Managing the capabilities, availability, and performance of all theater units' systems.
- Coordinating distribution of system patches and notifying the NEC, enterprise operations center, TNCC, GCC, and all units in the theater of impending patches.
- Synchronizing the global address list for all units in the theater.
- Managing e-mail hubs to support all units in the theater.
- Providing theater-level technical support for problems escalated from the NEC.
- Overseeing operation, management, and security of the DODIN-A throughout the theater.
- Assessing mission impact of network events for the SC(T) commander.
- Operating and managing all Army-controlled items on the public or Defense Information Systems Agency side of the installation network infrastructure.
- Operating and managing selected systems and networks on the installation.
- Enforcing cybersecurity policies and reporting violations on Army networks.
- Identifying physical or logical property to address through the configuration management process.
- Providing guidance to Army NOSCs in theater through technical channels.
- Supervising Army use of DODIN operations tools in theater.
- Operationally managing communications assets provided by external organizations and agencies.
- Managing signal interfaces with joint and multinational forces, including host-nation support interfaces.
- Managing and controlling network transport and information services from the generating force to the operational force.
- Performing network management and enterprise systems management activities required to manage the information systems infrastructure and multi-organizational networks supporting the operational mission.
- Ensuring the cybersecurity tools are in place to for network protection and integrity, and to support secure access controls and connectivity.
- Performing change and release management to support units in the theater.
- Providing public key infrastructure support for Army users in theater.

The RCC coordinates with the ACOIC to oversee defensive cyberspace operations-internal defensive measures, network vulnerability assessment, and incident management within the theater. Defensive cyberspace operations-internal defensive measures are operations in which authorized defense actions occur within the defended portion of cyberspace (JP 3-12). In coordination with the ACOIC, the RCC—

- Coordinates threat-based vulnerability assessments with the supporting counterintelligence element.
- Conducts attack signature sensing and warning analysis.
- Develops mitigation strategies to support network defense and prevent data loss, including loss due to spillage. Spillage is a security incident that results in the transfer of classified information onto an information system not authorized to store or process that information (CNSSI 4009).
- Conducts the computer defense assistance program (penetration testing, network assistance visits, and network damage assessments) to support commanders and theater units.

I. Installation-Level Department of Defense Information Network Operations Infrastructure

Installation-level infrastructure delivers network services and automated information systems support at fixed locations. As the joint information environment matures, many of these functions (such as Defense Enterprise E-mail) will take place at the enterprise level.

The network enterprise center is the facility that provides and acquires telecommunications and information management services on Army installations. The NEC implements and manages enterprise services (including e-mail, user storage, office automation, collaboration, and cybersecurity) according to current policy, procedural guidance, and management procedures.

Centralized access to communications and DISN services at the NEC—

- Reduces congestion in the electromagnetic spectrum—each unit does not have to use separate transmission systems for network access.
- Reduces redundant efforts, since each unit does not have to maintain networking equipment and network administration. This still allows the flexibility of a unit performing its own network management and cybersecurity compliance activities using installation as a docking station.
- Improves cybersecurity posture by standardizing cybersecurity implementation across the installation with fewer potential points of failure.
- Reduces the cyberspace threat surface by maintaining fewer access points to the DODIN.

The NEC provides overall DODIN operations on its post, camp, or station, or within a designated geographic area. NECs plan and budget for network and information systems upgrades or replacements to meet customer requirements. NECs work with external organizations to ensure proper operation of installation-level components of DOD or Army-level networks and information systems. The NEC's DODIN operations responsibilities include—

- Providing customer access to the installation campus area network and information systems infrastructure.
- Providing service desk support and problem resolution for networks and information systems on the installation, or in the service area for which the NEC is directly responsible.
- Sharing information with other network managers about lessons learned and innovative ideas to support users.
- Implementing DODIN operations best business practices according to DOD, Army, NETCOM, and SC(T) policy and guidance.

- Coordinating with the strategic signal brigade to manage inter-installation networks and information systems affecting their supported organizations.
- Establishing and managing the cybersecurity program for the installation campus area network.
- Providing an installation as a docking station connection so units can connect their tactical systems to the installation campus area network.
- Managing network and information system resources in its area of operations under the direction of the supporting RCC, in coordination with the SC(T).
- Assessing mission impact of outages, network defense incidents, and other network issues for the RCC and strategic signal brigade.
- Responding to RCC direction to support problem resolution, change requests, and information assurance vulnerability management (IAVM).

The Army National Guard treats each state and territory as an installation (post, camp, or station). The state G-6 Directorate of Information Management provides NEC services to the units and members of that state or territory National Guard. Each state has an installation processing node and data center that hosts unique applications and data relevant to only that state or territory.

III. Corps and Below Units

Combat and combined arms units at echelons corps and below have adequate organic signal capabilities to conduct their standard missions without requiring outside signal support. The theater army may place additional signal assets, such as a theater tactical signal brigade, ESB, or other signal elements under OPCON of the corps or division, as required. The RCC has overall responsibility for Army DODIN operations in the theater. All subordinate NOSCs take direction from the RCC and report status to the RCC through technical channels. The RCC maintains comprehensive situational understanding of the theater network.

For detailed information about available signal assets and requests for signal support, refer to FM 6-02.

A. Corps and Division

The corps and division are headquarters organizations able to exercise command and control over land forces or serve as a joint task force headquarters. Both the corps and division have organic headquarters elements that command assigned or attached maneuver and support elements to meet mission objectives. Corps and division organic signal assets consist of the G-6 section and the signal portion of the signal, intelligence, and sustainment company. Having the signal support and DODIN operations capabilities organic to the headquarters enables unity of command and unity of effort. Commanders can readily leverage their portion of the network as a warfighting platform and align signal and network support to mission priorities.

Figure 2-3 (facing page) shows the corps and below DODIN operations relationships and corresponding technical channels.

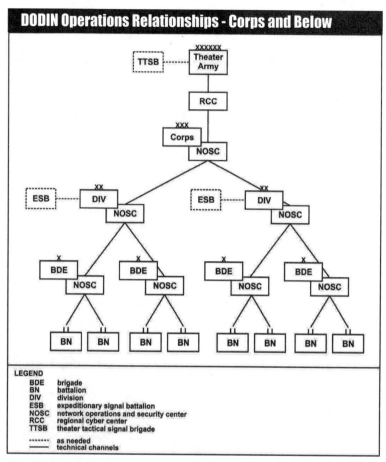

Ref: ATP 6-02.71 (Apr '19), fig. 2-3. Department of Defense information network operations relationships-corps and below.

Corps and Division G-6

G-6 sections in the corps and division are organized the same, except the grade structure. The G-6 controls DODIN operations within the unit's area of operations in compliance with joint, Army, and theater policies. The G-6 works closely with the higher headquarters G-6 or J-6, subordinate G-6, battalion or brigade signal staff officer (S-6), OPCON theater tactical signal brigade or ESB elements, and the organic signal, intelligence, and sustainment company to achieve integrated DODIN operations supporting the commander's intent. The G-6 staff plans and designs DODIN operations capabilities and support for command posts and subordinate units. The staff also provides training and readiness oversight for assigned and attached units.

The G-6 controls and monitors the network situational awareness view, including subordinate networks. The G-6 also helps integrate the network situational awareness view with that of the higher headquarters, for example the one controlled and maintained by a joint task force J-6. The situational awareness view consists of the status of all network components within the unit's area of operations, as well as the status of WAN links to theater, adjacent, and subordinate units. 2-69. Commanders have the authority to delay directed changes to their portion of the network. The commander may receive a network directive from higher headquarters that could adversely impact the unit's mission. In this case, issues will be resolved through command channels. Issue resolution requires close coordination between the commander, the G-6, and the higher headquarters commander and G-6 or J-6. The commander carefully considers the potential impact of delayed compliance with network directives and coordinates with higher headquarters and affected organizations to resolve compatibility issues and comply with the directed changes as soon as the tactical situation allows.

The G-6 section provides DODIN operations support to the main, tactical, and support area command posts and the mobile command group. G-6 DODIN operations activities integrate geographically separated units into the DODIN-A. Subordinate units' DODIN operations provide another level of management, which the G-6 coordinates as part of the overall DODIN operations plan.

Units without organic signal assets, such as functional support brigades, sometimes augment corps and divisions. The supported headquarters provides communications support for augmenting units, either with elements of their organic signal company or by requesting external, pooled assets through the request for forces process. The G-6 integrates the supporting brigade and other signal assets into the network and provides DODIN operations for the supporting unit. The expanded DODIN operations mission may require augmenting the DODIN operations section with external capabilities from the supporting unit S-6 or from a theater tactical signal brigade or ESB.

The corps or division G-6 has these DODIN operations responsibilities—

- Recommending communications system and DODIN operations priorities for networks and systems to support the commander's priorities.
- Establishing procedures for relevant information and information systems to develop the common operational picture, in coordination with the assistant chief of staff, operations.
- Managing IT infrastructure to follow theater and Army-wide policies and standards, in coordination with the SC(T).
- Serving as the Army component G-6 in a joint task force, when designated. This mission may require equipment and personnel augmentation and support from the SC(T) and RCC.
- Serving as the joint task force J-6, if designated. This mission may require equipment and personnel augmentation.
- Advising the commander, staff, and subordinate commanders on communications networks and information services.
- Supervising DODIN operations in the area of operations.
- Monitoring, and making recommendations for, communications networks and information services.
- Preparing, maintaining, and updating communications systems operation estimates, plans, and orders. These orders often require configuration management changes across multiple organizations.
- Providing signal units with direction and guidance for plans and diagrams to establish the information network.
- Providing signal units with unit locations, organizational status, and communications requirements.
- Planning the integration of information systems.
- Developing, updating, and distributing signal operating instructions.
- Coordinating with signal elements of higher, adjacent, subordinate, and multinational units.
- Preparing and publishing communications system standard operating procedures for command posts.

- Coordinating, planning, and conducting spectrum management operations in the area of operations.
- Planning and coordinating with higher and lower headquarters for information system upgrades, replacement, elimination, and integration.
- Performing network vulnerability and risk assessments, in coordination with the assistant chief of staff, intelligence and the information operations officer, and according to Army and theater cybersecurity policies and procedures.
- Monitoring and disseminating information that changes warfighting function priorities and control measures.
- Coordinating, planning, and directing cybersecurity activities.
- Ensuring the command complies with Army and theater automation and systems administration policies, procedures, and standards.
- Validating user information requirements to support the mission.
- Establishing and disseminating the electronic battle rhythm, in coordination with the chief of staff, or assistant chief of staff, operations.
- Establishing policies and procedures for using and managing information tools and resources.
- Planning DODIN operations support for the corps or division command posts, and those of subordinate units.

Corps and Division G-6 Signal Operations

The corps and division G-6 signal operations sections consist of network management elements, cybersecurity and communications security (COMSEC) cells, plans elements, and signal systems support elements.

The network management element performs these DODIN operations functions—

- Manages the unit's portion of the DODIN-A, from the applications through the connections to the theater network.
- Identifies, validates, establishes, plans, and manages communications requirements, including tracking the headquarters and subordinate units' communications requirements within the area of operations.
- Installs, operates, maintains, and secures communications networks across the unit, including subordinate units, within the area of operations.
- Executes deliberate network modifications to meet the commander's requirements.
- Installs, operates, and maintains COMSEC and transmission security devices to maintain confidentiality, integrity, availability, and authentication for transmission over private and public communications and media.

  For more information on COMSEC, refer to ATP 6-02.75.

- Performs fault, configuration, accounting, performance, and security management of network system components and services to ensure systems and applications meet the commander's operational requirements.
- Manages the quality of service of the network services, including those provided by systems the G-6 does not directly control, for example, Global Broadcast System and combat service support very small aperture terminal.
- Conducts spectrum management operations, including frequency allocation and deconfliction with signal and non-signal emitters.
- • Advises the commander, staff, and subordinate commanders on DODIN operations and network priorities to support the commander's intent.

(DODIN) II. Roles and Responsibilities 6-27

- Conducts information dissemination management and content staging so users can locate and retrieve voice and data information over SIPRNET, NIPRNET, and mission partner environment.
- Produces and distributes signal operating instructions.
- Prepares and publishes DODIN operations-related annexes and standard operating procedures.
- Prepares network reports for submission to the network management technician at G-6 and higher headquarters NOSCs.
- Provides network monitoring data to higher headquarters NOSCs and the network situational awareness view to local or subordinate users.
- Plans, integrates, and synchronizes network management with the cybersecurity, COMSEC, and information dissemination management cell.

The cybersecurity, COMSEC, and information dissemination management cell performs these DODIN operations functions:
- Manages cybersecurity compliance.
- Implements access controls for—
  - Information.
  - Information systems.
  - Networks.
- Manages software copies, updates, and security patches.
- Monitors for, detects, and analyzes anomalies that could cause network disruption, degradation, or denial.
- Executes response and restoration activities to resolve incidents that interrupt normal operations.
- Identifies threats against, and vulnerabilities of, the organization's information assets.
- Implements physical security from the outside perimeter to the inside operational space, including information system resources.
- Ensures the effectiveness of cybersecurity infrastructure (for example, firewalls and intrusion prevention system), and tools (Host Based Security System, antivirus, and software update service).
- Provides training and instruction on cybersecurity awareness, observations, insights, and lessons learned.
- Coordinates with higher and subordinate headquarters units to provide network defense-in-depth.
- Collaborates with the assistant chief of staff, intelligence to gain awareness of current threats and disseminate awareness of network anomalies.
- Conducts COMSEC management—
  - Plans and manages COMSEC operations, including subordinate units.
  - Implements procedures for detecting and reporting COMSEC insecurities.
  - Receives, transfers, accounts for, safeguards, and destroys COMSEC material.
  - Ensures users employ COMSEC key only for its intended purpose w/i the network.
- Conducts information dissemination management and content staging.
- Implements, manages, and maintains user services (web services, e-mail, database, collaboration tools, and mass storage).
- Plans and coordinates procedures for contingency operations, including continuity of operations and data recovery.

- Provides collaboration, messaging, and storage services to support information advantage.
- Prioritizes information resources.
- Monitors information delivery status and integrates it into the network situational awareness view.
- Manages information system and network support activities.
- Disseminates the common operational picture and executive information.

The plans element performs these DODIN operations functions:
- Prepares, maintains, and updates command information management estimates, plans and orders, including—
    - Mission analysis products.
    - Annex H (signal) to corps or division plans and orders.
    - Telecommunications plan.
    - Network management plan.
    - Information management plan.
- Establishes procedures to employ relevant information and information systems to develop the common operational picture, in coordination with the assistant chief of staff, operations.
- Coordinates local network capabilities and services.
- Conducts spectrum management operations.
- Plans combat visual information documentation.
- Coordinates future network connectivity, information dissemination management, and network interface with joint and multinational forces, including those of the host nation.
- Coordinates, plans, and directs development of the network situational awareness view for the main command post.
- Plans the transition of responsibility for the tactical network from the corps or division to permanent theater signal assets.

The signal systems support element performs these DODIN operations functions:
- Installs, operates, maintains, and secures servers for SIPRNET, NIPRNET, and mission partner environment to support the main, tactical, and support area command posts.
- Manages installation and operation of main, tactical, and support area command post local area networks, including cable and wire installation and troubleshooting.
- Establishes and operates the corps or division service desk to provide user assistance for voice, video teleconferencing, and e-mail services.

Corps and Division Network Operations and Security Centers (NOSC)

The NOSC performs the DODIN operations activities required to operate and secure the network within the corps or division area of operations. The NOSC responds to shifting network priorities to support the tactical plan and extend the DODIN's strategic capabilities to tactical formations. Signal elements coordinate with the NOSC to install, operate, maintain, and secure the network.

The NOSC establishes the network and provides operational and technical support to all signal elements in the area of operations. It also provides network status and running estimates to operations planners to support tactical operations.

B. Brigade Combat Team and Multifunctional Support Brigade

The BCT has organic signal assets that provide network transport and information services to support the commander's information requirements. The BCT's signal elements provide 24-hour communications networks to its formations, and install, operate, maintain, and secure these systems.

Each multifunctional support brigade (maneuver enhancement, field artillery, combat aviation, sustainment, or security force assistance) has an organic signal company. This company provides the brigade's tactical communications support.

Brigade Combat Team and Multifunctional Support Brigade S-6 Responsibilities

The brigade S-6 maintains DODIN operations in the brigade area of operations in compliance with joint, Army, and theater policies. The brigade S-6 may also serve as the Army component S-6 in a joint task force. The brigade S-6 works closely with its higher headquarters G-6 or J-6 and the brigade signal company to integrate DODIN operations while meeting the commander's intent. The brigade S-6 controls and monitors the status of the brigade portion of the DODIN-A, including subordinate units, to maintain network situational understanding. The S-6 also helps integrate the brigade network situational awareness view with that of the higher headquarters, such as corps or division G-6 or joint task force J-6.

The brigade commander has the authority to delay directed changes to the brigade portion of the network. The brigade commander may receive a network directive from higher headquarters that could adversely impact the brigade's mission. In this case, issues are resolved using command channels. Resolving issues requires close coordination between the brigade commander, the brigade S-6, and the higher headquarters commander and G-6 or J-6. To resolve compatibility issues and comply with the directed changes, the commander carefully considers the potential impact of delayed compliance with network directives and coordinates with higher headquarters and affected organizations as soon as the tactical situation allows.

If units without organic signal assets augment the brigade, or if the brigade must establish a network beyond its organic capabilities, the brigade S-6 defines communications and network support requirements, based on the situation and mission. The operational chain of command validates requests for signal support they cannot source internally and forwards them to United States Army Forces Command for approval and resourcing. The brigade S-6 assumes DODIN operations responsibility for the augmenting elements.

Refer to FM 6-02 for detailed information about available signal assets and requests for signal support.

The brigade S-6 staff plans DODIN operations capabilities and network support for brigade command posts and subordinate units. The S-6 section personnel are part of the brigade command post staffing to support the commander's critical information requirements.

Figure 2-4 (facing page) depicts the brigade DODIN operations relationships and technical channels with higher and lower headquarters.

Brigade Combat Team and Multifunctional Support Brigade Network Operations and Security Center (NOSC)

The brigade NOSC is the brigade's network control center that plans and directs DODIN operations. Designated personnel from the S-6 section staff the brigade NOSC with the same responsibilities as higher-level NOSCs, scaled to the size of the unit and operation.

Brigade DODIN Relationships

Ref: ATP 6-02.71, Techniques for Department of Defense Information Network Operations (Apr '19), pp. 2-20 to 2-21.

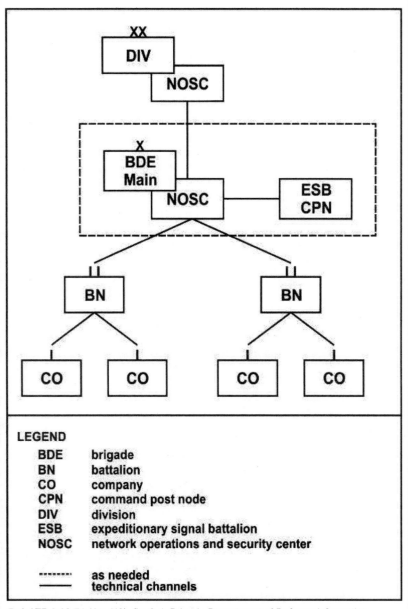

Ref: ATP 6-02.71 (Apr '19), fig. 2-4. Brigade Department of Defense information network operations relationships

The brigade NOSC reports directly to the brigade S-6. The NOSC uses the brigade's organic WIN-T network management capability to configure, monitor, and manage the brigade WAN. The brigade NOSC supports the S-6 section in installing, operating, maintaining, and securing the command post local area networks, and prioritizes information dissemination across the WAN. Under the direction of the brigade S¬6, the NOSC—

- Conducts brigade spectrum management operations.
- Plans and manages the brigade information network.
- Plans and manages cybersecurity, including—
  - Firewalls.
  - Intrusion detection systems.
  - Access control lists.
  - Key management and distribution.
  - Cybersecurity compliance.
- Plans and manages information dissemination management and content staging for the brigade— user profiles, file and user priorities, and dissemination policies—in coordination with higher headquarters NOSCs and the supporting RCC.
- Evaluates the brigade's network and communications relay requirements.
- Conducts DODIN operations to support the unit's mission.
- Advises operation planners of current network status, and provides estimates to support tactical operations.

Network Management Cell

The network management cell operates, maintains, and sustains networked systems to provide the desired level of quality and guarantee availability. The network management cell—

- Manages the brigade network, from the applications on brigade platforms through the connections to the division network.
- Identifies, validates, and manages communications requirements, including the headquarters and subordinate units' requirements within the area of operations.
- Monitors network performance and quality of service, including interoperability of the brigade network with external networks.
- Installs, operates, and maintains communications networks, including subordinate units within the area of operations.
- Executes deliberate network modifications to meet the commander's requirements.
- Manages quality of service for brigade network services, including those systems not directly controlled by the S-6, for example, Global Broadcast System and combat service support very small aperture terminal.
- Installs and maintains COMSEC and transmission security devices for secure transmission over private and public communications networks and media.
- Performs fault, configuration, accounting, performance, and security management of network system components and services (situational understanding, voice, video, data, and imagery) to ensure systems and software applications meet the commander's operational requirements.
- Conducts spectrum management operations.
- Advises the commander, staff, and subordinate commanders on DODIN operations and network priorities to support the commander's intent.

- Prepares and publishes DODIN operations related annexes and standard operating procedures.
- Develops, produces, updates, and distributes signal operating instructions.
- Plans, coordinates, integrates, and synchronizes network management with the cybersecurity and COMSEC cell and the signal system integration and oversight and information dissemination management cell.
- Reports network status to the network management technician, brigade S-6, and division NOSC.
- Provides the network situational awareness view to the division NOSC and authorized recipients in the brigade.

Cybersecurity and Communications Security Cell

The cybersecurity and COMSEC cell monitors and manages activities to provide data confidentiality, integrity, availability, and protection against unauthorized access. The brigade S-6 cybersecurity and COMSEC cell—

- Implements access controls for—
    - Information.
    - Information systems.
    - Communications networks.
- Manages software copies, updates, and security patches.
- Executes response and restoration activities to resolve incidents that interrupt normal operations.
- Conducts risk assessments to identify threats against, and vulnerabilities of, the organization's network.
- Manages cybersecurity compliance.
- Implements physical security, from the outside perimeter to the inside operational space, including all information system resources.
- Ensures the effectiveness of cybersecurity infrastructure, using firewalls and intrusion prevention system and tools, for example, host-based prevention, antivirus, and software update service.
- Monitors for, detects, and analyzes anomalies that may disrupt, degrade, or deny network service.
- Plans and manages brigade COMSEC operations, including subordinate units, and implements COMSEC incident detection and reporting procedures.
- Receives, transfers, accounts for, safeguards, and destroys COMSEC material.
- Ensures users employ COMSEC key only for its intended purpose on the network.
- Provides cybersecurity awareness, observations, insights, and lessons learned training.
- Coordinates with higher headquarters, and directs subordinates, to provide network defense-in-depth.
- Collaborates with the intelligence staff officer to gain threat awareness and disseminate awareness of network anomalies.
- Plans, coordinates, integrates, and synchronizes cybersecurity activities with the network management cell and the signal systems integration and oversight and information dissemination management cell.

Signal Systems Integration and Oversight and Information Dissemination Management Cell

The signal systems integration and oversight and information dissemination management cell provides training and maintenance oversight for the brigade signal company. It monitors and manages information storage and dissemination. The brigade S-6 signal systems integration and oversight and information dissemination management cell—

- Plans and manages the tactical radio network.
- Installs, operates, and maintains automated information systems.
- Enables the discovery of information, services, and applications.
- Provides collaboration, messaging, and information storage to support information advantage.
- Prioritizes information resources.
- Monitors information delivery status and integrates with overall situational understanding.
- Supports IT life cycle management.
- Integrates systems across the unit and with Army and joint, inter-organizational and multinational elements.
- Helps install user information systems.
- Implements the tactical intranet at brigade and below.
- Provides system administration and local area network management.
- Performs service desk functions and problem tracking.
- Plans, coordinates, integrates, and synchronizes signal systems integration and oversight and information dissemination management with the network management cell and the cybersecurity and COMSEC cell.

Maneuver and Support Battalions and Companies

Maneuver battalions and companies receive tactical internet support using organic WIN-T assets, managed by the brigade NOSC. Support battalions and companies receive their tactical internet support from elements of the brigade signal company, which also provide DODIN operations for the supported unit. Tactical battalion and below units have no inherent DODIN operations roles in the upper tier tactical internet.

Refer to ATP 6-02.60 for more information about WIN-T, and ATP 6-02.53 for information on DODIN operations in the lower tier tactical internet.

Functional Support Brigades

Theater-level commands may receive specialized support from functional support brigades. Examples of functional support brigades include—

- Military police.
- Engineer.
- Air and missile defense.
- Medical.
- Chemical, biological, radiological, and nuclear.
- Civil affairs.

Functional support brigades may be attached or OPCON to a corps or division. These brigades do not have organic signal companies or assets. They receive their signal support from pooled assets, such as an ESB, or from the internal assets of the supported unit. The supported unit assumes DODIN operations responsibility for supporting units' requirements. This expanded DODIN operations role may require DODIN operations augmentation to manage the network.

III. DODIN Network Operations Components

Ref: ATP 6-02.71, Techniques for Department of Defense Information Network Operations (Apr '19), pp. A-1 to A-7.

Department of Defense information network operations consist of enterprise management, cybersecurity, content management, network situational understanding, and their underlying principles and components.

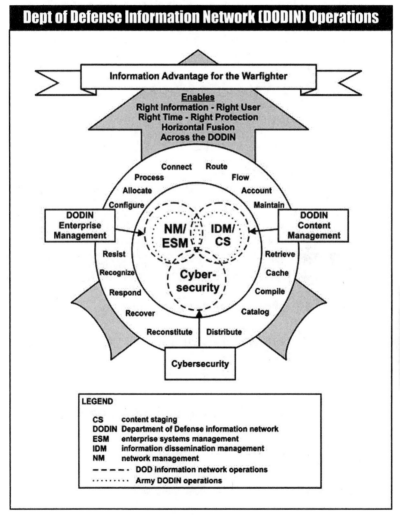

ATP 6-02.71, fig. A-1. Department of Defense information network operations operational construct.

I. DODIN Operations Operational Construct

DODIN enterprise management, DODIN content management, cybersecurity, and network situational understanding guide the installation, management, and protection of communications networks and information services to support operational forces. DODIN operations provide users and systems at all levels with end-to-end network and information system availability, information protection, and prompt information delivery.

Figure A-1 (previous page) depicts how Army DODIN operations tasks nest within, and correspond with, the wider joint DODIN operations functions. Network management and enterprise systems management (DODIN enterprise management) consist of steps to configure, assign, process, connect, route, flow, account for, and maintain network capabilities. Information dissemination management and content staging (DODIN content management) allow users to retrieve, cache, compile, catalog, and distribute information to support planning and decision making. Cybersecurity provides the means to resist and recognize intrusions and to recover and reconstitute network capabilities. The net effect of integrating the three tasks is information advantage. The right user gets the right information at the right time, with the right protection and in a usable format.

II. DODIN Enterprise Management

DODIN enterprise management is the technology, processes, and policies necessary to effectively and efficiently install, operate, maintain, and sustain communications networks, information systems, and applications. DODIN enterprise management merges IT services with DODIN operations critical capabilities.

A. Functional Services

The five major functional services of DODIN enterprise management foster the installation, operation, maintenance, and sustainment of communications networks and information services technologies to ensure their performance, availability, and security. These services are inherent at the strategic, operational, and tactical levels across all warfighting functions. The five functional services are—

Functional Services

1. Enterprise Systems Management
2. Systems Management
3. Network Management
4. Satellite Communications Management
5. Frequency Assignment

See facing page for an overview and further discussion.

6-36 (DODIN) III. Network Operations Components

Functional Services

Ref: ATP 6-02.71, Techniques for Department of Defense Information Network Operations (Apr '19), p. A-3.

Enterprise Systems Management
Enterprise systems management for end-user and system applications focuses on the availability, performance, and responsiveness of enterprise service capabilities. Enterprise services are IT services that span an entire large organization, or enterprise. In the case of the DODIN, enterprise refers to the DOD, including all of its organizational entities (DODD 8115.01). Some enterprise services are—
- DOD Enterprise E-mail
- DOD Enterprise Portal Service
- Defense Collaboration Services
- Enterprise search
- Enterprise file share
- Identity and access management

Systems Management
Systems management provides day-to-day administration of computer-based systems, elements of systems, and services including software applications, operating systems, databases, and hosts of end-users. Systems management is comprised of all measures necessary to operate enterprise systems and services effectively and efficiently.

Network Management
Network management provides a network infrastructure with the desired level of quality and guaranteed service. Networks included in enterprise management are located on all transmission media and tiers of communication: terrestrial, aerial, and satellite communications. They include digital telephony, packet routing, and cell-switched networks using fiber optic or wireless transport media.

Satellite Communications Management
Satellite communications management is the day-to-day operational control of satellite communications resources. In the Army, the operational authority for satellite communications management is United States Army Space and Missile Defense Command/Army Forces Strategic Command. Satellite communications management includes—
- Appropriate support when service is disrupted
- Providing satellite communications system status
- Maintaining situational understanding, including current and planned operations, as well as space, control, and earth segment asset and operational configuration management
- Satellite anomaly resolution and management
- Satellite communications interference to the network

Refer to FM 3-14 and ATP 6-02.54 for information about satellite communications mgmt.

Frequency Assignment
Frequency assignment involves ensuring frequency resources are available to support effective and efficient frequency use for planning, managing, and operating the wireless portion of the network. Ensuring frequency availability requires determining accurate spectrum requirements and coordinating with spectrum managers, who continually deconflict frequencies used for the network, enabling dynamic operations. The frequency assignment function of spectrum management operations enables frequency availability. In the Army, the operational authority for frequency assignment is the Army Spectrum Management Office.

Refer to ATP 6-02.70 for more information about spectrum management operations.

B. Critical Capabilities

DODIN enterprise management involves several critical capabilities associated with IT services. Enterprise managers must achieve these capabilities at the strategic, operational, and tactical levels across all warfighting functions. Enterprise management has five critical capabilities—

Critical Capabilities

1. Fault Management
2. Configuration Management
3. Accounting Management
4. Performance Management
5. Security Management

See facing page for an overview and further discussion.

C. Enabled Effects

DODIN enterprise management enables network and information system availability and information delivery. These effects are achieved by—

- Maintaining robust network capabilities in the face of component or system failure or adversary attack.
- Configuring and allocating network and information system resources.
- Rapidly and flexibly deploying network resources.
- Ensuring effective, efficient, and prompt processing.
- Ensuring connectivity, routing, and information flow.
- Planning for increased network use.

D. Objective

The objective of DODIN enterprise management is to provide network control for Army communications systems and enable interoperability with joint networks. Army DODIN operations managers conduct enterprise management at all levels of military operations.

Critical Capabilities

Ref: ATP 6-02.71, Techniques for Department of Defense Information Network Operations (Apr '19), pp. A-3 to A-4.

DODIN enterprise management involves several critical capabilities associated with IT services. Enterprise managers must achieve these capabilities at the strategic, operational, and tactical levels across all warfighting functions. Enterprise management has five critical capabilities—

Fault Management
Fault management is associated with failure of the network or information systems that affects connectivity and functionality. Fault management is a five-step process:
- Detect faults
- Locate faults
- Restore service
- Identify the cause of the fault
- Establish solutions so similar faults do not occur in the future

Configuration Management
Configuration management applies technical and administrative direction and surveillance to—
- Identify and document the functional and physical characteristics of a configuration item
- Control changes to those characteristics
- Record and report changes to processing and implementation status

Accounting Management
Accounting management helps effectively allocate internal and external resources. The goal of accounting management is to identify true requirements based on network monitoring and system use. The desired result is a network and information systems configuration that provides the most effective, efficient use of resources. Planners also use monitoring data to identify future resource requirements.

Performance Management
Performance management is monitoring and managing network and information systems performance. Performance management involves—
- Data monitoring
- Problem isolation
- Performance tuning
- Statistical analysis for trend recognition
- Resource planning

Security Management
Security management is implementing technical and administrative measures to secure access to the information transmitted over the network or processed and stored on information systems. Security management integrates enterprise management with cybersecurity.

III. Enterprise Management Activities

Specific enterprise management functions and tasks may vary, depending on the organization's mission and capabilities. All NOSCs share some common enterprise management activities. These activities occur during predeployment, deployment, and redeployment. Enterprise management consists of seven activities. Each activity is a different step in the enterprise management cycle. Each activity has associated network and information systems management resources identified to create controllable enterprise management. Each enterprise management activity involves specific functions and associated tasks, whether the activity applies to user communications networks, or to networks provided by support elements. These activities are—

- Physical and operational management
- Service delivery
- Service support
- Mission planning
- Capability design and engineering
- Sustainment
- Administration

DODIN enterprise management supports the commander's information requirements via physical and operational management, service delivery, and service support. The enterprise management cycle begins at the beginning of the operations process. The cycle is a continual process of identifying requirements (plan), determining courses of action (prepare), and execution (execute), with assessment integrated at every step.

Mission planning, and capability design and engineering, are centralized activities that design networks to meet users' service requirements. Sustainment support is required to maintain existing services and acquire equipment to meet new service requirements.

I. Cybersecurity Fundamentals

Ref: ATP 6-02.71, Techniques for Department of Defense Information Network Operations (Apr '19), pp. A-7 to A-15.

The Army depends on reliable networks and systems to access critical information and supporting information services to accomplish their missions. Threats to the DODIN exploit the increased complexity and connectivity of Army information systems and place Army forces at risk. Like other operational risks, cyberspace risks affect mission accomplishment. They can increase the needed time and space to conduct operations, or decrease a unit's performance or effectiveness. DOD networks experience adversary cyberspace attacks every day. Robust cybersecurity measures prevent adversaries from accessing the DODIN through known vulnerabilities. The cybersecurity measures apply to general threats and known vulnerabilities, as opposed to specific attacks.

Cybersecurity ensures IT assets provide mission owners and operators confidence in the confidentiality, integrity, and availability of information systems and information, and their ability to make choices based on that confidence. The DOD cybersecurity framework (see DODI 8500.01) provides the foundation for cybersecurity.

Cybersecurity supports effective operations in cyberspace where—
- Missions and operations continue under any cyberspace threat situation or condition.
- IT components of weapons systems and other defense platforms function as designed and adequately meet operational requirements.
- The DODIN collectively, consistently, and effectively defends itself.
- The information network securely and seamlessly extends to mission partners.
- U.S. forces and mission partners can access their information and command and control channels, but their adversaries cannot.

DOD cybersecurity complies with National Institute of Standards and Technology security and risk management publications to ensure mission partner interoperability. These publications are available online at the National Institute of Standards and Technology Computer Security Resource Center.

The cybersecurity framework consists of—
- Cybersecurity risk management
- Operational resilience
- Integration and interoperability
- Cyberspace defense
- Cybersecurity performance
- DOD information
- Identity assurance
- IT
- Cybersecurity workforce
- Mission partners

I. Cybersecurity Fundamental Attributes

Cybersecurity ensures the confidentiality, integrity, availability, authentication, and non-repudiation of friendly information and information systems while denying adversaries access to the same information and information systems. These attributes are—

- **Confidentiality** is assurance that sensitive information is not disclosed to unauthorized individuals, processes, or devices.
- **Integrity** is the reliability of an information system; the logical correctness and reliability of the operating system; the logical completeness of the hardware and software implementing the protection mechanisms; and the consistency of the data structures and occurrence of the stored data. A formal security model defines integrity more narrowly to protect against unauthorized modification or destruction of information.
- **Availability** is timely, reliable access to data, and information services by authorized users.
- **Authentication** is a security measure designed to—
  - Protect a communications system against acceptance of a fraudulent transmission or simulation by establishing the validity of a transmission, message, or originator.
  - Provide a means of identifying individuals and verifying their eligibility to receive specific categories of information.
- **Nonrepudiation** is assurance that the sender of data receives proof of delivery and the recipient receives proof of the sender's identity to create a record of the parties that processed the data.

Cybersecurity incorporates those actions taken to protect, monitor, analyze, detect, and respond to unauthorized activity on DOD information systems and computer networks. It incorporates protection, detection, and response while facilitating restoration of information systems. Cybersecurity provides end-to-end protection to ensure data quality and protection against unauthorized access and inadvertent damage or modification.

II. Cybersecurity Risk Management

Cybersecurity risk management identifies and analyzes threats against, and vulnerabilities of, networks and information systems; assesses the threat level; and determines how to deal with risks. Risk management also includes identifying vulnerabilities created by design weaknesses, ineffective security procedures, or faulty internal controls, which are susceptible to exploitation.

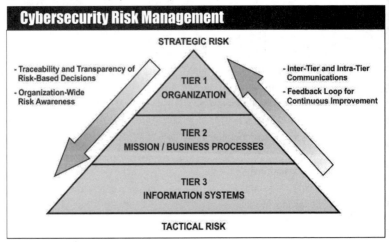

Ref: ATP 6-02.71 (Apr '19), A-2. Cybersecurity risk management.

III. Cybersecurity Principles

Ref: ATP 6-02.71, Techniques for Department of Defense Information Network Operations (Apr '19), pp. A-25 to A-26.

The principles of cybersecurity are not a checklist and may not apply the same way in every situation. These principles offer cybersecurity professionals a context for implementing the cybersecurity framework, developing strategies, and allocating resources.

Cybersecurity Principles

- Full Dimenion
- Layered
- Redundant
- Integrated
- Enduring

Full Dimension
Cybersecurity is not a linear activity, but a continuous process. Cybersecurity efforts and activities account for cyberspace risks in all directions, in all environments, at all times. Cybersecurity planning, coordination, and implementation occur anywhere the protection against, detection of, response to, and recovery from anomalous network activity is required. Network situational understanding supports, and leads to proper implementation of, full dimension cybersecurity.

Layered
Layered cybersecurity capabilities provide defense-in-depth. Layering reduces the destructive effects of a cyberspace attack. Layering may also provide time to focus response efforts.

Redundant
Redundancy ensures critical activities, systems, efforts, and capabilities have secondary or auxiliary efforts of equal or greater capability. Redundancy in this context is not simple duplication of effort. It emphasizes overlapping capabilities for seamless protection. Cybersecurity professionals identify critical points of failure or critical paths for each cybersecurity function, system, effort, and capability to apply redundancy. Cybersecurity efforts often overlap where there is an identified or expected vulnerability, weakness, or failure.

Integrated
Cybersecurity integrates with all other cyberspace operations, systems, efforts, and capabilities. This adds strength and structure to the overall cybersecurity effort. Integration occurs throughout the DODIN operations hierarchy in all operations. Cybersecurity integration supports and complements other cyberspace operations.

Enduring
Cybersecurity's enduring nature differentiates it from defensive cyberspace operations. Defensive cyberspace operations continue only until cyberspace forces can resume normal operation of the network, performing security to maintain freedom of action. Cybersecurity's persistent character preserves critical assets to enable mission assurance. Enduring cybersecurity affects freedom of action and resource allocation.

Risk management is a holistic activity integrated into every aspect of the organization. Figure A-2 above illustrates the three tiers of cybersecurity risk management. This tiered approach addresses risk-related concerns at the organization, mission and business process, and information systems levels.

Tier 1

Risk management at tier 1 addresses organizational risk. Tier 2 and 3 risk decisions inform and influence tier 1 risk management as part of the feedback loop.

A comprehensive information systems security governance structure aligns information systems security strategies to support mission and business objectives, ensures they are consistent with applicable laws and regulations, and assigns responsibilities.

The DOD information security risk management committee is comprised of the four mission area principal authorizing officials and other major DOD and intelligence community stakeholders. The information security risk management committee provides tier 1 risk management governance for the DOD.

Tier 2

Tier 2 addresses mission and business process risks. Tier 1 risk decisions guide, and tier 3 risk decisions inform and influence, tier 2 risk management.
- The activities at tier 2 begin with designing, developing, and implementing the mission and business processes defined at tier 1.
- The principal authorizing officials for each DOD mission area provide tier 2 governance for their respective mission areas.

Tier 3

Tier 3 addresses information systems and platform IT system risk. Tier 1 and 2 risk decisions guide tier 3 risk management.
- Requirement identification for specific protective measures takes place at tiers 1 and 2. Tier 3 includes applying the protective measures identified at tiers 1 and 2.
- Selecting and implementing appropriate security controls from National Institute of Standards and Technology Special Publication 800-53 satisfies information protection requirements.

See following pages (pp. 7-6 to 7-7) for a listing and discussion of Cybersecurity Risk Management Framework steps.

IV. Enabled Effects

Ref: ATP 6-02.71, Techniques for Department of Defense Information Network Operations (Apr '19), p. A-26.

Cybersecurity enables information protection, and network and system availability. DODIN operations personnel achieve these by—

- Instituting agile capabilities, such as firewalls, password protection, intrusion detection, and intrusion prevention, to resist adversary attacks by recognizing such attacks as they begin or progress.
- Detecting and analyzing anomalies or intrusions and reporting incidents to all NOSCs and ARCYBER.
- Implementing efficient, effective responses to reduce the effects of an attack, and to recover from attacks safely and securely.
- Informing others across the DODIN of local actions to counter intrusions or correct other incidents.
- Certifying, accrediting, and reporting on all networks, peripherals, and edge devices in their portion of the network, and enforcing information security controls.
- Evaluating subordinate units' security readiness and vulnerability for compliance with communications tasking orders and IAVM, and reporting compliance to higher echelons.
- Ensuring network management and defense training, awareness, and certification program compliance according to established policies and directives.
- Developing and deconflicting local contingency plans to defend against malicious activity and providing copies to higher-level commands and DODIN operations authorities.
- Conducting network risk assessments.
- Sharing cybersecurity information according to formal agreements and national disclosure policies, except where limited by law, policy, or security classification.
- Submitting reports, as directed by higher commands and DODIN operations authorities.
- Developing and maintaining remediation, mitigation, and reconstitution plans for critical infrastructure protection criteria.
- Reconstituting capabilities from reserve or reallocated assets when original capabilities are destroyed.
- Coordinating between user elements to distinguish between hostile cyber incidents and other system outages or degradations.

Risk Management Framework

Ref: ATP 6-02.71, Techniques for Department of Defense Information Network Operations (Apr '19), pp. A-9 to A-11.

The risk management framework (formerly the DOD Information Assurance Certification and Accreditation Process) provides a disciplined and structured process for combining information systems security and risk management into the system development life cycle. The DOD risk management framework complies with National Institute of Standards and Technology guidelines to align with federal civilian agencies. The risk management framework has six steps—

1. Categorize System
- Describe the system, including the system boundary, and document the description in the security plan.
- Register the system with the DOD Component cybersecurity program.
- Assign qualified personnel to risk management framework roles.

2. Select Security Controls
- Identify common controls.
- Identify the security control baseline for the system and document in the security plan.
- Develop and document a system-level strategy for continuously monitoring the effectiveness of security controls and proposed or actual changes to the system and its operating environment.
- Develop and implement processes whereby the authorizing official reviews and approves the security plan and system-level continuous monitoring strategy.

3. Implement Security Controls
- Implement security controls specified in the security plan in accordance with DOD implementation guidance.

4. Assess Security Controls
- Develop, review, and approve a plan to assess security controls using a methodology consistent with National Institute of Standards and Technology Special Publication 800-30.
- Assess security controls in accordance with the security assessment plan and DOD assessment procedures.
- Record the compliance status of security controls.
- Assign vulnerability severity value for security controls.
- Determine risk level for security controls.
- Assess and characterize the aggregate level of risk to the system.

5. Authorize System
- Prepare the program of action and milestones based on the vulnerabilities identified during the security control assessment.
- Assemble the security authorization package and submit to the authorizing official for adjudication.
- Determine the risk to organizational operations (including mission, functions, image, or reputation), organizational assets, individuals, other organizations, or the Nation.

- Decide whether the risk to organizational operations, organizational assets, individuals, other organizations, or the Nation is acceptable.
- If the risk is determined to be unacceptable, issue a denial of authorization to operate. If the system is already operational, the authorizing official will issue a denial of authorization to operate and stop operation of the system immediately.

6. Monitor Security Controls
- Determine the security impact of proposed or actual changes to the information system or platform IT system and its environment of operation.
- Assess a subset of the security controls employed within and inherited by the information system or platform IT system in accordance with the system-level continuous monitoring strategy.
- Conduct remediation actions based on the results of ongoing monitoring activities, risk assessment, and outstanding items in the program of action and milestones.
- The program manager or system manager updates the security plan and program of action and milestones, based on the results of the system-level continuous monitoring process. The information system security manager may recommend changes or improvements to the implementation of assigned security controls, the assignment of additional security controls, or changes or improvements to the design of the system to the security control assessor and authorizing official.
- Report the security status of the system, including the effectiveness of security controls, to the authorizing official and other appropriate organizational officials, in accordance with the monitoring strategy.
- The authorizing official continues to review the reported security status of the system, including the effectiveness of security controls, in accordance with the monitoring strategy, to determine whether the risk to organizational operations, organizational assets, individuals, other organizations, or the Nation remains acceptable.
- Implement a system decommissioning strategy, when needed. The decommissioning strategy defines the actions required when removing an information system or platform IT system from service.

Cybersecurity Reciprocity
Cybersecurity reciprocity aids rapid, efficient IT capability development and fielding. Reciprocity reduces redundant testing, assessment, and documentation, and the associated costs in time and resources. The risk management framework presumes acceptance of existing test and assessment results and authorization documentation from other Services and federal agencies.

The Services share security authorization packages and agree to accept other Services' test and assessment results and authorization to support cybersecurity reciprocity. Reciprocal acceptance of DOD and other federal agency and department security authorizations ensures interoperability and reduces redundant testing. It is important that each Service exercises due diligence in assessing, documenting, and approving systems, software, and configurations, since all Services share a risk accepted by one Service.

Refer to DODI 8510.01 for detailed, authoritative guidance on implementing the risk management framework.

V. Operational Resilience

DODIN operations personnel install, operate, maintain, and secure IT systems to—
- Ensure information and services are available to authorized users when and where they are required according to mission needs, priorities, and changing roles and responsibilities.
- Sense and correlate security posture, from individual devices or software objects to systems, and make it visible to mission owners and network operators across the DODIN.
- Ensure that whenever possible, hardware and software can reconfigure, optimize, self-defend, and recover with minimal human intervention. Attempts to reconfigure, self-defend, and recover should produce an incident audit trail.

Operational resilience requires three conditions be met:
- Information resources are trustworthy.
- Missions are ready for information resource degradation or loss.
- DODIN operations have the means to prevail in the face of adverse events.

Operational resilience is achieved by—
- Using trusted system and network requirements and best practices to protect mission-critical functions and components and manage risks.
- Performing developmental cybersecurity test and evaluation to inform acquisition and fielding decisions. This includes testing the ability of systems to detect and react to penetrations and exploitations and to protect and restore data and information.
- Including cybersecurity as a key element of program planning activities.
- Planning for mission continuation in the face of degraded or unavailable information resources.
- Conducting periodic exercises or evaluations of the ability to operate during loss of all information resources and connectivity.
- • Preserving trust in the security of DOD information during transmission.
- Protecting transmission of DOD information through established COMSEC and transmission security controls.
- Conducting COMSEC monitoring and cybersecurity readiness testing.
- Applying compromising emanations (TEMPEST) countermeasures.

VI. Cybersecurity Integration and Interoperability

Cybersecurity is integral to system life cycles. Adhering to the DOD IT architecture, a standards-based approach, and risk sharing among DOD Components helps achieve interoperability.

Cybersecurity personnel manage IT interconnections to reduce shared risk. Reducing vulnerabilities of each managed system protects the security posture of other interconnected systems.

See facing page for an overview and further discussion.

Cybersecurity Integration and Interoperability
Ref: ATP 6-02.71, Techniques for Department of Defense Information Network Operations (Apr '19), p. A-12.

Net-Centric Operations
A net-centric model provides personnel, services, and platforms the ability to discover one another and connect to form new capabilities or teams without being constrained by geographic, organizational, or technical barriers. The net-centric model allows collaboration to achieve shared ends. Cybersecurity design, organization, and management ensure systems can work together in any combination and maintain an expected level of readiness.

Integration
Cybersecurity is a visible element of organizational, joint, and DOD component architectures, capability identification and development processes, integrated testing, IT portfolios, acquisition, operational readiness assessments, supply chain risk management, security system engineering, and operations and maintenance.

Interoperability
Cybersecurity products—firewalls, file integrity checkers, virus scanners, intrusion detection systems, anti-malware—operate in a net-centric manner to enhance data exchange and shared security.

Semantic, technical, and policy interoperability integrate a wide range of cybersecurity products into a net-centric enterprise. This integration creates new information about the network and speeds up decision making and decision implementation.

Interoperability support products provide security for communications between different IT systems. The goal is seamless and secure exchange of critical classified or sensitive information.

Standards-Based Approach
One goal of the DOD cybersecurity strategy is interoperability through a standards-based approach. These standards conform to government, industry, and academic best business practices, and are available online from the National Institute of Standards and Technology Computer Security Resource Center.

Department of Defense Architecture Principles
Adhering to established DOD cybersecurity architectures enables interoperability and effective security management. All DOD Components use the same architectures to facilitate information sharing while managing the risk inherent in interconnecting systems.

Knowledge Repositories
Cybersecurity knowledge repositories enable sharing of best practices, benchmarks, standards, templates, checklists, tools, guidelines, rules, and principles. Examples include the National Vulnerability Database the Open Vulnerability and Assessment Language Repository and the Risk Management Framework Knowledge Service. Knowledge repositories enable policy and process interoperability and allow information sharing among cybersecurity professionals.

VII. Cyberspace Defense

Ref: ATP 6-02.71, Techniques for Department of Defense Information Network Operations (Apr '19), pp. A-12 to A-14.

Cyberspace defense is actions normally created within Department of Defense cyberspace for securing, operating, and defending the Department of Defense information network. Specific actions include protect, detect, characterize, counter, and mitigate (DODI 8500.01).

Cyberspace defense protects against, detects, characterizes, counters, and mitigates unauthorized activity and vulnerabilities on the DODIN. Sharing cyberspace defense information across the enterprise supports shared situational understanding. Cyberspace defense actions create desired effects inside the DODIN and other specified cyberspace.

Cyberspace defense uses architectures, cybersecurity, intelligence, counterintelligence, other security programs, law enforcement, and other military capabilities to—

- Make the DODIN more resistant to penetration and disruption.
- Facilitate response to unauthorized activity.
- Defend information and networks against cyberspace risks.
- Recover quickly from cyber incidents.

Department of Defense Information Technology
USCYBERCOM controls access to and defense of DOD IT systems and information networks. Cyberspace defense integrates with other elements of DODIN operations to secure IT systems.

Continuous Monitoring Capability
Continuous network and information systems monitoring provide consistent collection, transmission, storage, aggregation, and presentation of current operational status to affected DOD stakeholders. A common monitoring framework, terminology, and workflow across DOD components ensure interoperability and shared situational understanding.

Penetration and Exploitation Testing
Penetration and exploitation testing are part of developmental and operational test and evaluation. This evaluation includes independent threat representative (cyber protection team or cyber red team) penetration and exploitation testing and evaluation of all cyberspace defenses. This testing includes the controls and protection provided by network defense service providers. For more information, see CJCSM 6510.03.

As part of the risk assessment process, cybersecurity professionals should periodically request penetration and exploitation testing. Testing measures the level of performance and effectiveness of cyberspace defense actions. For more information, refer to AR 380-53.

Law Enforcement and Counterintelligence
The DOD Cyber Crime Center provides digital and multimedia forensics and specialized cyberspace investigative training and services. The DOD Cyber Crime Center coordinates working relationships across the law enforcement, intelligence, and homeland security communities.

Individual Service law enforcement and counterintelligence agencies deploy their investigative capabilities on DOD networks to identify and investigate human threats to IT systems and information. Cybersecurity supports counterespionage, counterterrorism, and counterintelligence insider threat detection.

Network administrators accommodate lawful deployment of law enforcement and counterintelligence tools. DOD law enforcement and counterintelligence organizations coordinate law enforcement and counterintelligence efforts with their respective authorizing officials, consistent with service level agreements and change management processes. Coordination helps avoid disrupting mission-critical systems and networks.

Insider Threat

An insider threat is the threat that an insider will use her/his authorized access, wittingly or unwittingly, to do harm to the security of the United States. This threat can include damage to the United States through espionage, terrorism, unauthorized disclosure, or through the loss or degradation of departmental resources or capabilities (CNSSI 4009). Trusted insiders with legitimate access to systems pose one of the most difficult threats to counter. Insiders are the most dangerous threat to operations security because they can readily access sensitive information. Whether recruited or self-motivated, insiders can access systems normally protected against attack, usually without raising suspicion. For this reason, operationally sensitive and critical information should only be shared with personnel who have both an appropriate security clearance and a valid need to know. While insiders can attack at almost any time, systems are most vulnerable during the design, production, transport, and maintenance stages. Risks from insiders may be intentional and malicious, or may cause damage unintentionally through negligence or inaction. Operations security awareness and learning to recognize threat indicators help identify and mitigate risks from insider threats. Reportable cyber indicators of a potential insider threat include—

- Excessive probing or scanning from either an internal or external source.
- Tampering with or introducing unauthorized data, software, or hardware into information systems.
- Hacking or password cracking activities.
- Unauthorized network access or unexplained user account.
- Social engineering, electronic elicitation, e-mail spoofing, or spear phishing.
- Use of DOD account credentials by unauthorized parties.
- Downloading, attempting to download, or installing non-approved computer applications.
- Key logging.
- Rootkits, remote access tools, and other backdoors.
- Unauthorized account privilege escalation.
- Account masquerading—changing credentials to look like another user's credentials.
- Unexplained storage of encrypted data.
- Encryption or steganography (hiding a coded message within an ordinary message) data propagation internally.
- Unauthorized use of USB removable media or other transfer devices.
- Denial of service attacks or suspicious network communication failures.
- Exfiltration of data to unauthorized domains or cross-domain violations.
- Unauthorized e-mail traffic to foreign destinations.
- Unauthorized downloads or uploads of sensitive data.
- Use of malicious code or blended threats such as viruses, worms, logic bombs, malware, spyware, or browser hijackers, especially those used for clandestine data exfiltration.
- Data or software deletion.
- Log manipulation.
- Unauthorized use of intrusion detection systems.

All personnel need annual threat awareness and reporting program training to maintain awareness of extremist, espionage, and insider threat indicators.

Refer to AR 381-12 for more information about the threat awareness and reporting program.

VIII. Cybersecurity Performance

Ref: ATP 6-02.71, Techniques for Department of Defense Information Network Operations (Apr '19), pp. A-14 to A-15.

Cybersecurity personnel measure, assess, and manage systems' performance relative to their contributions to mission outcomes and strategic goals and objectives. They collect data to support reporting and cybersecurity management across the system life cycle. Standardized IT tools, methods, and processes prevent duplicate costs and focus resources on technologically mature and verified solutions.

Services implement processes and procedures to accommodate three conditions necessary for consistent cybersecurity across the DOD—

- **Organization direction**—organizational mechanisms for establishing and communicating priorities and objectives, principles, policies, standards, and performance measures.
- A **culture of accountability**—aligning internal processes; maintaining accountability; and informing, making, and following through on cyberspace protection and defense decisions.
- **Insight and oversight**—measuring, reviewing, verifying, monitoring, facilitating, and remediating to ensure coordinated and consistent cybersecurity compliance without impeding local missions.

Department of Defense Information

The DOD cybersecurity program provides the mechanisms to measure, monitor, and enforce information security and information sharing policies and procedures as they relate to information in an electronic form, primarily by implementing security controls.

Information security guidance establishes the standards for protecting classified and controlled unclassified information. Information systems protect classified and controlled unclassified information from unauthorized access by requiring user authentication.

Identity Assurance

Identity assurance ensures strong identification and authentication and eliminates anonymity in DOD information systems and platform IT systems. The DOD uses department-wide public key infrastructure and public key infrastructure-enabled information systems. Cybersecurity personnel manage and safeguard biometric data supporting identity assurance according to applicable DOD and Army policies.

Information Technology

IT systems that receive, process, store, display, or transmit DOD information are acquired, configured, operated, maintained, and disposed of consistent with established cybersecurity policies, standards, and architectures. Planners identify cybersecurity requirements and consider them throughout the system life cycle.

Cybersecurity Workforce

Security configuration errors make DOD networks and information systems vulnerable to attack or failure. For this reason, cybersecurity personnel require careful screening and appropriate technical qualifications, according to DODD 8140.01. Cybersecurity managers ensure cybersecurity personnel have the required certifications base on their roles and integrate them across the range of military operations.

Mission Partners

Standards-based cybersecurity enables seamless collaboration with mission partners. The decision structures and processes in DODI 8500.01 govern shared cybersecurity. DOD information residing on mission partner information systems requires adequate safeguards. Documented interagency and multinational agreements specify the required levels of protection.

II. Cybersecurity Functions

Ref: ATP 6-02.71, Techniques for Department of Defense Information Network Operations (Apr '19), pp. A-16 to A-25.

Cybersecurity includes both technical and non-technical measures, such as risk management, personnel training, audits, and continuity of operations planning. Cybersecurity factors in all cyber incidents that occur through malicious or accidental activity by enemy, adversary or friendly entities.

Cybersecurity functions help an organization manage risk by organizing information, enabling risk management decisions, mitigating threats, and improving security by learning from earlier activities. These functions align with existing incident management methodologies and help show the impact of cybersecurity measures.

Cybersecurity Functions

- **I. Identify**
- **II. Protect**
- **III. Detect**
- **IV. Respond**
- **V. Recover**

Cybersecurity personnel perform these functions concurrently and continuously to mitigate the dynamic cyberspace risk.

I. Identify

The identify function develops situational understanding to manage cybersecurity risks to systems, assets, data, and capabilities. This function helps cybersecurity personnel understand the mission, the resources supporting critical functions, and related cybersecurity risks. This understanding allows an organization to focus and prioritize its efforts, consistent with its risk management strategy and mission needs.

A. Identify Mission-Critical Assets

Mission-critical assets are those resources without which the unit's key missions would significantly degrade or cease to function. The steps are—
- Inventory the organization's physical devices, systems, and software applications.
- Map the associated communication and data flows.
- Understand cybersecurity roles and responsibilities of higher & subordinate units.
- Identify the security categories for resources.

Information systems have assigned security categories, based on the potential impact of a breach to the security objectives of confidentiality, integrity, and availability. The security category (low, moderate, or high) determines the necessary cybersecurity controls.

Cybersecurity professionals and system owners identify mission-critical assets by determining the potential impact if there is a loss of—

- Confidentiality
- Integrity
- Availability

Identifying mission-critical assets and their security categories is a continual process. The process is mission-dependent and synchronized with the military decision-making process. Mission-critical assets may change rapidly based on operational phases, outcomes of running estimates, refined commander's intent, commander's critical information requirements, and essential elements of friendly information. Refer to Federal Information Processing Standards Publication 199 for more information on security categorization.

B. Identify Laws, Regulations, and Policies

Cybersecurity professionals must understand and follow applicable Army, DOD, and national laws, policies, and processes to meet regulatory, legal, risk, environmental, and operational requirements. They also develop internal policies and procedures for their organizations to mitigate cybersecurity gaps or leader requirements the existing laws, policies, and processes do not cover.

C. Identify Threat Activities

Threat actors can exploit system vulnerabilities and cause a loss of confidentiality, integrity, or availability of communications networks. These threat actors use many methods to disrupt, degrade, destroy, exploit, alter, or otherwise adversely affect the Army's use of cyberspace.

See following pages (pp. 7-16 to 7-17) for an overview and further discussion of cyber threat activities and pp. 7-18 to 7-19 for an overview and discussion of cyber attack tools. See also p. 2-24 for discussion of threats in cyberspace from FM 3-12.

D. Identify Vulnerabilities

Deploying forces require secure video, database connectivity, and the ability to send and receive data to enable reach operations, access to intelligence, and other essential support. Successful operations require reachback to access information residing outside the operational area. Soldiers' mobility and sustainment requirements may rely on commercial reach telecommunications, including international telecommunications and public switched networks.

Soldiers' increased reliance on reachback information capabilities creates vulnerabilities to attack from various sources. Adversaries can quickly exploit design weaknesses, ineffective or lax security, or insufficient internal controls to attack networks and information systems. Even an adversary who is not a technological equal could launch a covert or overt attack using inexpensive, commercial off-the-shelf products and readily available hacking tools. An adversary can attack from any location with Internet access. Recent trends that increase vulnerability include using commercial services, commercial off-the-shelf hardware and software; moving toward an open systems environment; and extensive interfacing with U.S. Government, industry, and public networks.

Vulnerabilities are flaws, loopholes, oversights, or errors a threat source can exploit. Vulnerability classification is based on the type of asset.

- Hardware and physical sites are vulnerable to environmental factors and uncontrolled access.

Cybersecurity Functional Services

Ref: ATP 6-02.71, Techniques for Department of Defense Information Network Operations (Apr '19), pp. A-25 to A-26.

The ten functional services of cybersecurity help protect friendly information, networks, and information systems while denying adversaries access to the same information, networks, and information systems. These functional services are—

- **Access control** provides the capability to restrict access to resources and protect them from unauthorized modification or disclosure. Access control measures can be technical, physical, or administrative. Access control uses hardware tools, software tools, or operational procedures.
- **Application security** protects software applications and software solution development. Some examples of application security solutions are software update service and patch management.
- A **continuity of operations plan** provides preservation and post-disaster recovery of information, network, or information system resources in case of incidents that interrupt or may interrupt normal operations.
- **COMSEC** provides the principles, means, and methods of encrypting voice, video, data, and imagery to ensure confidentiality, integrity, authentication, and nonrepudiation.
- **Risk analysis** identifies information assets, identifies risks, quantifies the possible damages that can occur to those information assets, and determines the most cost-effective way to mitigate the risks. Risk mitigation may include developing and implementing policies, standards, procedures, and guidelines.
- **Legal and regulatory compliance** fulfill the requirement for individuals to know and understand cybersecurity based on U.S. laws and DOD or Army regulations. Compliance also assists investigations to detect defensive breaches.
- **Development of cybersecurity policies and procedures** related to organizational personnel, hardware, software, and media. Cybersecurity policies and procedures identify security guidelines for data, media, telecommunications equipment, and information systems. These policies and procedures also help ensure the security of the users' activities. Examples of required activities are monitoring activity logs and analyzing audit trails.
- **Physical (environmental) security** protects the network facility from the outside perimeter to the inside operational space, including all information system resources. Physical security safeguards the network and information systems against damage, loss, and theft. Physical security includes determining and integrating site selection criteria and implementing effective perimeter and interior security for those facilities. Site selection also includes measures to enable adequate temperature, humidity, and fire controls.
- **Security in development and acquisition** implements principles, structures, and standards for hardware and software acquisition to enforce confidentiality, integrity, and availability. The key is integrating the common security criteria across DOD, Army, and international standards, including the trusted computing base and reference monitor models.
- **Telecommunications and network security** include implementing network architectures; transmission methods; transport formats; measures to provide confidentiality, integrity, and availability; and authentication for transmission over private and public communications networks and media. Secure networks incorporate cross domain solutions, remote access protocols, IP security, virtual private networking, and access control lists. Some common solutions include intrusion detection and prevention systems, antivirus solutions, web caches, and firewalls.

Cyber Threat Activities

Ref: ATP 6-02.71, Techniques for Department of Defense Information Network Operations (Apr '19), pp. A-25 to A-26.

Threat actors can exploit system vulnerabilities and cause a loss of confidentiality, integrity, or availability of communications networks. These threat actors use many methods to disrupt, degrade, destroy, exploit, alter, or otherwise adversely affect the Army's use of cyberspace.

Threat agent categories, methodologies, and intents include—

- **Unauthorized users**, such as hackers, are the source of most attacks against information systems in peacetime. They mostly target personal computers but have also targeted network communications, mainframes, and local area network-based computers.
- **Trusted insiders** with legitimate access to systems pose one of the most difficult threats to counter. Whether recruited or self-motivated, insiders can access systems normally protected against attack without leaving any indicators of malicious or unusual activity.
- **Terrorist groups** with access to commercial information systems, including the Internet, may access an information network without authorization, or direct physical attacks against the infrastructure. Organized terrorist groups pose serious threats to the information infrastructure and U.S. national security.
- **Non-state groups**, such as drug cartels and social activists, can take advantage of the information age to acquire the capabilities to strike at their foes' commercial, security, and communications infrastructures at low cost. Moreover, they can strike from any distance with near impunity.
- **Foreign intelligence entities**, which are active during both peacetime and conflict, take advantage of the anonymity offered by computers, bulletin boards, and the Internet. They hide organized collection or disruption activities behind the facade of unorganized attackers. Their primary targets are often commercial, scientific, and university networks. Foreign intelligence entities may also directly attack military and government networks and systems.
- **Opposing militaries or political opponents** are more traditionally associated with open conflict or war, but these attackers may invade U.S. computer and telecommunications networks during peacetime. Such strikes may seek to help frame situations to their advantage preceding the onset of hostilities.

Risks to the DODIN and Army networks are natural or man-made, worldwide in origin, technically multifaceted, and growing. They may come from individuals or groups motivated by military, political, cultural, ethnic, religious, personal, or industrial gain.

To understand intentional, malicious cyberspace attacks, cybersecurity professionals request information on the latest threat tactics, techniques, and procedures from supporting intelligence elements. Cybersecurity professionals monitor external data sources, for example vendor sites and computer emergency response teams, to maintain awareness of threat conditions and identify which security issues may affect their network.

Adversaries use cyberspace attacks against the DODIN to deny, degrade, disrupt, destroy, manipulate, or otherwise adversely affect friendly forces' ability to use cyberspace. The DODIN consists of several segments spanning multiple domains, with many means of communicating and different levels of interconnectivity and isolation. For this reason, enemies may employ a wide spectrum of capabilities to conduct offensive operations in the Army's portion of cyberspace. These capabilities may target any part of the DODIN, from specific nodes and links to the data traversing those nodes and links.

Friendly use of cyberspace requires control of the physical, logical, and cyber persona layers and the electromagnetic spectrum. Threat cyberspace attacks may target any of these to deny friendly cyberspace use. The cyber persona layer is vulnerable to social engineering and phishing. The logical network layer may be subject to cyberspace attack. The physical network layer may experience a physical or lethal attack. An adversary may try to deny access to the electromagnetic spectrum through an electronic attack, such as jamming.

Cyberspace Attacks
Cyberspace attacks are actions that create various direct denial effects in cyberspace (degradation, disruption, or destruction) or manipulation that leads to hidden denial, or that manifests in the air, land, maritime, or space domain.

Electronic Attack
Electronic attack is the division of electronic warfare involving the use of electromagnetic energy, directed energy, or antiradiation weapons to attack personnel, facilities, or equipment with the intent of degrading, neutralizing, or destroying enemy combat capability and is considered a form of fires (JP 3-13.1).

Physical Attacks
Physical attacks use measures to physically destroy or otherwise adversely affect a target. Because cyberspace network enclaves can be isolated, this may involve lethal attacks on network nodes. A physical attack can create effects within and outside cyberspace to help control the domain. Regardless of the degree of isolation, an adversary may decide a direct physical attack is the best option depending on the situation, the desired effect, and availability and suitability of other capabilities or options.

Social Engineering
Social engineering describes a non-technical intrusion that relies on human interaction, Social engineering involves tricking others into divulging information or violating security procedures. For example, a person using social engineering to break into the DODIN might try to gain the confidence of an authorized user and get them to reveal information, such as a password, that compromises network security. Social engineers rely on the natural helpfulness of people, as well as their weaknesses. Virus writers use social engineering to persuade people to run malware-laden e-mail attachments. Phishers use social engineering to convince people to divulge sensitive information. Scam software vendors may use social engineering to frighten people into running software that is either useless or malicious.

The outcome of adversary cyberspace activities is either an incident or event:
- **Cyber Incident**—Actions taken through the use of computer networks that result in an actual or potentially adverse effect on an information system, network, and/or the information residing therein (CNSSI 4009).
- **Event**—Any observable occurrence in a network or system (CNSSI 4009). Events sometimes indicate that a cyber incident is occurring.

Adversary cyberspace attacks can cause effects to communications, command and control capabilities, and other operational missions, such as fires. An adversary's ability to access cyberspace can result in a change to the information in Army systems. This change can influence future friendly actions or lead to reduced confidence in DOD information. This reduced confidence degrades situational understanding of the information environment.

See pp. 0-2 to 0-5 for an overview of the global cyber threats and cyber attacks against the U.S. See following pages (pp. 7-18 to 7-19) for an overview and discussion of common cyber attack tools. See also p. 2-24 for discussion of threats in cyberspace from FM 3-12.

Tools of Cyber Attacks
Ref: DCSINT Handbook No. 1.02, Critical Infrastructure (Aug '06), pp. IV-9 to IV-11.

Backdoor
This is used to describe a back way, hidden method, or other type of method of by passing normal security in order to obtain access to a secure area. It is also referred to as a trapdoor. Sometimes backdoors are surreptitiously planted on a network element; however, there are some cases where they are purposely installed on a system.

Denial of Service Attacks (DOS)
A DOS attack is designed to disrupt network service, typically by overwhelming the system with millions of requests every second causing the network to slow down or crash. An even more effective DOS is the distributed denial of service attack (DDOS). This involves the use of numerous computers flooding the target simultaneously. Not only does this overload the target with more requests, but having the DOS from multiple paths makes backtracking the attack extremely difficult, if not impossible. Many times worms are planted on computers to create zombies that allow the attacker to use these machines as unknowing participants in the attack. To highlight the impact of these type attacks, in February 2000, DOS attacks against Yahoo, CNN, eBay and other e-commerce sites were estimated to have caused over a billion dollars in losses. DOS attacks have also been directed against the military. In 1999, NATO computers were hit with DOS attacks by hactivists protesting the NATO bombing in Kosovo.

E-mail Spoofing
E-mail spoofing is a method of sending e-mail to a user that appears to have originated from one source when it actually was sent from another source. This method is often an attempt to trick the user into making a damaging statement or releasing sensitive information (such as passwords). For example, e-mail could be sent claiming to be from a person in authority requesting users to send them a copy of a password file or other sensitive information.

IP Address Spoofing
A method that creates Transmission Control Protocol/Internet Protocol (TCP/IP) packets using somebody else's IP address. Routers use the "destination IP" address to forward packets through the Internet, but ignore the "source IP" address. This method is often used in DDOS attacks in order to hide the true identity of the attacker.

Keylogger
A software program or hardware device that is used to monitor and log each of the keys a user types into a computer keyboard. The user who installed the program or hardware device can then view all keys typed in by that user. Because these programs and hardware devices monitor the actual keys being typed, a user can easily obtain passwords and other information the computer operator may not wish others to know.

Logic Bomb
A program routine that destroys data by reformatting the hard disk or randomly inserting garbage into data files. It may be brought into a computer by downloading a public-domain program that has been tampered with. Once it is executed, it does its damage immediately, whereas a virus keeps on destroying.

Physical Attacks
This involves the actual physical destruction of a computer system and/ or network. This includes destroying transport networks as well as the terminal equipment.

Sniffer
A program and/or device that monitors data traveling over a network. Although sniffers are used for legitimate network management functions, they also are used during cyber attacks for stealing information, including passwords, off a network. Once emplaced, they are very difficult to detect and can be inserted almost anywhere through different means.

Trojan Horse
A program or utility that falsely appears to be a useful program or utility such as a screen saver. However, once installed performs a function in the background such as allowing other users to have access to your computer or sending information from your computer to other computers.

Viruses
A software program, script, or macro that has been designed to infect, destroy, modify, or cause other problems with a computer or software program. There are different types of viruses. Some of these are:

- **Boot Sector Virus:** Infects the first or first few sectors of a computer hard drive or diskette drive allowing the virus to activate as the drive or diskette boots.
- **Companion Virus:** Stores itself in a file that is named similar to another program file that is commonly executed. When that file is executed the virus will infect the computer and/or perform malicious steps such as deleting your computer hard disk drive.
- **Executable Virus:** Stores itself in an executable file and infects other files each time the file is run. The majority of all computer viruses are spread when a file is executed or opened.
- **Overwrite Virus:** Overwrites a file with its own code, helping spread the virus to other files and computers.
- **Polymorphic Virus:** Has the capability of changing its own code allowing the virus to have hundreds or thousands of different variants making it much more difficult to notice and/or detect.
- **Resident Virus:** Stores itself within memory allowing it to infect files instantaneously and does not require the user to run the "execute a file" to infect files.
- **Stealth Virus:** Hides its tracks after infecting the computer. Once the computer has been infected the virus can make modifications to allow the computer to appear that it has not lost any memory and or that the file size has not changed.

Worms
A destructive software program containing code capable of gaining access to computers or networks and once within the computer or network causing that computer or network harm by deleting, modifying, distributing, or otherwise manipulating the data.

Zombie
A computer or server that has been basically hijacked using some form of malicious software to help a hacker perform a Distributed Denial of Service attack (DDOS).

See pp. 0-2 to 0-5 for an overview of the global cyber threats and cyber attacks against the U.S. See previous pages (pp. 7-16 to 7-17) and 7-26 to 7-27 for an overview of the cyber threat activities. See also p. 2-24 for discussion of threats in cyberspace from FM 3-12.

- Software is vulnerable to security flaws, software bugs, and poor password management.
- Insecure network architecture and complexity make networks vulnerable.
- Personnel are vulnerable due to inadequate security practices (visiting malicious websites).
- The lack of continuity plans, lack of audits, and failure to implement lessons learned make organizations vulnerable.

Vulnerability assessment includes systematically identifying and mitigating software, hardware, and procedural vulnerabilities. Auditing or penetration testing may identify some vulnerabilities. Assessment tools and techniques will search for and discover most vulnerabilities.

A major element of a vulnerability assessment is vulnerability scanning. Cybersecurity personnel conduct scheduled and unscheduled scans throughout all phases of operations. Operational scanning occurs in every layer of classified and unclassified networks. Both scheduled and no-notice scans are part of security policy and compliance enforcement.

For improved interoperability, preferred assessment tools express vulnerabilities in the common vulnerabilities and exposures naming convention and use the Open Vulnerability and Assessment Language. Cybersecurity professionals use one or more of these tools to discover as many vulnerabilities as possible:

- **Host Scanning tools** scan critical system files, active processes, file shares, and the configuration and patch level of a particular system. The results produced from this type of tool are usually very detailed because they run on the host system at the same permission level as the user conducting the scan.
- **Network scanning tools** scan available network services for vulnerabilities through banner grabbing, port status, protocol compliance, service behavior, or exploitation.
- **Web application scanning tools** are a specialized form of network or host scanner that interrogates web servers or scans web source code for known vulnerabilities. These tools search for the presence of default accounts, directory traversal attacks, form validation errors, unsecure cgi-bin files, demonstration web pages, and other vulnerabilities.
- **Database application scanning tools** are specialized network scanners, which interrogate database servers for known vulnerabilities.
- **Vulnerability and patch management tools** incorporate many aspects of vulnerability management. These tools apply to vulnerabilities, policy compliance, patch management, configuration management, and reporting. These solutions make managing large, complex networks more efficient and reduce manpower requirements.

System administrators and network managers need the consent of the information system security officer and the G-6 or S-6, who consider operational or mission status and bandwidth constraints before scanning. Table A-1 details the actions undertaken when scanning.

Sharing vulnerability scan results freely among appropriate personnel throughout the organization helps eliminate similar vulnerabilities in other systems. Vulnerability analysis for custom software and applications may require specialized approaches. These may include vulnerability scanning tools for applications, source code reviews, or static analysis of source code. If analysis identifies and verifies unauthorized activity, cybersecurity personnel follow the established incident and vulnerability reporting procedures, as outlined in CJCSM 6510.01 and AR 25-2.

Another major element of vulnerability assessment is managing vulnerabilities. Vulnerability management is a comprehensive process for notifying Services, DOD agencies and field activities, and joint and combatant commands about vulnerability alerts, bulletins, technical advisories, and countermeasures. Vulnerability management requires combatant commands, Services, and DOD agencies and field activities to acknowledge receipt, and provides specific time limits for implementing countermeasures, depending on the criticality of the vulnerability (CJCSM 6510.01). For the Army, ARCYBER is the lead agent for implementing IAVM. ARCYBER issues alerts, bulletins, technical tips, and system administrator reports based on mandatory USCYBERCOM IA vulnerability alerts and Army-generated IAVM requirements using DOD Enterprise E-mail and other vulnerability management systems, including—

- Army Network Operations Reporting Tool.
- Microsoft SYSMAN.
- Assured Compliance Assessment Solution.
- Continuous Monitoring and Risk Scoring.

Cybersecurity personnel perform routine vulnerability assessments and IAVM procedures to manage system and network vulnerabilities and maintain their remediation skills. DODIN operations personnel apply remediation actions specified in IAVM messages immediately. If they cannot implement an IAVM action, they must submit a mitigation plan in the Army Cyber Vulnerability Tracking databases for approval or disapproval.

II. Protect Function

Cybersecurity professionals, DODIN operations personnel, and users work together to develop and implement the appropriate safeguards to assure critical infrastructure services. The protect function supports the ability to limit or contain the impact of cybersecurity events. Each of the cyberspace layers (physical, logical, and cyber persona) and the electromagnetic spectrum have specific protection requirements.

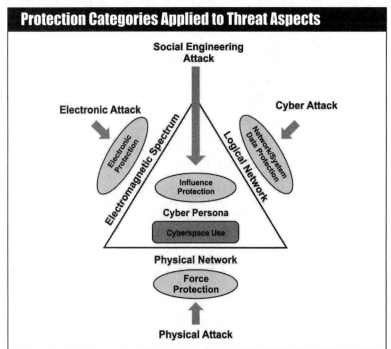

Ref: ATP 6-02.71 (Apr '19), fig. A-4. Protection categories applied to cyberspace threat aspects.

Information Operations Condition (INFOCON)

The information operations condition (INFOCON) system establishes a uniform process for posturing and defending against malicious activity that targets DOD information systems and networks (Strategic Instruction 527-1). The DOD identifies the threat level against its networks and information systems by using INFOCON status levels. The INFOCON system provides coordinated, structured defense against, and reaction to, attacks on DOD computers, networks, and information systems. The INFOCON system outlines countermeasures to scanning, probing, unauthorized access, data browsing and general threats at DOD computers, networks, and information systems. Refer to Strategic Instruction 527-1 for more information on the INFOCON system.

Protect Networks, Systems, and Data

Protection of networks, information systems, and data is achieved using—

- Identity and access control ensures strong identification and authorization and eliminates anonymity in the network. Identity and access control restricts access to resources and protects them from unauthorized modification or disclosure. Access control measures may be technical, physical, or administrative (hardware or software tools or procedures). Within the DODIN-A, Army forces use public key infrastructure. DOD-approved identity credentials (common access card or SIPRNET hardware token) authenticate users. This requirement extends to mission partners using Army systems. Cybersecurity personnel manage authorized devices and user identities and determine whether to apply permissions based on information and knowledge management plans. Cybersecurity personnel also manage and protect remote access points.

- COMSEC provides the principles, means, and methods of encrypting voice, video, data, and imagery to ensure confidentiality, integrity, authentication, and non-repudiation. Cybersecurity professionals and network or system administrators ensure users employ COMSEC appropriately to protect data at rest and in transit.

- Security in development and acquisition implements principles, structures, and standards for hardware and software acquisition and life cycle management to enforce confidentiality, integrity, and availability. Cybersecurity personnel formally manage assets throughout procurement, implementation, transfer, removal, and disposition. Integrity checking mechanisms verify software and firmware in commercial products do not include malicious code.

- Telecommunications and network security includes implementing network architectures; transmission methods; transport formats; measures to provide confidentiality, integrity, and availability; and authentication for transmission over private and public communications networks and media. Secure networks incorporate cross domain solutions, firewalls, intrusion prevention and protection systems, web caches, IP security, virtual private networking, and access control lists. As part of telecommunications and network security, cybersecurity professionals verify the network's engineering ensures integrity and provides the network capacity needed to ensure availability.

- Continuity of operations plans preserve, and provide post-disaster recovery of, information, networks, or information system resources in case of incidents that interrupt or may interrupt normal operations.

Force Protection

Physical and environmental security measures support force protection. Environmental security protects the network facility from the outside perimeter to the inside operational space, including all information system resources. Physical security safeguards the network and information systems against damage, loss, and theft. Physical security planning includes determining and integrating site selection criteria and implementing effective perimeter and interior security. Site selection also includes measures to control environmental factors (temperature, humidity, and fire controls).

Electronic Protection

Electronic protection is the division of electronic warfare involving actions taken to protect personnel, facilities, and equipment from any effects of friendly or enemy use of the electromagnetic spectrum that degrade, neutralize, or destroy friendly combat capability (JP 3-13.1). Electronic protection involves activities to address system hardening, electromagnetic compatibility, electromagnetic interference, reprogramming spectrum dependent devices, and emission control. Electronic protection focuses on both lethal and non-lethal electronic warfare activities. Electronic protection actions related to cybersecurity focus mainly on non-lethal electronic warfare. Electronic protection requires coordinating and integrating actions with electronic warfare and spectrum management personnel.

Influence Protection

The first line of defense against social engineering is the user. Cybersecurity professionals should develop internal standard operating procedures for managing potential social engineering attacks. They should ensure users complete periodic cybersecurity awareness training. Cybersecurity awareness training prepares users to perform information security-related duties consistent with applicable policies, procedures, and agreements.

Through social engineering attacks, adversaries seek out exploitable information to influence others. For this reason, operations security measures help achieve cybersecurity objectives. Operations security is a capability that identifies and controls critical information, indicators of friendly force actions attendant to military operations, and incorporates countermeasures to reduce the risk of an adversary exploiting vulnerabilities (JP 3-13.3). Operations security helps prevent adversaries gaining critical information through cyberspace. Critical information includes indicators that are sensitive, but unclassified, such as passwords. Operations security aims to identify unclassified activity or information that, when analyzed with other activities and information, can reveal protected and important friendly operations, information, or activities. Cybersecurity professionals integrate their activities with operations security efforts to enhance the effectiveness of both.

III. Detect Function

Detection involves activities such as anomaly and event handling and continuous monitoring. Detection uses intrusion detection systems and other sensor and logging devices to discover and report anomalies. Intrusion detection systems require continuous monitoring to provide timely warning of incidents and attacks. Monitoring also requires both automated and manual analysis. When a potential attack is detected, tracing and monitoring determine the severity and extent of the attack, gather evidence, limit the attack's effects, identify the potential for escalation, coordinate responses, and determine the effectiveness of countermeasures.

Continuous Monitoring

NOSCs monitor networks and systems in near real-time to detect anomalies and take preliminary defensive actions. Continuous network and information systems monitoring identifies unauthorized network connections, devices, and software, using both automated and manual processes. Monitoring includes not only the network, but also the physical environment and personnel activities.

Anomalies and Events

The ability to detect an event depends on established baseline network and information system configurations and behaviors. The baseline serves a basis for comparison to discover and analyze changes to identify their cause and potential impact.

Once cybersecurity personnel discover an incident and it reaches alert thresholds, they categorize and report it to the commander and higher-level DODIN operations authorities. Reporting allows DODIN operations personnel to aggregate and correlate events with other activities occurring in the cyberspace and physical domains.

Certain anomalies and events may require notifying law enforcement or counterintelligence agencies. Law enforcement and counterintelligence agencies can deploy capabilities on Army networks to identify and investigate the human element posing a threat to Army IT and DOD information. Cybersecurity personnel may support counterespionage, counterterrorism, and counterintelligence insider threat mitigation according to DODI 5240.26. Network administrators accommodate legitimate deployment of law enforcement and counterintelligence tools. Law enforcement and counterintelligence organizations, in turn, make all reasonable attempts to employ their solutions consistent with established change control processes to avoid disrupting mission-critical systems.

IV. Respond Function

Response involves steps to limit and mitigate the effects of unauthorized network activity. Effective response requires well-defined processes and procedures for handling incidents in an organized and disciplined manner, including coordination with internal and external stakeholders, such as law enforcement and counterintelligence. Damage containment and control prevents the spread of malicious code, minimizes the effects of an attack, and reduces exposure of interconnected networks and systems. The response preserves evidence and remnant files to enable forensic analysis. Technical support, including the analysis of logs and related activities, aids in incident response.

Mitigation

Cybersecurity professionals work with network operators, leaders, and users to counter potential threats by recommending and implementing activities to prevent an event's spread, minimize its effects, and create conditions that allow its elimination from the network. This may involve changing network configurations, creating new policies, or adopting new tactics, techniques, and procedures.

Analysis

Once cybersecurity personnel mitigate an event to a level that allows for mission assurance, they collect relevant information logs for forensic analysis. Forensic analysis validates the incident type, the intrusion method used, and the impacted system's shortcomings. This analysis helps cybersecurity professionals understand the incident's technical details, root cause, and operational impact. This understanding helps determine other information to gather, coordinate information sharing with others, and develop a course of action for response. Cybersecurity professionals report incidents to higher-level DODIN operations authorities, law enforcement, and counterintelligence, describing the threat event in detail. These authorities correlate this information with other incidents across the DODIN to identify relationships and trends between incidents in the short term and patterns across threat activities in the long term.

V. Recover Function

Recovery reestablishes normal operation of the network. The recover function includes activities to remove the vulnerability from the network and information systems

Recovery activities strive to improve operational, technical, and management controls to prevent recurrence of threat activity. After DODIN operations personnel follow the detailed recovery steps, cybersecurity professionals conduct a post-incident analysis to review the effectiveness of incident handling. Lessons learned from this analysis help in developing follow-up strategies that support prevention goals.

Chap 7
III. Protection, Detection, & Reaction

Ref: ATP 6-02.71, Techniques for Department of Defense Information Network Operations (Apr '19), pp. A-28 to A-32.

Securing Networks and Information Systems

 Protection

 Detection

 Reaction

Information systems and networks are critical to the military's ability to conduct operations. Adequately securing networks and information systems against attack requires the ability to—
- **Protect** the information computer systems and data networks pass and store.
- **Detect** intrusions into the network or information systems as they happen.
- **React** to limit or reduce damage, and repair the network or information system.

I. Protection

Information protection consists of active and passive measures to secure and defend friendly information and information systems to ensure timely, accurate, and relevant friendly information. Information protection denies enemies, adversaries, and threat actors the opportunity to exploit friendly information and information systems for their own purposes. Information protection includes cybersecurity and electronic protection.

Information protection applies to any data medium or form, including hardcopy, electronic, magnetic, video, imagery, voice, telegraph, computer, and human. Information protection involves determining appropriate security measures, based on the value of information protected. Protection measures reflect the changing value of the information about each operational phase of the mission. Leaders, information producers, processors, and users ensure information protection.

Continuity of operations plans, operation plans, and operation orders specify the priorities for protecting networks and information systems. Protection measures consist of the firewalls, intrusion protection systems, and software that harden these systems against intruders. Protecting information stored on U.S. computers and flowing through the networks is vital.

Army network and system managers devise and implement comprehensive plans for a full range of security measures. These plans include external and internal perimeter protection. External perimeter protection consists of COMSEC, router filtering, access control lists, security guards, and physical isolation as barriers to outside networks, such as the Internet. Internal perimeter protection consists of firewalls and router filtering. These serve as barriers between echelons of interconnected

Cyber Attacks

Ref: ATP 6-02.71, Techniques for Department of Defense Information Network Operations (Apr '19), pp. A-26 to A-28.

Some attacks have delayed effects while others are immediate. Both delayed and immediate attacks may corrupt databases and control programs and may degrade or physically destroy the system attacked. Prompt attack detection is essential to initiating intrusion response and network restoration.

Computer attacks generally target software or data in either end-user computers or platform information technology systems. Platform information technology is information technology, both hardware and software, that is physically part of, dedicated to, or essential in real-time to the mission performance of special purpose systems (DODI 8500.01). Adversaries may attempt to unobtrusively access information, modify software and data, or destroy software and data. These activities may target one or more computers connected to a local area network or WAN. Computer attacks may occur during routine operations and may disrupt major military missions. These attacks can happen during both wartime and peacetime. Attacks can be part of a major nation-state attempt to cripple U.S. information infrastructure or come from mischievous or vengeful insiders, criminals, political dissidents, terrorists, or foreign intelligence entities.

Attackers may design attacks to unleash computer viruses, trigger future attacks, or install software that compromises or damages information and information systems. Malicious attacks may also involve unauthorized file exfiltration or deletion, or malicious software or data introduction. Malicious software is executable software code secretly introduced into a computer, including viruses, Trojan horses, and worms. Malicious data insertion, also known as spoofing, seeks to mislead users or disrupt systems operation. For example, an attack may disrupt a packet data network by introducing false routing table data into one or more routers. An attacker who denies service, or corrupts data on a wide scale may weaken user confidence in the information on the network by corrupting or sending false data.

Physical Attacks

Physical attacks generally deny service and involve destruction, damage, overrun, or capture of system components. These components may include end-user computers, communications devices, and network infrastructure components. Another form of physical attack is theft of cryptographic keys or passwords. This is a major concern, since these items can support subsequent electronic or computer attack or analysis activities.

Electronic Attacks

Electronic attacks may focus on specific or multiple targets across a wide area. Attacks against communications links include jamming and signals intelligence exploitation. Jamming overwhelms friendly signals, corrupts data being transmitted, and may cause denial of service. Two types of signals intelligence operations are signal interception and analysis to compromise data and emitter direction findings; and geo-location to support signal analysis and physical attacks.

Information Systems Security

Cybersecurity programs include the full range of security measures. Information systems security succeeds only when all assets connected to the local area network and throughout the WAN adhere to common technical standards. Protection from intrusions into, or via, a WAN begins with coordinating information systems security between all of the Services and the Defense Information Systems Agency. All measures to detect, respond to, and report attacks and intrusions must adhere to public laws, applicable DOD directives, and Army regulations. All users, system administrators and DODIN operations managers require cybersecurity awareness and certification training. Training prepares users to reduce vulnerabilities and risks by taking proper actions, depending on their DODIN operations role.

Protection Levels

Protection levels apply only to confidentiality requirements. Protection levels are based on the required clearance, formal access approval, and need-to-know of direct and indirect users who receive information from information systems without manual intervention and reliable human review. Protection levels indicate the level of trust placed in the system's technical capabilities. Service providers and users cooperate to implement the required protection level. Soldiers need assurance that their information systems have the level of protection or trust needed for mission success.

Mitigating Insider Threats

The insider is anyone with current or past authorized access to a DOD information system. Potential insider threats include military members, DOD civilians, employees of other federal agencies, and contractors.

The insider threat is real and significant. A DOD Inspector General investigation determined 87 percent of identified intruders into DOD information systems were either employees or others internal to the organization. Insiders may cause security risks through—

- Malicious intent.
- Disdain for security practices.
- Carelessness.
- Ignorance of security policy, security practices, and proper information system use.

An effective insider threat mitigation strategy implements best practices across multiple disciplines. Key elements of this strategy include—

- Determine which assets are critical to the mission.
- Establish trustworthiness—seek to reduce the threat by establishing a high level of assurance in the trustworthiness of people, practices, systems, and programs thorough personnel security, cybersecurity awareness training, and cybersecurity best practices.
- Strengthen and enforce personnel security and management practices.

Protect information assets by—

- Controlling asset-sharing through cybersecurity.
- Isolating information and capabilities, based on security clearance and need-to-know, through compartmentalization and system architecture.
- Identifying and reducing known vulnerabilities through cybersecurity.
- Employing and enforcing effective physical security policies.
- Detect problems through cybersecurity.
- React or respond to cyber incidents and events.
- Maintain command emphasis on the Army counterintelligence Threat Awareness and Reporting Program.

See pp. 0-2 to 0-7 for discussion of global cyber threats. See pp. 7-16 to 7-17 for dicussion of cyber threat activities and pp. 7-18 to 7-19 for common cyber attack tools. For more information on cyberspace threat from FM 3-12, see p. 2-24.

networks and information systems. Internal COMSEC barriers are also required. Local workstation protection consists of individual access controls, configuration audit capability, protection and intrusion detection tools, and security procedures.

Protection against intrusions into friendly computer networks denies unauthorized entry and access and protects networks and systems. Operations security procedures allow commanders to identify actions that adversary intelligence systems and intruders observe and provides awareness of the indicators adversary intelligence systems might collect. Operations security identifies and selects information adversaries could exploit, and countermeasures to mitigate threats. Since most vulnerabilities result from human error, operations security training helps protect against network intrusions. Many measures affect operations security, including information security, transmission security, COMSEC, and emission control.

Commercial capabilities, such as imaging, positioning, and cellular systems, allow adversaries to access significant information about U.S. forces. The ability of Army and other Service personnel to send information directly from the battlefield via e-mail to points around the world presents an attractive target for potential adversary exploitation. These e-mails may contain sensitive or classified information. Improper disclosure of this information could endanger friendly personnel and compromise missions.

Information on Army web pages is also a security concern. Operations security guidelines for web pages are the same as for any other information within the Army. Sensitive and classified information requires protection against disclosure to unauthorized personnel.

Security measures actively and passively preserve the confidentiality, integrity, availability, and functionality of information systems. Protection includes real-and near real-time measures to prevent intrusions and restore affected devices or systems. These security measures include—

- Vigorous cybersecurity protection programs.
- Denial of unauthorized access.
- Hardening of programs and gateways using software and hardware tools.
- Quality assurance procedures in all program and hardware acquisition.
- Strict access controls for networked computers and other devices.

Transmission security secures information across the various networks. Trunk encryption devices, in-line encryption devices, frequency hopping, and time division multiplexing and modulation techniques usually secure transmissions. Transmission security helps ensure information security. Any nonsecure system or device connected or entering into a secure network must use in-line encryption between the network entry point and the entering equipment.

COMSEC protects information on networks and system devices. Keying variables enable encryption of voice and data passing through transmission devices and computers. The National Security Agency controls most encryption keys and governs local key generation, distribution, and storage.

Information security is the protection of information and information systems from unauthorized access, use, disclosure, disruption, modification, or destruction in order to provide confidentiality, integrity, and availability (CNSSI 4009). Information security policies deny unauthorized access to classified or sensitive information. These policies establish measures to prevent disclosure of valuable information from other aspects of communications, such as traffic flow or message analysis, and to enhance the authentication of communications.

II. Detection

Routine DODIN operations activities include security management and intrusion detection to detect violations of security policies. Selected events or occurrences, such as several failed login attempts in a defined period, are monitored using conventional protection and detection tools and devices. When DODIN operations managers detect violations, they act to prevent further violations and report the event to the commander, information system security officer, and next higher NOSC in the DODIN operations technical channels.

NOSCs monitor networks and systems in near real-time to detect anomalies and take preliminary defensive actions. Prompt defensive actions mitigate the damage and reduce the operational impact of insecurities.

III. Reaction

Reaction to a network or information system intrusion includes restoring essential information services. A detailed continuity of operations plan includes procedures for various levels of restoration and addresses various potential disasters. Immediate restoration may rely on backup or redundant network links or system components, backup databases, or alternate means of network transport.

DODIN operations managers do not need permission to react to attacks or intrusions if their activities conform to appropriate regulations, statutes, and public law. DODIN operations managers or system administrators take these emergency steps when they verify an intrusion—

- Stop the breach, if possible, and restore any destroyed or compromised data from backups or other identified continuity of operations capabilities.
- Follow cyber incident policy, as outlined in the standard operating procedure and applicable regulations.
- Report the incident to the commander, cybersecurity manager, or information system security officer.
- Report the incident to next higher level NOSC.

Security management devices and IAVM messages warn DODIN operations personnel of intrusion attempts, attacks, and other network and systems anomalies. The appropriate response to these alerts depends on the severity of the attack, intrusion, or breach. These alerts trigger appropriate response measures. DODIN operations managers need to consider operational or mission status before responding to alerts. Protecting information systems requires real-time security management as a component of DODIN operations. When detection occurs, DODIN operations managers may need to take one or more of these actions—

- Change boundaries and perimeters.
- Reconfigure firewalls, guards, and routers.
- Reroute traffic.
- Change encryption levels or re-key COMSEC devices.
- Zeroize suspected compromised communications.
- Re-establish a network without compromised members.
- Change passwords and authentication.

Response begins immediately upon anomaly detection. The objective of a response is to restore services to a level that supports acceptable, if reduced, operations. Restoration returns services to the same level as before to the event. Responses may be offensive or defensive. Defensive responses include all measures and countermeasures available to a commander to limit an adversary's attack, exploitation, military deception, or electronic protection capabilities to protect against further attacks.

Information Assurance Vulnerability Management (IAVM)

Information assurance vulnerability management is the comprehensive distribution process for notifying combatant commands, Services, agencies and field activities about vulnerability alerts, bulletins, technical advisories, and countermeasures information. The IAVM program requires combatant commands, Services, agencies, and field activities to receipt acknowledgment and provides specific time parameters for implementing appropriate countermeasures depending on the criticality of the vulnerability (CJCSM 6510.01).

The IAVM program helps mitigate vulnerabilities. The ACOIC is the Army's lead agent for implementing IAVM. The Army Knowledge Online Knowledge Management Center mail service issues alerts, bulletins, technical tips, and system administrator reports for the ACOIC, based on mandatory ACOIC IA vulnerability alerts and Army-generated IAVM requirements.

IAVM is the program to identify and resolve discovered vulnerabilities on DOD systems and platforms. This program includes IA vulnerability alerts, IAVM messages, and technical advisories

Scanning and Remediation

Scanning is automated or semi-automated polling for information system and device configuration data. Scans help in system identification; maintenance; security assessment and investigation; verification of vulnerability compliance; or discovery of compromised systems. Scanning includes network port scanning and vulnerability scanning, whether wired or wireless, and classified or unclassified. Cybersecurity personnel conduct scheduled and unscheduled scans throughout all phases of operations.

Operational scanning occurs in every layer of the enterprise management structure and on all classifications of networks. Both scheduled and no-notice scans are part of security policy and compliance enforcement. NETCOM maintains scanning tool software licenses.

New vulnerabilities require proactive management. Assessors use a five-step methodology for assessment scanning—

- Identify assets
- Determine vulnerabilities
- Review vulnerabilities
- Remediate vulnerabilities
- Validate remediation measures

The system administrator or network manager ensures confidentiality of information by preventing unauthorized access to computer equipment. The system administrator, network manager, and operators patch security vulnerabilities on all Army platforms. The NEC and tactical system administrators validate patches, whether the system is operating on the installation network or stored. Operation orders and other command directives should include these requirements.

Continuity of Operations

A continuity of operations plan is a plan for emergency response, backup operations, transfer of operations, and post-disaster recovery maintained by an activity as a part of its cybersecurity program. A continuity of operations plan ensures the organization can continue to function after a catastrophic event and defines procedures to protect and restore the organization's vital data and resume operations. Units conduct continuity of operations exercises at least annually.

Refer to AR 500-3 and DA Pam 25-1-2 for more information on continuity of operations.

Chap 8

I. Acronyms & Abbreviations

Ref: JP 3-12, Cyberspace Operations (Jun '18) and FM 3-12, Cyberspace Operations and Electromagnetic Warfare (Aug '21).

A
AOR — area of responsibility
ARCYBER — U.S. Army Cyber Command

B
BDA — battle damage assessment

C
C2 — command and control
CCDR — combatant commander
CCMD — combatant command
CCMF — Cyber Combat Mission Force
CEMA — cyberspace electromagnetic activities
CERF — cyber effects request format
CEWO — cyber electromagnetic warfare officer
CI — counterintelligence
CI/KR — critical infrastructure and key resources
CIO — chief information officer
CMF — Cyber Mission Force
CMT — combat mission team
CNMF — Cyber National Mission Force
CNMF — cyber national mission force
CNMF-HQ — Cyber National Mission Force Headquarters
CO — cyberspace operations
COA — course of action
COCOM — combatant command (command authority)
CO-IPE — cyberspace operations-integrated planning element
CONOPS — concept of operations
CONPLAN — concept plan
COP — common operational picture
CPF — Cyber Protection Force
CPT — cyberspace protection team
CSA — combat support agency
CSSP — cybersecurity service provider
CST — combat support team

D
D3A — decide, detect, deliver, and assess
DACO — directive authority for cyberspace operations
DC3 — Department of Defense Cyber Crime Center
DCI — defense critical infrastructure
DCO — defensive cyberspace operations
DCO-IDM — defensive cyberspace operations-internal defensive measures
DCO-RA — defensive cyberspace operations-response actions
DCO-RA — defensive cyberspace operations-response actions
DHS — Department of Homeland Security
DIA — Defense Intelligence Agency
DIB — defense industrial base
DISA — Defense Information Systems Agency
DOD — Department of Defense
DODIN — Department of Defense information network
DODIN-A — Department of Defense information network-Army
DSCA — defense support of civil authorities

E
EA — electromagnetic attack
EMI — electromagnetic interference
EMOE — electromagnetic operational environment
EMS — electromagnetic spectrum
EMSO — electromagnetic spectrum operations
EP — electromagnetic protection
ES — electromagnetic support
EW — electromagnetic warfare
EXORD — execute order

I. Abbreviations & Acronyms 8-1 *

G
GCC	geographic combatant commander
GFMIG	Global Force Management Implementation Guidance

I
I2CEWS	intelligence, information, cyber, electromagnetic warfare and space
IAW	in accordance with
IC	intelligence community
IGL	intelligence gain/loss
IJSTO	integrated joint special technical operations
IO	information operations
IP	Internet protocol
IPB	intelligence preparation of the battlefield
IR	intelligence requirement
IRC	information-related capability
ISP	Internet service provider
ISR	intelligence, surveillance, and reconnaissance
IT	information technology

J
JEMSO	joint electromagnetic spectrum operations
JEMSOC	joint electromagnetic spectrum operations cell
JFC	joint force commander
JFHQ-C	Joint Force Headquarters-Cyber
JFHQ-C	joint force headquarters-cyberspace
JIACG	joint interagency coordination group
JOA	joint operations area
JP	joint publication
JPP	joint planning process
JS	Joint Staff
JTF	joint task force
JTL	joint target list

L
LE	law enforcement
LOC	line of communications

M
MILDEC	military deception
MISO	military information support operations
MNF	multinational force
MOE	measure of effectiveness
MOP	measure of performance
MTFP	mission-tailored force package
NCO	noncommissioned officer
NETCOM	United States Army Network Enterprise Technology Command

N
NIPRNET	Non-classified Internet Protocol Router Network
NMT	national mission team
NST	national support team

O
OA	operational area
OCO	offensive cyberspace operations
OE	operational environment
OPCON	operational control
OPLAN	operation plan
OPORD	operation order
OPSEC	operations security
OSC	offensive space control
OSD	Office of the Secretary of Defense
OSINT	open-source intelligence

P
PIT	platform information technology
PN	partner nation
PPD	Presidential policy directive

R
RFI	request for information
RFS	request for support
ROE	rules of engagement

S
SATCOM	satellite communications
SCC	Service cyberspace component
SecDef	Secretary of Defense
SIGINT	signals intelligence
SIPRNET	SECRET Internet Protocol Router Network

T
TACON	tactical control
TCPED	tasking, collection, processing, exploitation, and dissemination
TSS	targeting sensing software
TST	time-sensitive target

U
USC	United States Code
USCYBERCOM	U.S. Cyber Command

(CYBER1-1)
II. Glossary

Ref: JP 3-12, Cyberspace Operations (Jun '18) and FM 3-12, Cyberspace Operations and Electromagnetic Warfare (Aug '21). This combined glossary lists acronyms and terms with Army, multi-Service, or joint definitions, and other selected terms. The proponent publication for a term is listed in parentheses after the definition.

A

Adversary. A party acknowledged as potentially hostile to a friendly party and against which the use of force may be envisaged. (JP 3-0)

Army design methodology. A methodology for applying critical and creative thinking to understand, visualize, and describe problems and approaches to solving them. Also called ADM. (ADP 5-0)

Assessment. 1) A continuous process that measures the overall effectiveness of employing capabilities during military operations. 2) Determination of the progress toward accomplishing a task, creating a condition, or achieving an objective. 3) Analysis of the security, effectiveness, and potential of an existing or planned intelligence activity. 4) Judgment of the motives, qualifications, and characteristics of present or prospective employees or "agents." (JP 3-0)

C

Chaff. Radar confusion reflectors, consisting of thin, narrow metallic strips of various lengths and frequency responses, which are used to reflect echoes for confusion purposes. (JP 3-85)

combat power. The total means of destructive, constructive, and information capabilities that a military unit or formation can apply at a given time. (ADP 3-0)

constraint. A restriction placed on the command by a higher command. (FM 6-0)

countermeasures. That form of military science that, by the employment of devices and/or techniques, has as its objective the impairment of the operational effectiveness of enemy activity. (JP 3-85)

cyberspace attack. Actions taken in cyberspace that create noticeable denial effects (i.e., degradation, disruption, or destruction) in cyberspace or manipulation that leads to denial that appears in a physical domain, and is considered a form of fires. (JP 3-12)

cyberspace capability. A device or computer program, including any combination of software, firmware, or hardware, designed to create an effect in or through cyberspace. (Approved for inclusion in the DOD Dictionary.)

cyberspace defense. Actions taken within protected cyberspace to defeat specific threats that have breached or are threatening to breach cyberspace security measures and include actions to detect, characterize, counter, and mitigate threats, including malware or the unauthorized activities of users, and to restore the system to a secure configuration. (Approved for inclusion in the DOD Dictionary.)

cyberspace defense. Actions taken within protected cyberspace to defeat specific threats that have breached or are threatening to breach cyberspace security measures and include actions to detect, characterize, counter, and mitigate threats, including malware or the unauthorized activities of users, and to restore the system to a secure configuration. (JP 3-12)

cyberspace electromagnetic activities. The process of planning, integrating, and synchronizing cyberspace operations and electromagnetic warfare operations in support of unified land operations. Also called CEMA. (ADP 3-0)

cyberspace exploitation. Actions taken in cyberspace to gain intelligence, maneuver, collect information, or perform other enabling actions required to prepare for future military operations. (JP 3-12)

cyberspace operation. The employment of cyberspace capabilities where the primary purpose is to achieve objectives in or through cyberspace. Also see CO. (JP 3-0)

cyberspace security. Actions taken within protected cyberspace to prevent unauthorized access to, exploitation of, or damage to computers, electronic communications systems, and other information technology, including platform information technology, as well as the information contained therein, to ensure its availability, integrity, authentication, confidentiality, and nonrepudiation. (JP 3-12)

cyberspace superiority. The degree of dominance in cyberspace by one force that permits the secure, reliable conduct of operations by that force and its related land, air, maritime, and space forces at a given time and place without prohibitive interference. (Approved for incorporation into the DOD Dictionary.)

cyberspace. A global domain within the information environment consisting of the interdependent networks of information technology infrastructures and resident data, including the Internet, telecommunications networks, computer systems, and embedded processors and controllers. (JP 3-12)

D

defeat. To render a force incapable of achieving its objectives. (ADP 3-0)

defensive cyberspace operations. Missions to preserve the ability to utilize blue cyberspace capabilities and protect data, networks, cyberspace-enabled devices, and other designated systems by defeating on-going or imminent malicious cyberspace activity. Also called DCO. (JP 3-12)

defensive cyberspace operations-internal defensive measures. Operations in which authorized defense actions occur within the defended portion of cyberspace. Also called DCO-IDM. (JP 3-12)

defensive cyberspace operations-response actions. Operations that are part of a defensive cyberspace operations mission that are taken external to the defended network or portion of cyberspace without permission of the owner of the affected system. Also called DCO-RA. (JP 3-12)

Department of Defense information network. The set of information capabilities and associated processes for collecting, processing, storing, disseminating, and managing information on demand to warfighters, policy makers, and support personnel, whether interconnected or stand-alone. Also called DODIN. (JP 6-0)

Department of Defense information network operations. Operations to secure, configure, operate, extend, maintain, and sustain Department of Defense cyberspace to create and preserve the confidentiality, availability, and integrity of the Department of Defense information network. Also called DODIN operations. (JP 3-12)

Department of Defense information network-Army. An Army-operated enclave of the Department of Defense information network that encompasses all Army information capabilities that collect, process, store, display, disseminate, and protect information worldwide. Also called DODIN-A. (ATP 6-02.71)

directed energy. An umbrella term covering technologies that relate to the production of a beam of concentrated electromagnetic energy or atomic or subatomic particles. Also called DE. (JP 3-85)

directed-energy warfare. Military actions involving the use of directed-energy weapons, devices, and countermeasures. Also called DEW. (JP 3-85)

directed-energy weapon. A weapon or system that uses directed energy to incapacitate, damage, or destroy enemy equipment, facilities, and/or personnel. (JP 3-85)

direction finding. A procedure for obtaining bearings of radio frequency emitters by using a highly directional antenna and a display unit on an intercept receiver or ancillary equipment. Also called DF. (JP 3-85)

directive authority for cyberspace operations. The authority to issue orders and directives to all Department of Defense components to execute global Department of Defense information network operations and defensive cyberspace operations internal defensive measures. Also called DACO. (Approved for inclusion in the DOD Dictionary.)

dynamic targeting. Targeting that prosecutes targets identified too late or not selected for action in time to be included in deliberate targeting. (JP 3-60)

E

electromagnetic attack. Division of electromagnetic warfare involving the use of electromagnetic energy, directed energy, or antiradiation weapons to attack personnel, facilities, or equipment with the intent of degrading, neutralizing, or destroying enemy combat capability and is considered a form of fires. Also called EA. (JP 3-85)

electromagnetic compatibility. The ability of systems, equipment, and devices that use the electromagnetic spectrum to operate in their intended environments without causing or suffering unacceptable or unintentional degradation because of electromagnetic radiation or response. Also called EMC. (JP 3-85)

electromagnetic hardening. Actions taken to protect personnel, facilities, and/or equipment by blanking, filtering, attenuating, grounding, bonding, and/or shielding against undesirable effects of electromagnetic energy. (JP 3-85)

electromagnetic intrusion. The intentional insertion of electromagnetic energy into transmission paths in any manner. The objective of electromagnetic intrusion is to deceive threat operators or cause confusion. (JP 3-85)

electromagnetic jamming. The deliberate radiation, reradiation, or reflection of electromagnetic energy for the purpose of preventing or reducing an enemy's effective use of the electromagnetic spectrum, and with the intent of degrading or neutralizing the enemy's combat capability. (JP 3-85)

electromagnetic masking. The controlled radiation of electromagnetic energy on friendly frequencies in a manner to protect the emissions of friendly communications and electronic systems against enemy electromagnetic support measures/signals intelligence without significantly degrading the operation of friendly systems. (JP 3-85)

electromagnetic probing. The intentional radiation designed to be introduced into the devices or systems of adversaries to learn the functions and operational capabilities of the devices or systems. (JP 3-85)

electromagnetic protection. Division of electromagnetic warfare involving actions taken to protect personnel, facilities, and equipment from any effects of friendly or enemy use of the electromagnetic spectrum that degrade, neutralize, or destroy friendly combat capability. Also called EP. (JP 3-85)

electromagnetic pulse. A strong burst of electromagnetic radiation caused by a nuclear explosion, energy weapon, or by natural phenomenon, that may couple with electrical or electronic systems to produce damaging current and voltage surges. (JP 3-85)

electromagnetic reconnaissance. The detection, location, identification, and evaluation of foreign electromagnetic radiations. (JP 3-85)

electromagnetic security. The protection resulting from all measures designed to deny unauthorized persons information of value that might be derived from their interception and study of noncommunications electromagnetic radiations (e.g., radar). (JP 3-85)

electromagnetic spectrum superiority. That degree of control in the electromagnetic spectrum that permits the conduct of operations at a given time and place without prohibitive interference, while affecting the threat's ability to do the same. (JP 3-85)

electromagnetic support. Division of electromagnetic warfare involving actions tasked by, or under the direct control of, an operational commander to search for, intercept, identify, and locate or localize sources of intentional and unintentional radiated electromagnetic energy for immediate threat recognition, targeting, planning, and conduct of future operations. Also called ES. (JP 3-85)

electromagnetic vulnerability. The characteristics of a system that cause it to suffer a definite degradation (incapability to perform the designated mission) as a result of having been subjected to a certain level of electromagnetic environmental effects. (JP 3-85)

electromagnetic warfare. Military action involving the use of electromagnetic and directed energy to control the electromagnetic spectrum or to attack the enemy. Also called EW. (JP 3-85)

electromagnetic warfare reprogramming. The deliberate alteration or modification of electromagnetic warfare or target sensing systems, or the tactics and procedures that employ them, in response to validated changes in equipment, tactics, or the electromagnetic environment. (JP 3-85)

enemy. An enemy is a party identified as hostile against which the use of force is authorized. (ADP 3-0)

essential task. A specified or implied task that must be executed to accomplish the mission. (FM 6-0)

execution. The act of putting a plan into action by applying combat power to accomplish the mission and adjusting operations based on changes in the situation. (ADP 5-0)

H

hazard. A condition with the potential to cause injury, illness, or death of personnel, damage to or loss of equipment or property, or mission degradation. (JP 3-33)

high-payoff target. A target whose loss to the enemy will significantly contribute to the success of the friendly course of action. Also called HPT. (JP 3-60)

high-value target. A target the enemy commander requires for the successful completion of the mission. (JP 3-60)

hybrid threat. A hybrid threat is the diverse and dynamic combination of regular forces, irregular forces, terrorists, or criminal elements acting in concert to achieve mutually benefitting effects. (ADP 3-0)

I

implied task. A task that must be performed to accomplish a specified task or mission but is not stated in the higher headquarters' order. (FM 6-0)

information assurance. None. (Approved for removal from the DOD Dictionary.)

information collection. An activity that synchronizes and integrates the planning and employment of sensors and assets as well as the processing, exploitation, and dissemination systems in direct support of current and future operations. (FM 3-55)

information operations. The integrated employment, during military operations, of information-related capabilities in concert with other lines of operation to influence, disrupt, corrupt, or usurp the decision-making of adversaries and potential adversaries while protecting our own. Also called IO. (JP 3-13)

intelligence operations. The tasks undertaken by military intelligence units through the intelligences disciplines to obtain information to satisfy validated requirements. (ADP 2-0)

intelligence preparation of the battlefield. The systematic process of analyzing the mission variables of enemy, terrain, weather, and civil considerations in an area of interest to determine their effect on operations. Also called IPB. (ATP 2-01.3)

intelligence. 1) The product resulting from the collection, processing, integration, evaluation, analysis, and interpretation of available information concerning foreign nations, hostile or potentially hostile forces or elements, or areas of actual or potential operations. 2) The activities that result in the product. 3) The organizations engaged in such activities. (JP 2-0)

K

knowledge management. The process of enabling knowledge flow to enhance shared understanding, learning, and decision making. (ADP 6-0)

N

named area of interest. The geospatial area or systems node or link against which information that will satisfy a specific information requirement can be collected. Also called NAI. (JP 2-01.3)

O

offensive cyberspace operations. Missions intended to project power in and through cyberspace. Also called OCO. (JP 3-12)

operational environment. A composite of the conditions, circumstances, and influences that affect the employment of capabilities and impact the decisions of the commander assigned responsibility for it. Also called OE. (JP 3-0)

operational initiative. The setting or tempo and terms of action throughout an operation. (ADP 3-0)

operations process. The major command and control activities performed during operations: planning, preparing, executing, and continuously assessing the operation. (ADP 5-0)

operations security. A capability that identifies and controls critical information, indicators of friendly force actions attendant to military operations, and incorporates countermeasures to reduce the risk of an adversary exploiting vulnerabilities. Also called OPSEC. (JP 3-13.3)

P

planning. The art and science of understanding a situation, envisioning a desired future, and laying out effective ways of bringing that future about. (ADP 5-0)

position of relative advantage. A location or the establishment of a favorable condition within the area of operations that provides the commander with temporary freedom of action to enhance combat power over an enemy or influence the enemy to accept risk and move to a position of disadvantage. (ADP 3.0)

preparation. Those activities performed by units and Soldiers to improve their ability to execute an operation. (ADP 5.0)

priority of fires. The commander's guidance to the staff, subordinate commanders, fires planners, and supporting agencies to employ fires in accordance with the relative importance of a unit's mission. (FM 3-09)

priority of support. A priority set by the commander to ensure a subordinate unit has support in accordance with its relative importance to accomplish the mission. (ADP 5-0)

R

radio frequency countermeasures. Any device or technique employing radio frequency materials or technology that is intended to impair the effectiveness of enemy activity, particularly with respect to precision-guided and sensor systems. (JP 3-85)

risk management. The process to identify, assess, and control risks and make decisions that balance risk cost with mission benefits. (JP 3-0)

S

scheme of fires. The detailed, logical sequence of targets and fire support events to find and engage targets to support commander's objectives. (JP 3-09)

specified task. A task specifically assigned to an organization by its higher headquarters. (FM 6-0)

T

target. An entity or object that performs a function for the adversary considered for possible engagement or other actions. See also objective area. (JP 3-60)

target area of interest. The geographical area where high-valued targets can be acquired and engaged by friendly forces. (JP 2-01.3)

targeting. The process of selecting and prioritizing targets and matching the appropriate response to them, considering operational requirements and capabilities. (JP 3-0)

W

warfighting function. A group of tasks and systems united by a common purpose that commanders use to accomplish missions and training objectives. (ADP 3-0)

wartime reserve modes. Characteristics and operating procedures of sensor, communications, navigation aids, threat recognition, weapons, and countermeasure systems that will contribute to military effectiveness if unknown to or misunderstood by opposing commanders before they are used, but could be exploited or neutralized if known in advance. (JP 3-85)

[CYBER1-1]
Index

1st Information Operations Command, 6-10

A
Accidents and Natural Hazards, 1-13
Acronyms, 8-1
Airborne Electronic Attack, 3-26
Airborne Electromagnetic Attack Support, 4-28
ANNEX C–OPERATIONS, 4-35
ANNEX H–SIGNAL, 4-35
Anonymity, 1-13
Anticipated Operational Environments, 0-7
Appendix 12 to Annex C, 4-35
Army Cyber Operations and Integration Center (ACOIC), 6-12
Army Design Methodology, 4-2
Army Enterprise Service Desk, 6-12
Army Information Warfare Operations Center, 2-27
Army Organizations, 2-27
Assess, 4-33
Assessment, 1-55, 2-39
Assignment of Cyberspace Forces, 1-23
Assistant Chief of Staff, Intelligence, 2-32
Assistant Chief of Staff, Signal, 2-33
Authorities, 1-30

B
Battalion Electronic Warfare Personnel, 3-13

C
CDRUSCYBERCOM, 1-33
CI/KR Protection, 1-30
Civil Considerations Data Files, Overlays, and Assessments, 4-h
Civil-Military Operations (CMO), 0-14, 4-48
Close Air Support (CAS), 3-24
Command and Control (C2), 1-48, 2-27
Commander's Communication Synchronization (CCS), 4-46
Commander's Role, 2-28
Common Operational Picture (COP), 5-14
Company CREW Specialists, 3-16
Competition Continuum, 2-16
Concealment, 3-32
Conflict and Competition, 2-16
Congested Environments, 2-10
Connectivity and Access, 1-7
Considerations When Targeting, 4-34
Contemporary Operational Environment, 0-6
Contested Environments, 2-10
Continuity of Operations, 7-30
Core Activities   1-15
Core Competencies, 2-12
Counter Radio-Controlled Improvised Device (CREW), 3-28
Critical Capabilities, 6-38
Critical Variables, 0-6
Cyber Attack, 7-26
Cyber Attack Tools, 7-18
Cyber Combat Mission Force (CCMF), 1-10
Cyber Effects Request Format (CERF), 4-9, 4-11
Cyber Electromagnetic Warfare Officer (CEWO), 2-30, 3-12
Cyber Kill Chain, 4-I
Cyber Mission Force (CMF), 1-10
Cyber National Mission Force (CNMF), 1-1
Cyber Protection Force (CPF), 1-10
Cyber Threat, 0-2
Cyber Warfare Officer or Cyber-Operations Officer, 2-31
Cyber-Persona Layer, 2-6
Cyber-Personal Layer, 1-3, 1-3
Cybersecurity, 7-1
Cybersecurity Functions, 7-13, 7-15
Cybersecurity Fundamentals, 7-1
Cybersecurity Performance, 7-12
Cybersecurity Principles, 7-3
Cybersecurity Risk Management, 7-2
Cyberspace, 0-1
Cyberspace Actions, 1-20, 2-21
Cyberspace and the Electromagnetic Spectrum, 2-1, 2-2
Cyberspace Attack, 1-21, 2-22
Cyberspace Defense, 1-21, 2-21, 7-10
Cyberspace Domain, 2-6
Cyberspace-Enabled Activities, 1-15
Cyberspace (CEMA) in Operations Orders, 4-35
Cyberspace (CEMA) Operations Planning, 4-1
Cyberspace Electromagnetic Activities (CEMA) Section, 2-29

Index-1 *

Cyberspace Electromagnetic Activities (CEMA) Working Group, 2-29
Cyberspace Electromagnetic Activities at Corps and Below, 2-28
Cyberspace Electromagnetic Activities Spectrum Manager, 2-31
Cyberspace Exploitation, 1-21, 2-21
Cyberspace Layer Model, 1-2
Cyberspace Missions, 0-1, 1-15, 2-19
Cyberspace Operations (CO), 0-1, 0-15, 1-1, 2-17, 4-48
Cyberspace Operations & EW Logic Chart, 2-3
Cyberspace Operations (Missions & Actions), 2-19
Cyberspace Operations Forces, 1-10
Cyberspace Security, 1-20, 2-21

D
DCO-IDM, 1-19, 2-7
DCO-RA, 1-19, 2-7
Decide, 4-32
Deconflicting the Electromagnetic Spectrum, 3-19
Deconfliction, 1-52
Defense Information Systems Agency (DISA), 6-9
Defense of Non-DOD Cyberspace, 1-19
Defensive Cyberspace Operations (DCO), 1-19, 2-18
Defensive Cyberspace Operations Internal Defensive Measures (DCO-IDM), 2-20
Defensive Cyberspace Operations Response Action (DCO-RA), 2-20
Defensive Electromagnetic Attack, 3-2, 3-28
Define the Operational Environment, 4-b
Deliver, 4-33
Deny, 1-22

Department of Defense Information Network (DODIN) Operations, 1-6, 2-6, 2-18, 6-1
Describe Environmental Effects on Operations, 4-e
Detect, 4-33
Detect Function, 7-23
Detection, 7-29
Determine Threat Courses of Action, 4-n
Direction Finding (DF), 3-36
DOD Information Network (DODIN), 1-6, 2-6, 2-18, 6-1
DOD Ordinary Business Operations, 1-17
DODIN Enterprise Management, 6-36
DODIN Network Operations Components, 6-35
DODIN Operations Operational Construct, 6-36
DODIN Operations, 1-18
DODIN, 1-6, 2-6, 6-1

E
Electromagnetic Attack (EA), 3-2
Electromagnetic Attack Request, 4-27
Electromagnetic Environment (EME), 5-2
Electromagnetic Environment (EME) Survey, 3-36
Electromagnetic Environmental Effects (E3), 5-8
Electromagnetic Interference (EMI), 3-10, 3-31
Electromagnetic Interference (EMI) Battle Drill, 3-33
Electromagnetic Jamming, 3-31
Electromagnetic Operational Environment (EMOE), 5-2
Electromagnetic Order of Battle (EOB), 5-15
Electromagnetic Protection (EP), 3-6
Electromagnetic Pulse (EMP), 5-8
Electromagnetic Spectrum (EMS), 2-1, 5-2

Electromagnetic Spectrum (EMS) Factors, 1-52
Electromagnetic Spectrum Operations (EMSO), 5-1
Electromagnetic Spectrum Superiority, 2-17
Electromagnetic Support (ES), 3-8
Electromagnetic Warfare (EW), 3-1
Electromagnetic Warfare (EW) Organizations, 2-36
Electronic Attack (EA), 5-6
Electronic Attack Effects, 3-21
Electronic Attack Techniques, 3-21
Electronic Protection Techniques, 3-29
Electronic Reconnaissance, 3-35
Electronic Warfare Assessment, 4-26
Electronic Warfare Configurations, 4-23
Electronic Warfare Control Authority, 3-16
Electronic Warfare Employment Considerations, 4-24
Electronic Warfare Execution, 3-20
Electronic Warfare Personnel, 3-11
Electronic Warfare Planning, 4-15
Electronic Warfare Preparation, 3-17
Electronic Warfare Running Estimate, 4-16
Electronic Warfare Support Techniques, 3-35
Electronic Warfare Technician, 3-12
EMOE Estimate, 5-16
EMS Superiority Approach, 5-20
Enabled Effects, 6-38, 7-5
Enterprise Management Activities, 6-40
Enterprise Operations Center, 6-17

* Index-2

Equipment and Communications Enhancements, 3-34
Evaluate the Threat, 4-i
Event Matrix, 4-q
Event Template, 4-p
Execution, 2-39
Exploitation, 5-6

F
Fires Support Element, 2-34
Frequency Interference Resolution, 3-10
Frequency Interference Resolution, 5-9
Friendly EMS-Use Requirements, 5-20
Functional Network Operations and Security Centers (NOSC), 6-13
Functional Services, 6-36
Fundamental Principles, 2-12

G
G-6 or S-6 Spectrum Manager, 2-34
Geographic Combatant Commander (GCC), 6-14
Geography Challenges, 1-13
Global Cyber Threat, 0-2
Glossary, 8-3

H
Hazards, 2-11
HERF, 5-8
HERO, 5-8
HERP, 5-8
High-Altitude Electromagnetic Pulse (HEMP), 5-8
High-Value Targets, 4-m

I
Identify Vulnerabilities, 7-14
Individuals or Small Group Threat, 1-12
Information, 0-10, 1-28
Information (Planning Considerations), 5-18
Information Assurance (IA), 4-49
Information Assurance Vulnerability Management (IAVM), 7-30
Information Collection, 2-40
Information Environment, 0-10, 1-8
Information Environment Operations (IEO), 0-9, 1-9
Information Function, 0-10
Information Function Activities, 0-12
Information Operations (IO), 0-11, 2-25, 4-44, 4-45
Information Operations Officer or Representative, 2-34
Information Operations Planning, 4-51
Information Systems Security, 7-26
Information-Influence Relational Framework, 4-45
Integrating / Coordinating Functions of IO, 4-45
Integrating Cyberspace Operations, 1-9
Integrating Processes, 2-40
Integration of Cyberspace Fires, 1-53
Integration through the Operations Process, 2-37
Intelligence, 4-49
Intelligence and Operational Analytic Support, 1-43
Intelligence Gain/Loss (IGL), 1-44
Intelligence Operations, 2-23
Intelligence Preparation of the Battlefield (IPB), 2-40, 4-a
Intelligence Requirements (IRs), 1-43, 4-44
Intelligence, Information, Cyber, EW, & Space (I2CEWS), 2-36
Interorganizational Considerations, 1-57
Interrelationship with Other Operations, 2-23
ISR in Cyberspace, 1-45

J
Jamming, 3-31
JEMSMO Cell Actions and Outputs, 5-17
JEMSO Actions, 5-6
JEMSO Staff Estimate, 5-20
Joint Cyberspace Center (JCC), 6-14
Joint Cyberspace Operations, 1-1
Joint Electromagnetic Spectrum Operations (JEMSO), 4-50, 5-5
Joint Functions, 0-10, 1-24
Joint Interagency Coordination Group (JIACG), 4-47
Joint Planning Group (JPG), 4-51
Joint Planning Process (JPP), 1-39, 4-41
Joint Restricted Frequency List (JRFL), 4-22

K
Key Leader Engagement (KLE), 0-14, 4-50
Key Terrain, 1-8
Knowledge Management, 2-41

L
Large Scale Combat Operations, 3-28
Legal Considerations, 1-38
Leveraging Information, 0-14
Live Spectrum Analysis, 5-14
Location and Ownership 1-6
Logical Network Layer, 1-3, 2-6

M
Manage, 5-7
Manipulate, 1-22
Measures of Effectiveness (MOEs), 1-56
Measures of Performance (MOPs), 1-56
Military Deception (MILDEC), 0-14, 4-49
Military Decision-Making Process (MDMP), 4-2
Military Information Support Operations (MISO), 0-14, 4-49
Mission Variables (METT-TC), 2-9
Mission-Tailored Force Package (MTFP), 1-49

Mitigating Insider Threats, 7-27
Modified Combined Obstacle Overlay, 4-g
Multi-Domain Extended Battlefield, 0-8, 2-4, 2-16
Multinational Considerations, 1-58
National Incident Response, 1-29
National Intelligence Operations, 1-17
Nation-State Threat, 1-12

N
Nature of Cyberspace, 1-2
Non-State Threats, 1-12

O
Offensive Cyberspace Operations (OCO), 1-18, 2-20
Offensive Electromagnetic Attack, 3-2
Open-Source Intelligence (OSINT), 1-45
Operational Environment (OE), 0-6, 1-7, 2-4
Operational Initiative, 2-4
Operational Resilience, 7-8
Operational Risks, 2-42
Operational Variables (PMESII-PT), 2-8
Operations Orders, 4-35
Operations Process, 2-37
Operations Security (OPSEC), 0-15, 4-50
Operations Security Risks, 2-43

P
Phasing, 4-54
Physical Network Layer, 1-3, 2-6
Planning, 2-38, 4-1
Planning Considerations, 1-39
Planning Insights, 4-44
Planning Joint EMS Operations (JEMSO), 5-15
Planning Timelines, 1-40
Policy Risks, 2-43
Positions of Relative Advantage, 2-16

Preparation, 2-38
Protect, 5-6
Protect Function, 7-21
Protection, 7-25
Protection Levels, 7-27
Public Affairs (PA), 0-14, 4-48

R
Reaction, 7-29
Recover Function, 7-24
Remedial Electronic Protection Techniques, 3-32
Requesting Cyberspace Effects, 4-9
Respond Function, 7-24
Risk Concerns, 1-53
Risk Management, 2-41
Risk Management Framework, 7-6
Risks In Cyberspace and the EMS, 2-42
Roles and Responsibilities, 1-30

S
Scanning and Remediation, 7-30
Sensing Activity Distinctions, 3-18
Services' Cyberspace Doctrine, 1-4
SMO inputs to the MDMP, 5-10
SMO Support to the Warfighting Functions, 5-12
Space Operations, 0-15, 4-49, 2-24
Special Technical Operations (STO), 0-15, 4-50
Spectrum Management, 3-10, 5-9
Spectrum Management Operations (SMO/JEMSO), 3-10, 5-4, 5-1
Spectrum Management Operations Core Functions, 5-5
Spectrum Manager, 3-13
Staff and Support at Corps and Below, 2-32
Staff Judge Advocate, 2-34

Staff Members and Electronic Warfare, 3-14
Strategic Communication (SC), 4-46
Synchronization, 1-52

T
Target Access, 1-46
Target Nomination and Synchronization, 1-46
Targeting (D3A), 1-46, 2-41, 4-29
Targeting Crosswalk, 4-31
Targeting Methodology, 4-30
Technical Risks, 2-42
Technology Challenges, 1-13
Terrain Effects Matrix, 4-h
Theater Network Operations Control Center (TNCC), 6-14
Threat Activities, 7-14
Threat Capabilities, 4-k
Threat Description Table, 4-f
Threat Detection and Characterization, 1-44
Threat Electronic Attack, 3-32
Threat Model, 4-l
Threat Overlay, 4-f
Threat Situation Template, 4-o
Threats, 1-12, 2-10
Time-Sensitive Targets (TSTs), 1-48
Tools of Cyber Attacks, 7-18
Trends and Characteristics, 2-10

U
U.S. Army Network Enterprise Technology Command (NETCOM), 6-10
United States Army Cyber Command (ARCYBER), 2-27, 6-10
United States Code, 1-31
United States Cyber Command (USCYBERCOM), 1-10, 6-9

W
Warfighting Functions, 2-14
Warning Intelligence, 1-45
Weather, Light, and Illumination Charts or Tables, 4-h

* Index-4

SMARTbooks
INTELLECTUAL FUEL FOR THE MILITARY

Recognized as a "**whole of government**" doctrinal reference standard by military, national security and government professionals around the world, SMARTbooks comprise a **comprehensive professional library** designed with all levels of Soldiers, Sailors, Airmen, Marines and Civilians in mind.

The SMARTbook reference series is used by **military, national security, and government professionals** around the world at the organizational/institutional level; operational units and agencies across the full range of operations and activities; military/government education and professional development courses; combatant command and joint force headquarters; and allied, coalition and multinational partner support and training.

Download FREE samples and SAVE 15% everyday at:
www.TheLightningPress.com

The Lightning Press is a **service-disabled, veteran-owned small business**, DOD-approved vendor and federally registered — to include the SAM, WAWF, FBO, and FEDPAY.

SMARTbooks
INTELLECTUAL FUEL FOR THE MILITARY

MILITARY REFERENCE: SERVICE-SPECIFIC

Recognized as a "whole of government" doctrinal reference standard by military professionals around the world, SMARTbooks comprise a comprehensive professional library.

MILITARY REFERENCE: MULTI-SERVICE & SPECIALTY

SMARTbooks can be used as quick reference guides during operations, as study guides at professional development courses, and as checklists in support of training.

JOINT STRATEGIC, INTERAGENCY, & NATIONAL SECURITY

The 21st century presents a global environment characterized by regional instability, failed states, weapons proliferation, global terrorism and unconventional threats.

  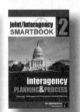 

 The Lightning Press is a **service-disabled, veteran-owned small business**, DOD-approved vendor and federally registered — to include the SAM, WAWF, FBO, and FEDPAY.

RECOGNIZED AS THE DOCTRINAL REFERENCE STANDARD BY MILITARY PROFESSIONALS AROUND THE WORLD.

THREAT, OPFOR, REGIONAL & CULTURAL

In today's complicated and uncertain world, the military must be ready to meet the challenges of any type of conflict, in all kinds of places, and against all kinds of threats.

HOMELAND DEFENSE, DSCA, & DISASTER RESPONSE

Disaster can strike anytime, anywhere. It takes many forms—a hurricane, an earthquake, a tornado, a flood, a fire, a hazardous spill, or an act of terrorism.

DIGITAL SMARTBOOKS (eBooks)

In addition to paperback, SMARTbooks are also available in digital (eBook) format. Our digital SMARTbooks are for use with Adobe Digital Editions and can be used on up to **six computers and six devices**, with free software available for **85+ devices and platforms**— including **PC/MAC, iPad and iPhone, Android tablets and smartphones, Nook, and more!** Digital SMARTbooks are also available for the **Kindle Fire** (using Bluefire Reader for Android).

Download FREE samples and SAVE 15% everyday at:
www.TheLightningPress.com

Purchase/Order

SMARTsavings on SMARTbooks! Save big when you order our titles together in a SMARTset bundle. It's the most popular & least expensive way to buy, and a great way to build your professional library. If you need a quote or have special requests, please contact us by one of the methods below!

View, download FREE samples and purchase online:
www.TheLightningPress.com

Order SECURE Online
Web: www.TheLightningPress.com
Email: SMARTbooks@TheLightningPress.com

24-hour Order & Customer Service Line
Place your order (or leave a voicemail)
at 1-800-997-8827

Phone Orders, Customer Service & Quotes
Live customer service and phone orders available
Mon - Fri 0900-1800 EST at (863) 409-8084

Mail, Check & Money Order
2227 Arrowhead Blvd., Lakeland, FL 33813

Government/Unit/Bulk Sales

The Lightning Press is a **service-disabled, veteran-owned small business**, DOD-approved vendor and federally registered—to include the SAM, WAWF, FBO, and FEDPAY.

We accept and process both **Government Purchase Cards** (GCPC/GPC) and **Purchase Orders** (PO/PR&Cs).

Keep your SMARTbook up-to-date with the latest doctrine! In addition to revisions, we publish incremental "**SMARTupdates**" when feasible to update changes in doctrine or new publications. These SMARTupdates are printed/produced in a format that allow the reader to insert the change pages into the original GBC-bound book by simply opening the comb-binding and replacing affected pages. Learn more and sign-up at: **www.thelightningpress.com/smartupdates/**